Graph Theory

Undergraduate Mathematics

Graph Theory

Undergraduate Mathematics

Khee Meng Koh
National University of Singapore, Singapore

Fengming Dong
Nanyang Technological University, Singapore

Kah Loon Ng
National University of Singapore, Singapore

Eng Guan Tay
Nanyang Technological University, Singapore

 World Scientific

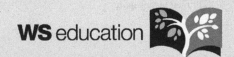 WS education

Published by

World Scientific Publishing Co. Pte. Ltd.
5 Toh Tuck Link, Singapore 596224
USA office: 27 Warren Street, Suite 401-402, Hackensack, NJ 07601
UK office: 57 Shelton Street, Covent Garden, London WC2H 9HE

Library of Congress Cataloging-in-Publication Data
Koh, K. M. (Khee Meng), 1944–
    Graph theory : undergraduate mathematics / by Khee Meng Koh (NUS, Singapore), Fengming Dong (NTU, Singapore), Kah Loon Ng (NUS, Singapore), Eng Guan Tay (NTU, Singapore).
      pages cm
    NUS represents National University of Singapore. NTU represents Nanyang Technological University, Singapore.
    Includes bibliographical references and index.
    ISBN 978-9814641586 (hardcover : alk. paper) -- ISBN 978-9814641593 (pbk. : alk. paper)
    1. Graph theory--Textbooks. 2. Mathematics--Study and teaching (Higher) I. Dong, Fengming
1962–   II. Ng, Kah Loon. III. Tay, Eng Guan. IV. Title.
    QA166.K635 2015
    511'.5--dc23

                                                    2014044827

British Library Cataloguing-in-Publication Data
A catalogue record for this book is available from the British Library.

The digraph on the cover, sometimes referred to as the Koh-digraph, is the smallest counterexample to a conjecture by Lovász. The Lion Head Symbol is inserted in the center of the digraph to commemorate the 50th anniversary of independence of Singapore in 2015.

Printed in Singapore

Preface

Discrete Mathematics is a branch of mathematics dealing with finite or countable processes and elements. Graph Theory is an area in Discrete Mathematics which studies configurations involving a set of vertices interconnected by edges (called graphs). From humble beginnings and almost recreational type problems, Graph Theory has found its calling in the modern world of complex systems and especially of the computer. Graph Theory and its applications can be found not only in other branches of mathematics, but also in scientific disciplines such as engineering, computer science, operational research, management sciences and the life sciences. Since computers require discrete formulation of problems, Graph Theory has become an essential and powerful tool for engineers and applied scientists, in particular, in the area of designing and analyzing algorithms for various problems which range from designing the itineraries for a shipping company to sequencing the human genome in the life sciences.

This book is an expansion of our first book Introduction to Graph Theory: H3 Mathematics. While the first book was intended for capable high school students and university freshmen, this version covers substantially more ground and is intended as a reference and textbook for undergraduate studies in Graph Theory. In fact, the topics cover a few modules in the Graph Theory taught at the National University of Singapore. The material in the book, and especially the variety and quantity of the problems, are derived very much from the enormous wealth of knowledge and experience gained from the thirty plus years of teaching and researching of the first author, Koh Khee Meng.

Certain features of this book are worth mentioning. Care is specially taken so that concepts are explained clearly and developed properly; it strives to be readable and at the same time be mathematically rigorous.

v

At suitable junctures, questions are inserted for discussion. This is to ensure that the reader understands the preceding section fully before proceeding on to new ideas and concepts. Each chapter includes applications of the concepts in real-life. They are added for general interest and as substantiation of the usefulness of Graph Theory concepts. There are many items in the Exercise component following most sections. Some are exercises intended for reinforcing what is learnt earlier while others test the full range of understanding and problem solving in the concepts acquired. Each chapter concludes with a large selection of interesting problems that cover all the sections in that chapter. Some of these problems are from research articles and these add to the depth and cutting-edge aspect of the endeavor. Proofs of most important theorems are given in their full mathematical rigour. Finally, we believe that "good research enlivens teaching, and good teaching encourages research" and so we have made a conscious effort to include recent research work at the frontiers of areas of graph theory into this book.

Chapter 1 covers the fundamental concepts and basic results in Graph Theory tracing its history from Euler's solution of the problem of the Bridges of Konigsberg. Fundamental concepts include those of graphs, multigraphs, vertex degrees, paths, cycles, distance, eccentricity of vertices, and the radius and diameter of a graph.

In Chapter 2, congruence is defined in terms of isomorphism rather than a vague notion of shape and this allows a 'handle' to compare graphs. Chapter 2 further fleshes out the concept of a graph by introducing the attendant concepts of subgraphs and the complement of a graph, and coding of a graph via graphic sequences.

In Chapter 3, we introduce two important families of graphs, namely trees and bipartite graphs. A tree, in some sense, forms the 'skeleton' of a connected graph and in general, a forest of trees forms the 'skeleton' of any graph. Thus, the structure and properties of trees are very important. Bipartite graphs are another family of graphs that have found applications in many real-life situations such as matching a group of job seekers with a set of potential jobs under certain conditions.

Chapter 4 returns the reader to Euler's seminal work on the Bridges of Konigsberg. Euler is memorialized for his contribution by having graphs with the property that one can have a walk that traverses all edges exactly once and then returns to the starting vertex named after him – Eulerian multigraphs. This chapter gives a fuller treatment of Eulerian multigraphs. This study includes the Chinese Postman Problem which is an optimization problem on multigraphs with weighted edges.

William Rowan Hamilton was a famous mathematical prodigy. He introduced the vertex analogous concept of the Eulerian circuit. Such a graph that admits a cycle that visits all the vertices in the graph was named Hamiltonian in his honor. In Chapter 5, we study Hamiltonian graphs and the Traveling Salesman Problem which is another important optimization problem on graphs with weighted edges.

Be it transportation systems, computer networks, or supply chains, the notion of the robustness of a system, in the sense of the connectedness of its components, is certainly of great importance. Since graphs are used often to model systems, in Chapter 6, we shall introduce an important parameter, called connectivity, which measures how 'strong' the connectedness of a given graph is. Two versions of the notion, namely, the vertex version and the edge version, will be presented.

Three important features of sets of vertices and sets of edges are discussed in Chapter 7. The first arises from the 'natural' relation of independence among vertices, giving rise to the notion of an independent set, intuitively seen as a set of vertices with no edges between them. Another 'natural' relation of independence among edges gives rise to the notion of a matching, intuitively seen as a set of edges with no common vertex between any pair of them. Thirdly, the analogous notions of a vertex cover and an edge cover are intuitively seen as a set of vertices which 'touch' every edge in the graph and a set of edges which 'touch' every vertex in the graph. Independence, matching and covering are graph features that are useful when modeling some real-life scenarios. For example, a complete matching can be used to find a system of distinct representatives (SDR). And in the next chapter, independence is a necessary condition for vertex-coloring, which in turn can be used in scheduling activities.

Can all maps be colored with at most four colors? Many people believed that the answer is affirmative, but no one could prove it for a long time. This is known as the Four Color Problem. In Chapter 8, we study vertex-colorings of graphs as an approach to tackle this problem. The chromatic number is introduced and an algorithm and some techniques to estimate or enumerate it are discussed. Interesting applications of vertex-coloring to scheduling problems are given in some detail.

A set of vertices in a graph is called a dominating set if any vertex not in the set is adjacent to a vertex in the set. This idea of dominating sets in graphs can be used to solve real world problems. For example, suppose transmitting stations are to be built in some cities in a country so that every city without a transmitting station can receive messages from some adjacent city with one. The problem for selecting the cities for the transmitting stations is that of finding a dominating set in the corresponding graph. Chapter 9 studies this very useful concept and its variants, including the very recent conception of Roman Domination.

Finally, Chapter 10 studies graphs with 'directions' indicated on the edges. Such are called directed graphs or digraphs. Such digraphs suitably model many situations where relationships between items are directional. The chapter covers some basic concepts and provides some detail for a special family of digraphs, called tournaments. Material in this chapter also reaches as far as the frontier area of optimal orientations of graphs.

Problems in a section are referenced as

Problem [Chapter].[Section].[Number];

for example, Problem 1.3.4 means Problem 4 at the end of Section 1.3. General problems related to all the concepts in a chapter are placed in a special section at the end of the chapter and are referenced as Problem [Chapter].[Number]; for example, Problem IV.3 means Problem 3 at the end of Chapter 4. While a list of notations is provided at the beginning of the book, a list of indices can be found at the end. References are cited in the book according to the format, for example, Dirac (1960) refers to the article authored by Dirac published in 1960.

We would like to thank Goh Chee Ying, Hang Kim Hoo, Ku Cheng Yeaw, Zeinab Maleki, Ng Boon Leong, Soh Kian Wee, Tay Tiong Seng, Chia Gek Ling, Tan Ban Pin, Ting Tao Siang and Anders Yeo for reading through the draft and checking through the problems – any mistakes that remain are ours alone.

Koh Khee Meng
Dong Fengming
Ng Kah Loon
Tay Eng Guan

Notations

$G \wedge H$	the graph given by $(G-wg) \cup (H-wh) + gh$ where w is a common vertex of G and H, wg is an edge in G and wh is an edge in H	336
$\Delta(G)$	maximum degree of G	19
$\delta(G)$	minimum degree of G	19
$\tau(G)$	number of spanning trees of G	128
$\kappa(G)$	vertex-connectivity of G	239
$\kappa'(G)$	edge-connectivity of G	240
$\rho(G)$	deficiency of G	275
$\alpha(G)$	independence number of G	293
$\alpha'(G)$	matching number of G	293
$\beta(G)$	vertex covering number of G	294
$\beta'(G)$	edge covering number of G	296
$\chi(G)$	chromatic number of G	313
$\omega(G)$	clique number of G	317
$\gamma(G)$	domination number of G	373
$i(G)$	independent domination number of G	378
$\gamma_R(G)$	Roman domination number of G	384
$i_R(G)$	independent Roman domination number of G	388
$d^-(v)$	in-degree of v	398
$d^+(v)$	out-degree of v	398
$s(v)$	score of a vertex v	416
$I(v)$	in-set of v	423
$O(v)$	out-set of v	423
$G(D)$	underlying graph of digraph D	401
$\operatorname{diam}(D)$	diameter of digraph D	403
$A(D)$	adjacency matrix of digraph D	411
$\overrightarrow{D}$	converse of digraph D	414
$K_r(D)$	set of all r-kings in digraph D	436
$k_r(D)$	number of r-kings in digraph D	436
$\overrightarrow{d}(G)$	orientation number of G	451
T_n	tournament of order n	416

Contents

Chapter 1

Fundamental Concepts and Basic Results

1.1 The Königsberg Bridge Problem

The River Pregel flowed through the old city of Königsberg in 18th century Eastern Prussia. Back then, there were seven bridges over the river connecting the two islands (B and D) and two opposite banks (A and C) as shown in Fig. 1.1.1.

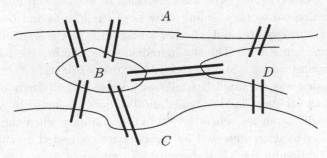

Fig. 1.1.1

It was said that the people in the city had always amused themselves with the following problem:

> *Starting with any one of the four places A, B, C or D as shown in Fig. 1.1.1, is it possible to have a walk which passes through each of the seven bridges once and only once, and return to where you started?*

No one could find such a walk; and after a number of tries, people believed that it was simply not possible, but no one could convincingly

1

prove it either.

Leonhard Euler, the greatest mathematician that Switzerland has ever produced, was told of the problem. He noticed that the problem was very much different in nature from the problems in traditional geometry, and instead of considering the original problem, he studied its much more general version which encompassed any number of islands or banks, and any number of bridges connecting them. His finding was contained in the article Euler (1736). As a direct consequence of his finding, he deduced the impossibiity of having such a walk in the Königsberg bridge problem. This was historically the first time a proof was given from the mathematical point of view.

How did Euler generalize the Königsberg bridge problem? How did he solve his more general problem? What was his finding?

1.2   Multigraphs and Graphs

Euler observed that the Königsberg bridge problem had nothing to do with traditional geometry where the measurements of lengths and angles, and relative locations of vertices count. How large the islands and banks are, how long the bridges are, and whether an island is at the south or north of a bank are immaterial. The key ingredients are whether the islands or banks are connected by a bridge, and by how many bridges.

Euler's idea was essentially as follows: represent the islands or banks by '**dots**', one for each island or bank, and two dots are joined by k '**lines**' (not necessarily straight), where $k \geq 0$, when and only when the respective islands or banks represented by the dots are connected by k bridges. Thus the situation for the Königsberg bridge problem is represented by the diagram in Fig. 1.2.1.

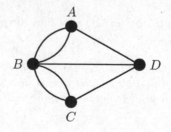

Fig. 1.2.1

The diagram in Fig. 1.2.1 is now known as a **multigraph**. Intuitively, a **multigraph** is a diagram consisting of 'dots' and 'lines', where each line joins some pair of dots, and two dots may be joined by no lines or any number of lines. More formally, we call a 'dot' a **vertex** (plural, vertices) and call a 'line' an **edge**.

For instance, in the multigraph of Fig. 1.2.1, there are four vertices and seven edges, where each edge joins some pair of vertices; vertices A and C are not joined by any edges, A and D are joined by one edge, and B and C are joined by two edges, etc.

Note that the sizes and the relative locations of the dots (vertices), and the lengths of the lines (edges) are immaterial. Only the 'linking relations' among the vertices and the number of edges that join any two vertices count. Thus, the situation for the Königsberg bridge problem can equally well be represented by the multigraph of Fig. 1.2.2.

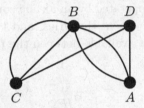

Fig. 1.2.2

Let us give more examples of multigraphs which represent situations of different natures.

Example 1.2.1. There were six people: A, B, C, D, E and F in a party and several handshakes among them took place. Suppose that

    A shook hands with B, C, D, E and F;

    B, in addition, shook hands with C and F;

    C, in addition, shook hands with D and E;

    D, in addition, shook hands with E;

    E, in addition, shook hands with F.

The situation can be clearly shown by the multigraph in Fig. 1.2.3, where people are represented by vertices and two vertices are joined by an edge whenever the corresponding persons shook hands.

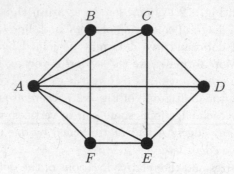

Fig. 1.2.3

Example 1.2.2.   The diagram in Fig. 1.2.4 is a multigraph which shows the availability of flights operated by an airline company between a number of cities. The vertices represent the cities, and two vertices are joined by an edge if there is a flight available between the two corresponding cities.

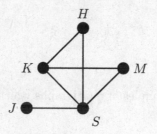

Fig. 1.2.4

Example 1.2.3.   The diagram in Fig. 1.2.5 is a multigraph which models a job application situation. The vertices are divided into two parts: X and Y, where the vertices in X (shaded grey) represent the applicants, while those in Y (shaded black) represent the jobs available. A vertex in X is joined to a vertex in Y by an edge if the corresponding applicant applies for the corresponding job.

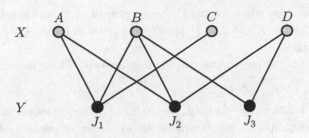

Fig. 1.2.5

Question 1.2.4. *Give three examples from our everyday life where the situations can be modeled by multigraphs.*

It is noted that in the three multigraphs shown in Fig. 1.2.3 - Fig. 1.2.5, every two vertices are joined by at most one edge (that is, either no edges or exactly one edge). These situations are different from the multigraph in Fig. 1.2.1 (or Fig. 1.2.2) where there are vertices joined by more than one edge. To distinguish them, we call the diagrams in Fig. 1.2.3 - Fig. 1.2.5 **simple graphs**, or simply, **graphs**. Thus the diagram in Fig. 1.2.1 (or Fig. 1.2.2) is a multigraph, but not a (simple) graph.

Let us consider another example.

Example 1.2.5. In the diagram shown in Fig. 1.2.6, there are

- four vertices: u, v, w and z, and
- eight edges: f_1 and f_2 joining u and v; e_1, e_2 and e_3 joining w and z; h_1 joining v and w; h_2 joining u and w; h_3 joining v to itself.

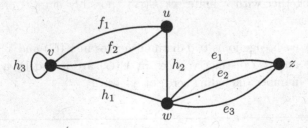

Fig. 1.2.6

Two or more edges joining the same pair of vertices are called **parallel edges**. Thus in Fig. 1.2.6, f_1 and f_2 are parallel edges; e_1, e_2 and e_3 are parallel edges.

Any edge joining a vertex to itself is a **loop**. Thus, in Fig. 1.2.6, h_3 is a loop.

Remark 1.2.6.   In this book, we shall not consider 'loops' in the diagram of any multigraph. A diagram with the existence of parallel edges is **not** a (**simple**) **graph**. Another example of a multigraph which is **not** a (simple) graph is shown in Fig. 1.2.7.

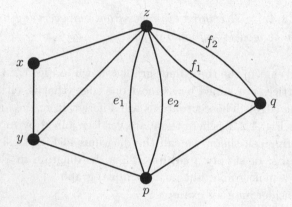

Fig. 1.2.7

We now give formal definitions of a 'graph' and a 'multigraph'.

A **multigraph** G consists of a non-empty finite set $V(G)$ of vertices together with a finite set $E(G)$ (possibly empty) of edges such that

    (i) each edge joins two distinct verties in $V(G)$ and

    (ii) any two distinct vertices in $V(G)$ are joined by a finite number (including zero) of edges.

The sets $V(G)$ and $E(G)$ are called the **vertex set** and the **edge set** of G respectively. The number of vertices in G, denoted by $v(G)$, is called the **order** of G (thus $v(G) = |V(G)|$). The number of edges in G, denoted by $e(G)$, is called the **size** of G (thus $e(G) = |E(G)|$).

A multigraph G is called a (simple) **graph** if any two vertices in $V(G)$ are joined by at most one edge (that is, either they are not joined by an edge or joined by exactly one edge).

Remark 1.2.7.

(1) It follows from the above definitions that

  (i) every graph is a multigraph but not vice versa; and
  (ii) no loops are allowed in any multigraph.

  When a concept is defined or a statement is made for multigraphs, they are also valid, in particular, for graphs.

(2) If e is the only edge joining two vertices u and v, then we may write $e = uv$ or $e = vu$. The ordering of u and v in the expression is immaterial.

(3) An (n, m)-multigraph is a multigraph of order n and size m.

Example 1.2.8. Let G be the multigraph shown in Fig. 1.2.7. Then
$V(G) = \{x, y, z, p, q\}$,
$E(G) = \{xy, xz, yz, yp, e_1, e_2, f_1, f_2, pq\}$,
$v(G) = 5$ and $e(G) = 9$, that is, G is a $(5, 9)$-multigraph.

Let H be the graph shown in Fig. 1.2.3. Then
$V(H) = \{A, B, C, D, E, F\}$,
$E(H) = \{AB, AC, AD, AE, AF, BC, BF, CD, CE, DE, EF\}$,
$v(H) = 6$ and $e(H) = 11$, that is, H is a $(6, 11)$-graph.

Question 1.2.9. *Let G be the multigraph shown in Fig. 1.2.8. Find $V(G)$, $E(G)$, $v(G)$ and $e(G)$.*

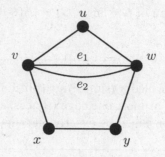

Fig. 1.2.8

Question 1.2.10. *Let H be the graph with*

$$V(H) = \{a, b, c, x, y, z\} \text{ and } E(H) = \{ab, ay, bx, by, cx, cz, xz, yz\}.$$

Find $v(H)$ and $e(H)$, and draw a diagram of H.

Matrices and multigraphs

As discussed earlier, a multigraph G can be represented by a diagram consisting of 'dots' and 'lines', and can be defined in terms of its vertex set $V(G)$ and edge set $E(G)$. Multigraphs can also be represented by matrices in various ways. In what follows, we introduce two of them.

Example 1.2.11. Let G be the multigraph shown in Fig. 1.2.9, where its four vertices are named v_1, v_2, v_3 and v_4.

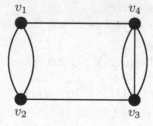

Fig. 1.2.9

Consider the following 4×4 matrix $\boldsymbol{A}$:

$$\boldsymbol{A} = \begin{pmatrix} 0\ 2\ 0\ 1 \\ 2\ 0\ 1\ 0 \\ 0\ 1\ 0\ 3 \\ 1\ 0\ 3\ 0 \end{pmatrix}.$$

Can you find any relation between G and $\boldsymbol{A}$?

What is the value of the $(1,2)$-entry in $\boldsymbol{A}$? It is '2'. How many edges are there in G joining v_1 and v_2? There are also '2'.

How many edges are there in G joining v_3 and v_4? There are '3'. What is the value of the $(3,4)$-entry in $\boldsymbol{A}$? It is also '3'.

Indeed, it is observed that the value of the (i,j)-entry in $\boldsymbol{A}$ is the number of edges in G joining v_i and v_j, where $i,j \in \{1,2,3,4\}$. Note that the value of each (i,i)-entry (that is, a diagonal entry) in $\boldsymbol{A}$ is '0' as there is no edge in G joining v_i to itself.

In general, for any multigraph G, if two vertices u and v are joined by an edge, we say that u and v are **adjacent**. This is why the 4×4 matrix in Example 1.2.11 is called the **adjacency matrix** of G. Evidently, the matrix is dependent on the labeling of the vertices.

Let G be a multigraph of order n with

$$V(G) = \{v_1, v_2, ..., v_n\}.$$

The **adjacency matrix** of G is the $n \times n$ matrix

$$\boldsymbol{A}(G) = (a_{ij})_{n \times n},$$

where a_{ij}, the (i,j)-entry in $\boldsymbol{A}(G)$, is the number of edges joining v_i and v_j for all $i,j \in \{1,2,...,n\}$.

Question 1.2.12.   *Let G be the multigraph shown in Fig. 1.2.10.*

Graph Theory

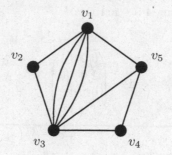

Fig. 1.2.10

(i) Find $A(G)$.

(ii) Is $A(G)$ symmetric (i.e., (i,j)-entry $= (j,i)$-entry)?

(iii) What is the sum of the values of the entries in each row (respectively, column)?

(iv) What is your interpretation of the 'sum' obtained in (iii)?

Question 1.2.13. *The adjacency matrix of a multigraph G is given below:*

$$A(G) = \begin{pmatrix} 0 & 2 & 1 & 0 & 1 \\ 2 & 0 & 1 & 0 & 0 \\ 1 & 1 & 0 & 3 & 2 \\ 0 & 0 & 3 & 0 & 0 \\ 1 & 0 & 2 & 0 & 0 \end{pmatrix}.$$

Draw a diagram of G.

Example 1.2.14. Let G be the multigraph discussed in Example 1.2.11, where its seven edges are named $e_1, e_2, ..., e_7$ as shown in Fig. 1.2.11:

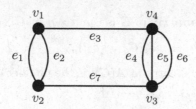

Fig. 1.2.11

Consider the following 4×7 matrix $\boldsymbol{B}$ in which the value of each entry is either '0' or '1':

$$\boldsymbol{B} = \begin{pmatrix} 1 & 1 & 1 & 0 & 0 & 0 & 0 \\ 1 & 1 & 0 & 0 & 0 & 0 & 1 \\ 0 & 0 & 0 & 1 & 1 & 1 & 1 \\ 0 & 0 & 1 & 1 & 1 & 1 & 0 \end{pmatrix}.$$

Can you find any relation between G and $\boldsymbol{B}$?

What is the order of G? How many rows are there in $\boldsymbol{B}$?

What is the size of G? How many columns are there in $\boldsymbol{B}$?

What is the value of the $(1,3)$-entry? What is the relation between vertex v_1 and edge e_3?

What is the value of the $(2,5)$-entry? What is the relation between vertex v_2 and edge e_5?

In general, for any multigraph G, if two vertices u and v are joined by an edge e, we say that u (and v) is **incident** with the edge e. That is why the 4×7 matrix in Example 1.2.14 is called the **incidence matrix** of G. Evidently, the matrix is dependent on the labeling of the vertices and edges.

Let G be an (n, m)-multigraph with

$$V(G) = \{v_1, v_2, ..., v_n\} \text{ and } E(G) = \{e_1, e_2, ..., e_m\}.$$

The **incidence matrix** of G is the $n \times m$ matrix

$$\boldsymbol{B}(G) = (b_{ij})_{n \times m},$$

where

$$b_{ij} = \begin{cases} 1 \text{ if } v_i \text{ is incident with } e_j, \\ 0 \text{ otherwise.} \end{cases}$$

Question 1.2.15. *Let G be the following $(5, 9)$-multigraph.*

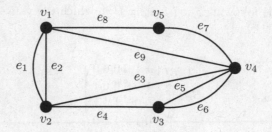

Fig. 1.2.12

(i) Find $B(G)$.

(ii) What is the sum of the values of the entries in each column of $B(G)$? Why?

(iii) What is the sum of the values of the entries in each row of $B(G)$? Why?

Question 1.2.16. *The incidence matrix of a multigraph G is given below:*

$$B(G) = \begin{pmatrix} 1 & 0 & 0 & 0 & 0 & 0 & 1 & 0 & 0 \\ 1 & 1 & 0 & 0 & 1 & 1 & 0 & 0 & 0 \\ 0 & 1 & 1 & 1 & 0 & 0 & 0 & 0 & 0 \\ 0 & 0 & 1 & 1 & 0 & 0 & 0 & 1 & 0 \\ 0 & 0 & 0 & 0 & 1 & 1 & 0 & 1 & 1 \\ 0 & 0 & 0 & 0 & 0 & 0 & 1 & 0 & 1 \end{pmatrix}.$$

Draw a diagram of G and find $A(G)$.

Exercise for Section 1.2

1. Let G be the multigraph shown in Fig. 1.2.13. Is G a simple graph? Determine $V(G), E(G), v(G)$ and $e(G)$.

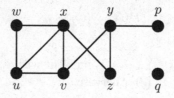

Fig. 1.2.13

2. Define a graph G such that $V(G) = \{2, 3, 4, 5, 11, 12, 13, 14\}$ and two vertices s and t are adjacent if and only if $gcd\{s, t\} = 1$. Draw a diagram of G and find its size $e(G)$.

3. Let G be a graph with $V(G) = \{1, 2, ..., 10\}$ such that two numbers i and j in $V(G)$ are adjacent if and only if $|i - j| \leq 3$. Draw the graph G and determine $e(G)$.

4. Let G be a graph with $V(G) = \{1, 2, ..., 10\}$ such that two numbers i and j in $V(G)$ are adjacent if and only if $i + j$ is a multiple of 4. Draw the graph G and determine $e(G)$.

5. Let G be a graph with $V(G) = \{1, 2, ..., 10\}$ such that two numbers i and j in $V(G)$ are adjacent if and only if $i \times j$ is a multiple of 10. Draw the graph G and determine $e(G)$.

6. Find the adjacency matrix and the incidence matrix of the following $(5, 7)$-graph G.

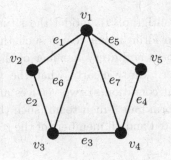

Fig. 1.2.14

7. The adjacency matrix of a multigraph G is shown below:

$$\begin{pmatrix} 0 & 1 & 0 & 2 & 3 \\ 1 & 0 & 1 & 2 & 2 \\ 0 & 1 & 0 & 1 & 1 \\ 2 & 2 & 1 & 0 & 1 \\ 3 & 2 & 1 & 1 & 0 \end{pmatrix}.$$

Draw a diagram of G.

8. Four teams of three specialist soldiers each (a scout, a signaller and a sniper) are to be sent into enemy territory. However, some of the soldiers cannot work well with some others. The following table shows the soldiers, their specializations and who they cannot work with.

Soldier	Specialization	Cannot cooperate with
1	Scout	5, 7, 10
2	Scout	–
3	Scout	5, 6, 8, 9, 11
4	Scout	8, 12
5	Signaller	1, 3, 9
6	Signaller	3, 10, 11
7	Signaller	1, 9, 12
8	Signaller	3, 4, 9, 10
9	Sniper	3, 5, 7, 8
10	Sniper	1, 6, 8
11	Sniper	3, 6
12	Sniper	4, 7

(i) Draw a multigraph to model the situation so that we may see how to form 3-men teams such that each specialization is represented and every member of the team can work with every other. State clearly what the vertices represent and under what condition(s) two vertices are joined by an edge.

(ii) Can you form four 3-men teams such that each specialization is represented and all members of the team can work with one another?

1.3 Vertex Degree

Let G be a multigraph. Recall that two vertices u and v are said to be adjacent if there is an edge e joining u and v. In this case, we write $e = uv$ and say that u is a **neighbor** of v (and vice versa), and e is incident with u and v. The two vertices u and v are also called the two **ends** of e.

The set of all neighbors of v in G is denoted by $N(v)$; that is,

$$N(v) = \{x \mid x \text{ is a neighbor of } v\}.$$

Example 1.3.1.   Let G be the multigraph shown in Fig. 1.3.1.

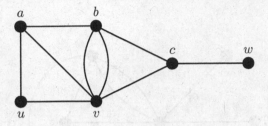

Fig. 1.3.1

Then

(i) the vertices a and b are adjacent, so are b and v, but not a and c;
(ii) the vertices a and u are the two ends of the edge au;
(iii) the edge av is incident with the vertices a and v;
(iv) the vertex a has three neighbors, namely, b, u and v; and
(v) $N(a) = \{b, u, v\}, N(b) = \{a, v, c\}, N(w) = \{c\}$, etc.

Let G be a multigraph. We now introduce a very useful and important number associated with each vertex in G.

Given a vertex v in G, the **degree** v in G, denoted by $d_G(v)$, is defined as the number of edges incident with v.

For simplicity, we shall replace $d_G(v)$ by $d(v)$ if there is no danger of confusion.

Question 1.3.2.

(i) *Find the degree of each vertex in G of Fig. 1.3.1.*
(ii) *Find $N(x)$ for each vertex x in G of Fig. 1.3.1.*
(iii) *By definition, is it necessarily true that $d(v) = |N(v)|$? Under what condition will $d(v) = |N(v)|$?*

Example 1.3.3. Let G be the multigraph shown in Fig. 1.3.2.

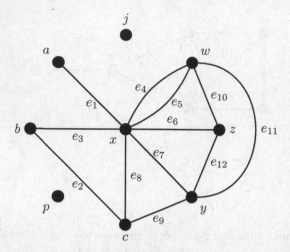

Fig. 1.3.2

Observe that there are seven edges incident with vertex x. Thus, $d(x) = 7$. There are no edges incident with the vertex j. Thus, $d(j) = 0$. The degrees of the vertices in G are shown in the table below.

Vertex	a	b	c	j	p	w	x	y	z
Degree	1	2	3	0	0	4	7	4	3

The degree of a vertex is also called the **valency** of the vertex as it is related to the valency of an atom in chemical compounds as shown in Fig. 1.3.3.

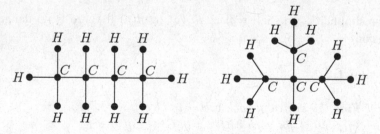

Fig. 1.3.3: Butane and Isobutane, C_4H_{10}

Two types of vertices having two smallest degrees have special names.

A vertex v is called an **isolated-vertex** if $d(v) = 0$; it is called an **end-vertex** if $d(v) = 1$.

Thus, in the multigraph of Fig. 1.3.2, the vertices j and p are isolated-vertices and the vertex a is an end-vertex.

Remark 1.3.4. An *end-vertex* and an *end* of an edge (see beginning of this section) are two different concepts. While an *end-vertex* is a vertex of degree one, an *end* of an edge has nothing to do with its degree.

In the multigraph G of Fig. 1.3.2, the total sum of the degrees of its vertices, as can be seen from the table, is 24. What is the size of G? The answer is: $e(G) = 12$. Observe that the total sum 24 is double the size 12 of G. Is this a coincidence?

Question 1.3.5. *Consider the multigraph G of Fig. 1.3.1. Find $e(G)$ and the sum of the degrees of the six vertices. Is the sum twice of $e(G)$?*

In general, is the sum of the degrees of the vertices in a multigraph always double its size?

There is a Chinese saying: *whenever you drink water, think of its source.* Where do the *degrees* of the vertices come from? The answer is: *from the existence of 'edges'.* No edges implies no degrees. How many degrees can each edge contribute? The answer is '2' as an edge is incident with its two ends. To *compute the sum of degrees of vertices, we count each edge twice, once for each end of the edge.* Thus, we have the following result due to Euler.

Theorem 1.3.6. *Let G be a multigraph with $V(G) = \{v_1, v_2, ..., v_n\}$. Then*

$$\sum_{i=1}^{n} d(v_i) = 2e(G).$$

$\square$

Remark 1.3.7.

    (1) If the vertices of G in Theorem 1.3.6 are not named as shown, the result can also be expressed as

$$\sum_{v \in V(G)} d(v) = 2e(G).$$

    (2) A group of n (≥ 2) persons were together and various handshakes took place among them (two persons might even shake hands more than once). Each person recorded the number of hands he or she had shaken. The sum of these n numbers would be double the number of handshakes that took place in the gathering. Theorem 1.3.6 is, therefore, also known as the **Handshaking Lemma**.

An important way to classify the vertices of a multigraph G is by means of the **parity** (i.e., being even or odd) of their degrees.

A vertex w in G is said to be **even** if $d(w)$ is even; and said to be **odd** if $d(w)$ is odd.

Thus, in the multigraph G of Fig. 1.3.2, there are five even vertices: b, j, p, w and y; and four odd vertices: a, c, x and z.

Question 1.3.8.

    (a) How many odd vertices are there in each of the multigraphs shown in the previous examples?

    (b) Can you construct a multigraph containing (i) exactly one odd vertex? (ii) exactly three odd vertices?

Instead of merely considering the multigraph of Fig. 1.2.1, which represents the Königsberg bridge problem, Euler (1736) studied a much more general problem:

Let G be a multigraph. Suppose that one starts with an arbitrary vertex in G, and finds that it is possible to have a walk which passes through each edge exactly once and then be able to end at the starting vertex. What can be said about such a multigraph?

Evidently, in order to have such a walk one must be able to enter and exit a vertex, hence each vertex must be even. Is the converse true? We will discuss this in Chapter 6.

In order to study this problem and its related issues, Euler introduced the notion of 'odd vertices' and determined the parity of the number of 'odd vertices'. To get to this, Euler first established Theorem 1.3.6, and then deduced from it the following consequence.

Corollary 1.3.9. *The number of odd vertices in any multigraph is even.*

Proof. Let G be a multigraph. Let A be the set of odd vertices in G, and B be the set of even vertices in G. Our aim is to show that $|A|$ is even. Indeed, as $V(G) = A \cup B$ and by Theorem 1.3.6, we have

$$\sum_{v \in A} d(v) + \sum_{v \in B} d(v) = \sum_{v \in V(G)} d(v) = 2e(G).$$

Since $\sum_{v \in B} d(v)$ and $2e(G)$ are even, $\sum_{v \in A} d(v)$ is also even. As $d(v)$ is odd for each v in A, it follows that $|A|$ must be even, as required. $\qquad\square$

Let us proceed to introduce two useful quantities pertaining to the degrees of vertices in a multigraph.

Let G be a multigraph. The **maximum degree** of G, denoted by $\Delta(G)$, is defined as the **maximum number** among all vertex degrees in G.

Likewise, the **minimum degree** of G, denoted by $\delta(G)$, is defined as the **minimum number** among all vertex degrees in G.

That is,

$$\Delta(G) = \max\{d(v) \mid v \in V(G)\} \text{ and}$$

$$\delta(G) = \min\{d(v) \mid v \in V(G)\}.$$

Thus, in the multigraph G of Fig. 1.3.2, $\Delta(G) = 7$ and $\delta(G) = 0$.

Example 1.3.10. Consider the graphs G and H shown in Fig. 1.3.4 (a) and (b) respectively.

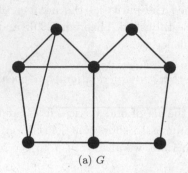
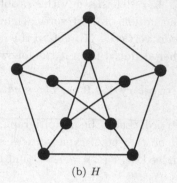

(a) G (b) H

Fig. 1.3.4

It can be checked that $\Delta(G) = 5$ while $\delta(G) = 2$, and $\Delta(H) = \delta(H) = 3$.

Remark 1.3.11. The graph H shown in Fig. 1.3.4(b) is a famous graph, known as the **Petersen graph**. It was named after Julius Petersen (1839 − 1910), a Danish mathematician, who discussed the graph in Petersen (1891). It should be remarked that the Petersen graph plays an important part in the disproval of several conjectures in the past as the graph is often used as a counter-example to the conjectures.

Notice that in the Petersen graph H, every vertex has the same degree, namely '3'. There are many graphs in which every vertex has the same degree. We now single out this special family of graphs by giving them a name.

A graph G is said to be **regular** if every vertex in G has the same degree. More precisely, G is said to be k-**regular** if $d(v) = k$ for each vertex v in G, where $k \geq 0$.

Thus, a graph G is k-regular if and only if $\Delta(G) = \delta(G) = k$. Note that the Petersen graph is 3-regular. For $k = 0, 1, 2, 3$ and 4, a k-regular graph is shown in Fig. 1.3.5.

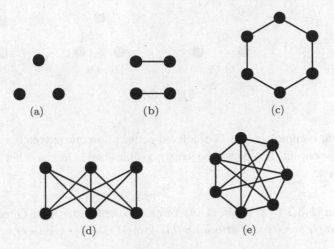

Fig. 1.3.5

Remark 1.3.12.   A 3-regular graph is also called a **cubic graph**.

Question 1.3.13.   *Construct a 5-regular graph of order* 10. *What is its size?*

To end this section, we introduce three important families of regular graphs: the '*empty graphs*', '*complete graphs*' and '*cycles*'.

The empty graphs

By the definition of a graph G, the vertex set $V(G)$ is never empty, but its edge set $E(G)$ may be empty. The graph in Fig. 1.3.5(a) is an example.

A graph G is called an **empty graph** (or a **null graph**) if $E(G)$ is empty. An empty graph of order n is denoted by O_n.

Clearly, each O_n is 0-regular and $e(O_n) = 0$. The four smallest empty graphs are shown in Fig. 1.3.6.

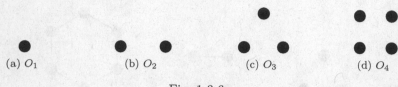

Fig. 1.3.6

Among (simple) graphs G of a fixed order n, on one extreme, the empty graph O_n contains no edges (the least possible size). On the other extreme, we may ask:

Question 1.3.14.   *What is the largest possible size that G can have? Which graph has its size attaining this largest possible number?*

The complete graphs

A graph is called a **complete graph** if every two of its vertices are adjacent. A complete graph of order n is denoted by K_n.

Clearly, each K_n is $(n-1)$-regular and $e(K_n) = \binom{n}{2}$. The four smallest complete graphs are shown in Fig. 1.3.7.

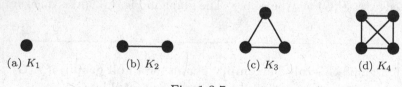

(a) K_1 (b) K_2 (c) K_3 (d) K_4

Fig. 1.3.7

The cycles
The 2-regular graph shown in Fig. 1.3.5(c) is called a **cycle**.

A graph G of order $n \geq 3$ is called a **cycle** if its n vertices can be named as $v_1, v_2, ..., v_n$ such that v_1 is adjacent to v_2, v_2 is adjacent to $v_3,...$, v_{n-1} is adjacent to v_n, v_n is adjacent to v_1, and no other adjacency exists; that is,

$$V(G) = \{v_1, v_2, ..., v_n\} \text{ and}$$

$$E(G) = \{v_1v_2, v_2v_3, ..., v_{n-1}v_n, v_nv_1\}.$$

A cycle of order n is denoted by C_n. We call C_n an n-**cycle**, and C_3 a **triangle**.

Clearly, every cycle is 2-regular and $e(C_n) = v(C_n) = n$. Three more examples of cycles are shown in Fig. 1.3.8.

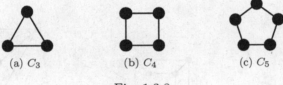

(a) C_3 (b) C_4 (c) C_5

Fig. 1.3.8

Remark 1.3.15. The graph C_n is defined for $n \geq 3$. For $n = 2$, as shown in Fig. 1.3.9, C_2 is also called a cycle, but it is not a graph (it is a multigraph).

Fig. 1.3.9

To end this section, we show in Fig. 1.3.10 the five famous regular polyhedra and their associated regular graphs in Fig. 1.3.11.

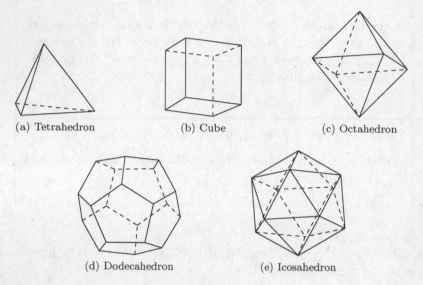

(a) Tetrahedron   (b) Cube   (c) Octahedron

(d) Dodecahedron   (e) Icosahedron

Fig. 1.3.10

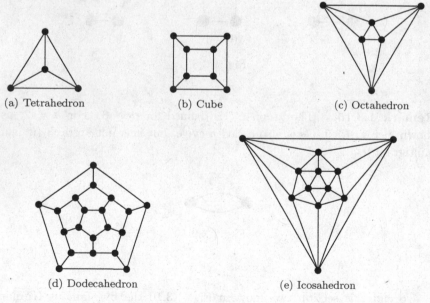

(a) Tetrahedron   (b) Cube   (c) Octahedron

(d) Dodecahedron   (e) Icosahedron

Fig. 1.3.11

Exercise for Section 1.3

1. In the following multigraph G, find

   (i) the size of G,
   (ii) the degree of each vertex,
   (iii) the sum $\sum_{v \in V(G)} d(v)$,
   (iv) the number of odd vertices,
   (v) $\Delta(G)$, and
   (vi) $\delta(G)$.

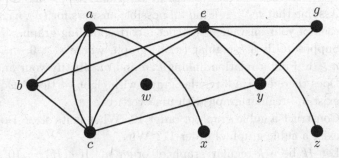

Fig. 1.3.12

   Is your answer for (iii) double your answer for (i)? Is your answer for (iv) an even number?

2. Construct a $(6, 7)$-multigraph in which every vertex is odd.

3. Let G be a multigraph with $V(G) = \{v_1, v_2, ..., v_n\}$. Prove that the sum of all the entries in the ith row of the adjacency matrix $A(G)$ is the degree of the vertex v_i for each $i = 1, 2, ..., n$.

4. Consider the multigraph G given in Question 1.2.16. Let $B(G)^T$ denote the transpose of $B(G)$. Find $B(G)B(G)^T$ and give an interpretation for the (i, j)-entry in $B(G)B(G)^T$.

5. Let G be a $(8, 15)$-graph in which each vertex is of degree 3 or 5. How many vertices of degree 5 does G have? Construct one such graph G.

6. Let H be a graph of order 10 such that $3 \leq d(v) \leq 5$ for each vertex v in H. Not every vertex is even. No two odd vertices are of the same degree. What is the size of H?

7. Let G be a $(14, 30)$-graph in which every vertex is of degree 4 or 5. How many vertices of degree 5 does G have? Construct one such graph G.

8. Does there exist a multigraph G of order 8 such that $\delta(G) = 0$ while $\Delta(G) = 7$? What if 'multigraph G' is replaced by 'graph G'?

9. Characterize the 1-regular graphs.

10. Draw all regular graphs of order n, where $2 \le n \le 6$.

11. (i) Does there exist a graph G of order 5 such that $\delta(G) = 1$ and $\Delta(G) = 4$?

    (ii) Does there exist a graph G of order 5 which has two vertices of degree 4 and $\delta(G) = 1$?

12. Let H be a $(8, 13)$-graph with $\delta(H) = 2$ and $\Delta(H) = 4$. Denote by n_i the number of vertices in H of degree i, where $i = 2, 3, 4$. Assume that $n_3 \ge 1$. Find all possible answers for (n_2, n_3, n_4). For each of your answers, construct a corresponding graph.

13. Suppose G is a k-regular (n, m)-graph, where $k \ge 0$, $m \ge 0$ and $n \ge 1$. Find a relation linking k, n and m. Justify your answer.

14. Does there exist a 3-regular graph with eight vertices? Does there exist a 3-regular graph with nine vertices?

15. Construct a cubic graph of order 12. What is its size? Does there exist a cubic graph of order 11? Why?

16. Let H be a k-regular graph of order n. If $e(H) = 10$, find all possible values for k and n; and for each case, construct one such graph H.

17. Let G be a 3-regular graph with $e(G) = 2v(G) - 3$. Determine the values of $v(G)$ and $e(G)$. Construct all such graphs G.

18. Find all integers n such that $100 \le e(K_n) \le 200$.

19. Let G be a multigraph of order 13 in which each vertex is of degree 7 or 8. Show that G contains at least eight vertices of degree 7 or at least seven vertices of degree 8.

20. Let G be a graph of order n in which there exist **no** three vertices u, v and w such that uv, vw and wu are all edges in G. Show that $n \ge \delta(G) + \Delta(G)$.

1.4   Paths, Cycles and Connectedness

Figure 1.4.1(a) shows a section of the street system of a town. It can be modeled as a graph as shown in Fig. 1.4.1(b), where a vertex represents a junction and two vertices are joined by an edge if and only if the corresponding junctions are linked by a street. For certain purposes, we may have to traverse the street system by passing through some junctions and streets. In order to show more precisely and succintly the way we traverse, in this section, we shall introduce some basic terms in general multigraphs which serve the purpose.

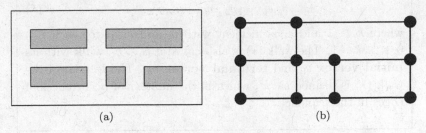

(a) (b)

Fig. 1.4.1

Consider the multigraph H of Fig. 1.4.2. If we start at vertex a, then we can reach vertex x via the edge f_1, and from x to y via the edge f_7. We can further proceed to reach z via f_{12}. This process can be conveniently expressed by the following sequence of vertices and edges:

$$a f_1 x f_7 y f_{12} z.$$

Such a sequence is called a **walk** or, more precisely, an $a - z$ **walk** as a and z are, respectively, the **initial** and **terminal** vertices of the walk.

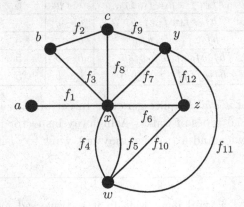

Fig. 1.4.2

A **walk** in a multigraph G is an alternating sequence of vertices and edges beginning and ending at vertices:

$$v_0 e_0 v_1 e_1 v_2 ... v_{k-1} e_{k-1} v_k, \qquad (*)$$

where $k \geq 1$ and e_i is incident with v_i and v_{i+1}, for each $i = 0, 1, ..., k-1$. The walk $(*)$ is also called a $v_0 - v_k$ **walk** with its **initial vertex** v_0 and **terminal vertex** v_k. The **length** of the walk $(*)$ is defined as 'k', which is the number of occurrences of edges in the sequence.

Remark 1.4.1. The vertices v_i's or edges e_i's in $(*)$ need not be distinct.

Question 1.4.2. *Is the following sequence:*

$$c f_9 y f_{11} w f_1 b$$

a walk in H of Fig. 1.4.2? Why?

Example 1.4.3. Some walks in H of Fig. 1.4.2 and their respective lengths are shown in the table below.

	sequence	walk	length
(1)	$b f_3 x f_4 w f_5 x f_4 w f_{10} z f_{10} w$	$b - w$	6
(2)	$b f_3 x f_4 w f_5 x f_7 y$	$b - y$	4
(3)	$b f_3 x f_4 w f_{11} y f_9 c$	$b - c$	4
(4)	$b f_2 c f_8 x f_4 w f_5 x f_3 b$	$b - b$	5
(5)	$b f_2 c f_9 y f_{12} z f_6 x f_3 b$	$b - b$	5

The definition of a *walk* is quite general. In certain circumstances, we will need some special types of walks. A highway inspector may not want to inspect a road twice and a traveler may not want to visit a city more than once.

A walk is called a **trail** if no *edge* in it is traversed more than once. A walk is called a **path** if no *vertex* in it is visited more than once.

Question 1.4.4. *Is every trail a path? Is every path a trail?*

In Example 1.4.3

(i) walk (1) is neither a trail nor a path (why?);
(ii) walk (2) is a trail but not a path (why?);
(iii) walk (3) is both a trail and a path (why?);
(iv) walks (4) and (5) are both trails but not paths (why?).

Remark 1.4.5. It is now clear that if some edge is traversed more than once, then at least one of its two ends is visited more than once. Thus, **every path must be a trail**. Its converse is, however, not true.

The walk (1) in Example 1.4.3 is a $b - w$ walk, but not a $b - w$ path as the vertices x and w are visited more than once. However, it can be cut short, say to bf_3xf_4w, to become a $b - w$ path. Likewise, the $b - y$ walk (2) is not a $b - y$ path, but it can be cut short to bf_3xf_7y to become a $b - y$ path.

Question 1.4.6. *Is it true that every $u - v$ walk always contains a $u - v$ path?*

A $u - v$ walk is said to be **closed** if $u = v$, that is, its initial and terminal vertices are the same; and **open** otherwise.

Thus, in Example 1.4.3, the walks (1), (2) and (3) are open while the walks (4) and (5) are closed.

A closed walk of length at least two in which no edge is repeated is called a **circuit**.

Thus, in Example 1.4.3, the closed walks (4) and (5) are circuits. Note that vertices are allowed to be repeated in a circuit.

Question 1.4.7. *In the multigraph H of Fig. 1.4.2, find a circuit of length 2 and a circuit of length 8.*

A circuit is called a **cycle** if no vertex is repeated (except the initial and terminal vertices).

Thus, in Example 1.4.3, while the circuit (4) is not a cycle (why?), the circuit (5) is a cycle.

Question 1.4.8. *For each* $k = 3, 4, 5, 6$, *find a cycle of length* k *which passes through the vertex* z *in the multigraph of Fig. 1.4.2.*

Question 1.4.9. *The circuit (4) in Example 1.4.3 is not a cycle, but it can be cut short to* $bf_2cf_8xf_3b$ *to become a cycle. Is it true that every circuit always contains a cycle?*

Remark 1.4.10.

(1) At the end of Section 1.3, a cycle is introduced as a graph. In the above discussion, however, a cycle is regarded as a special closed walk in a multigraph. What a cycle means should be clear from the context when it is used.

(2) We express a walk in a multigraph as an alternating sequence of vertices and edges. It is necessary to name the edges since two vertices may be joined by more than one edge and we want to know which edge is traversed to visit these two vertices. However, when we confine ourselves to graphs, as two adjacent vertices are joined by a unique edge, such an expression can be simplified by dropping the names of the edges. Thus, in the graph of Fig. 1.4.3, the $u - w$ walk $ue_1ve_7ye_4ze_3w$ can simply be denoted by $uvyzw$ without any ambiguity.

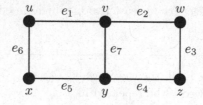

Fig. 1.4.3

The notion of walks or paths enables us to introduce a very important class of multigraphs, called **connected** multigraphs.

Let G be a multigraph and u, v be two vertices in G. We say that u and v are **connected** if they are joined by a path.

If every two vertices in G are connected, we say that G is **connected**.

Example 1.4.11. There are two graphs G and H in Fig. 1.4.4. It can easily be checked that every two vertices in G are connected. Thus G is a connected graph. However, the graph H is not connected since, for instance, the vertices r and u are not connected.

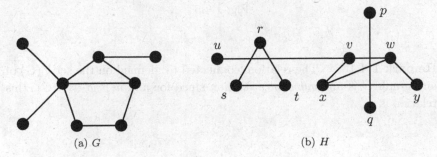

(a) G (b) H

Fig. 1.4.4

A mutigraph is said to be **disconnected** if it is not connected.

Thus the graph H is Fig. 1.4.4(b) is disconnected.

Question 1.4.12. *Consider the disconnected graph H in Fig. 1.4.4(b).*

(i) Which vertices are reachable from the vertex r via a path?
(ii) Which vertices are reachable from the vertex u via a path?
(iii) Which vertices are reachable from the vertex p via a path?

The answers to (i), (ii) and (iii) of the above question can be, respectively, found in Fig. 1.4.5(a), (b) and (c). Note that each of them is a connected 'piece', and is called a **connected component** of H. From now on, we simply call a connected component a **component**. Thus the disconnected graph H has three components. In general, we denote the number of components in a graph G by $c(G)$. Thus, a graph G is connected if and only if $c(G) = 1$.

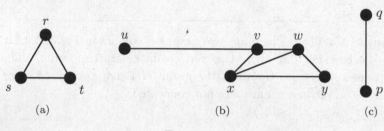

Fig. 1.4.5

Remark 1.4.13.   The relation 'connected to' defined on the set $V(G)$ of any graph G is an *equivalence relation* since for any $u, v, w \in V(G)$, this relation is

(1) **reflexive** (every vertex $v \in V(G)$ is connected to itself);
(2) **symmetric** (if u is connected to v, then v is also connected to u);
(3) **transitive** (if u is connected to v and v is connected to w, then u is connected to w).

This equivalence relation partitions $V(G)$ into **equivalence classes** where vertices are in the same equivalence class if and only if they are connected to each other. In fact, it is now easy to see that each equivalence class is a component of G.

Question 1.4.14.   *Consider the $(12,9)$-graph G in Fig. 1.4.6. Is the graph connected? If not, determine $c(G)$.*

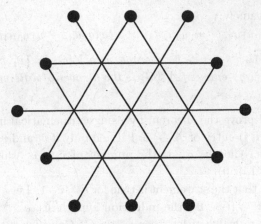

Fig. 1.4.6

To conclude this section, we discuss an interesting relationship between the adjacency matrix of a graph and the notion of walks introduced earlier. Recall that in a graph G with $V(G) = \{v_1, v_2, ..., v_n\}$, if two vertices v_i and v_j are joined by an edge e, the (i, j)- and (j, i)-entry in $A(G)$ will be '1'. Correspondingly, via the edge e, there is exactly '1' walk of length one from v_i to v_j and vice versa. Can $A(G)$ tell us more?

Example 1.4.15. Let G be the folowing graph.

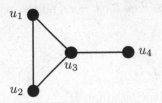

Fig. 1.4.7

Notice that

$$A(G) = \begin{pmatrix} 0 & 1 & 1 & 0 \\ 1 & 0 & 1 & 0 \\ 1 & 1 & 0 & 1 \\ 0 & 0 & 1 & 0 \end{pmatrix} \quad A^2(G) = \begin{pmatrix} 2 & 1 & 1 & 1 \\ 1 & 2 & 1 & 1 \\ 1 & 1 & 3 & 0 \\ 1 & 1 & 0 & 1 \end{pmatrix} \quad A^3(G) = \begin{pmatrix} 2 & 3 & 4 & 1 \\ 3 & 2 & 4 & 1 \\ 4 & 4 & 2 & 3 \\ 1 & 1 & 3 & 0 \end{pmatrix}.$$

The $(1, 3)$-entry in $A^3(G)$ is 4 and there are exactly 4 $u_1 - u_3$ walks of length 3 in G, namely:

$$u_1u_2u_1u_3, \qquad u_1u_3u_1u_3, \qquad u_1u_3u_2u_3, \qquad u_1u_3u_4u_3.$$

Theorem 1.4.16. *Let G be a graph with $V(G) = \{v_1, v_2, ..., v_n\}$. For $k = 1, 2, ...,$ the (i,j)-entry of $A^k(G)$ is the number of different $v_i - v_j$ walks of length k in G.*

Proof. We will prove the theorem by using mathematical induction on k. For $k = 1$, the (i,j)-entry of $A(G)$ is 1 if and only if v_i and v_j are adjacent, and 0 otherwise. Clearly, a_{ij} is the number of $v_i - v_j$ walks of length 1 (which is either 1 or 0).

Now assume that the statement is true for $k = r-1$. Let a_{ij}^{r-1} denote the (i,j)-entry of $A^{r-1}(G)$. By the induction hypothesis, a_{ij}^{r-1} is the number of different $v_i - v_j$ walks of length $r - 1$ in G. Consider the (i,j)-entry of $A^r(G) = A^{r-1}(G)A(G)$.

$$a_{ij}^r = \sum_{t=1}^{n} a_{it}^{r-1}a_{tj}$$

$$= \sum_{t=1}^{n}(\text{number of different } v_i - v_t \text{ walks of length } r - 1 \text{ in } G)a_{tj}. \qquad (*)$$

For a fixed t, $a_{tj} = 1$ if and only if v_t and v_j are adjacent in G, thus $a_{it}^{r-1}a_{tj}$ is the number of different $v_i - v_j$ walks of length r in G **that has v_t preceding** v_j. Summing over all t, expression $(*)$ gives the total number of different $v_i - v_j$ walks of length r in G.

Thus the statement is true for $k = r$ and the proof is complete.

$\square$

Exercise for Section 1.4

    1. Consider the following graph H.

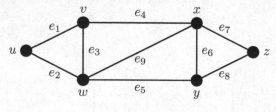

Fig. 1.4.8

(a) Which of the following sequences represents a $u - z$ walk in H?

   (i) $ue_2we_5xe_7z$;
   (ii) $ue_1ve_5ye_8z$;
   (iii) $ue_1ve_3we_3ve_4xe_7z$.

(b) Find a $u - z$ trail in H that is not a path.
(c) Find all $u - z$ paths in H which pass through e_9.

2. Consider the following multigraph G.

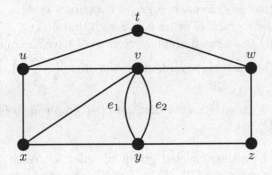

Fig. 1.4.9

(a) For $k = 2, 3, 4, 5, 6, 7$, find a cycle of length k in G.
(b) Find a circuit of length 6 in G that is not a cycle.
(c) Find a circuit of length 8 in G that does not contain t.
(d) Find a circuit of length 9 in G that contains t and v.

3. Is the following graph H disconnected? If so, determine the value of $c(H)$.

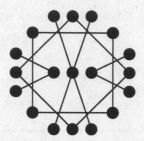

Fig. 1.4.10

4. Let G be a graph with $V(G) = \{1, 2, ..., n\}$, where $n \geq 5$, such that two numbers i and j in $V(G)$ are adjacent if and only if $|i - j| = 5$. Find $c(G)$.

5. Show that any $u - v$ walk in a graph contains a $u - v$ path.

6. Show that any circuit in a graph contains a cycle.

7. Let G be a graph of order $n \geq 2$ such that $\delta(G) \geq \frac{1}{2}(n - 1)$. Show that any two non-adjacent vertices in G have a common neighbor.

8. Let G be an (n, m)-graph such that $m > \binom{n-1}{2}$. Show that G is connected.

9. For $n \geq 2$, construct a disconnected graph of order n and size $\binom{n-1}{2}$.

10. Let G be a disconnected graph of order 5. What is the largest possible value of $e(G)$? If G is a disconnected graph of order $n \geq 2$, what is the largest possible value for $e(G)$? Construct one such extremal graph of order n.

11. Let G be a graph of order $n \geq 2$ and u, v be two non-adjacent vertices in G such that $d(u) + d(v) \geq n + r - 2$. Show that u and v have at least r common neighbors.

12. Let G be a connected graph that is not complete. Show that there exist three vertices x, y, z in G such that x and y, y and z are adjacent, but x and z are not adjacent in G.

13. Let G be an (n, m)-graph such that $\Delta(G) = n - 2$ and any two non-adjacent vertices have a common neighbor. Show that $m \geq 2n - 4$.

14. Let G be a graph such that $N(u) \cup N(v) = V(G)$ for every pair of vertices u, v in G. What can be said about G?

15. Let H be a graph of order $n \geq 2$. Suppose that H contains two

distinct vertices u, v such that

(i) $N(u) \cup N(v) = V(H)$ and
(ii) $N(u) \cap N(v)$ is non-empty.

What is the least possible value of $e(H)$?

16. Suppose G is a disconnected graph which contains exactly two odd vertices u and v. Must u and v be in the same component of G? Why?

17. Show that any two longest paths in a connected graph have a vertex in common.

18. Show that a graph G is connected if and only if for any partition of $V(G)$ into two non-empty sets A and B, there is an edge in G joining a vertex in A and a vertex in B.

1.5 Distance between Two Vertices

In this final section, we introduce an important quantity associated with a pair of vertices in a connected multigraph. The graph H shown in Fig. 1.5.1 is connected, and any two vertices in H are joined by at least one path.

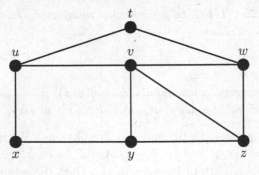

Fig. 1.5.1

Some paths joining the vertices x and w and their respective lengths are shown in the following table.

$x - w$ path	length
$xyzvutw$	6
$xuvyzw$	5
$xyzvw$	4
$xutw$	3
$xyzw$	3

We notice that '3' is the smallest length in the table. We ask: is there any $x - w$ path of length less than '3' in H? The answer is 'no'. That is, among all paths joining x and w in H, the smallest length is '3'. In this case, we say that the **distance from x to w** is '3', and we write $d(x, w) = 3$.

Let G be a connected multigraph, and u, v be any two vertices in G. The **distance from u to v**, denoted by $d(u, v)$, is defined as the *smallest length* of all $u - v$ paths in G.

Question 1.5.1. *In the graph H of Fig. 1.5.1, find $d(u, u)$, $d(x, y)$, $d(y, x)$, $d(x, z)$, $d(z, x)$, $d(u, z)$ and $d(y, t)$.*

As we shall see, the concept of distance is an important notion in studying the structure of a graph, and plays also a prominent role in applications of graphs. Four basic properties of distance are given below.

Theorem 1.5.2. *Let G be a connected multigraph, and x, y, z be any three vertices in G. Then*

    (i) $d(x, x) = 0$,
    (ii) $d(x, y) > 0$ if $x \neq y$,
   (iii) (the symmetric property) $d(x, y) = d(y, x)$,
   (iv) (the triangle inequality) $d(x, y) + d(y, z) \geq d(x, z)$.

Proof. The properties (i), (ii) and (iii) are obvious. We now prove (iv). Assume that $p = d(x, y)$ and $q = d(y, z)$. Let P be an $x - y$ path of length p and Q be a $y - z$ path of length q in G. Then the sequence beginning with P followed by Q is a $x - z$ walk of length $p + q$ in G. As every $x - z$ walk contains a $x - z$ path (see Problem 1.4.5), there exists a $x - z$ path of length at most $p + q$ in G. It follows by definition that

$$d(x, z) \leq p + q = d(x, y) + d(y, z).$$

$\square$

Question 1.5.3.

(i) If $d(x,y) = 1$, what is the relation between x and y?
(ii) If $d(x,y) > 1$, what is the relation between x and y?
(iii) Is it possible that $d(x,y) + d(y,z) = d(x,z)$?
(iv) Is it possible that $d(x,y) + d(y,z) > d(x,z)$?

Example 1.5.4. Consider the connected graph G of order 9 shown in Fig. 1.5.2. We wish to label the nine vertices $v_1, v_2, ..., v_8, v_9$ so that $v_9 = a$ and each of the rest has a higher-labeled neighbor in G. Is there any general way to do it?

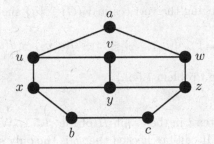

Fig. 1.5.2

We label the vertices starting from $a = v_9$ in descending order.

(1) There are 2 vertices, namely u and w, of distance 1 from a. Label them v_8 and v_7 arbitrarily, say $u = v_8$ and $w = v_7$.
(2) There are 3 vertices, namely v, x and z, of distance 2 from a. Label them v_6, v_5 and v_4 arbitrarily, say $v = v_6$, $x = v_5$ and $z = v_4$.
(3) The remaining 3 vertices are of distance 3 from a. Label $y = v_3$, $b = v_2$ and $c = v_1$.

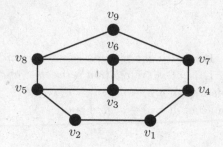

Fig. 1.5.3

The resulting labeling is shown in Fig. 1.5.3. It can be seen that every v_i $(i = 1, 2, ..., 8)$ is adjacent to a higher-labeled v_j in G.

In general, for any connected graph G of order $n \geq 2$ and any vertex 'a' in G, the n vertices in G can be labeled $v_1, v_2, ..., v_{n-1}, v_n$ so that $a = v_n$ and each vertex other than a has a higher-labeled neighbor in G. This can be achieved by arranging the vertices in $V(G) - \{a\}$ such that

$$d(a, v_{n-1}) \leq d(a, v_{n-2}) \leq ... \leq d(a, v_2) \leq d(a, v_1).$$

Why is this so? (See Problem 1.5.9.)

Consider the vertex t in the graph H of Fig. 1.5.1. Which vertices in H are furthest from t? It can be checked that y is the only such vertex. What is the value of $d(t, y)$? The answer is '3'. This '3', which measures the distance between t and a vertex furthest from t, is called the **eccentricity** of the vertex t in H.

Let G be a connected multigraph and $v \in V(G)$. The **eccentricity** of v, denoted by $e(v)$, is the distance between v and a vertex furthest from v in G. That is,

$$e(v) = \max\{d(v, x) \mid x \in V(G)\}.$$

As an example, the eccentricities of the six vertices in the graph G of Fig. 1.5.4 are shown in parentheses.

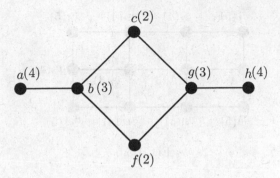

Fig. 1.5.4

Among the six eccentricities shown in the figure, we notice that '2' is the smallest while '4' is the largest. In this situation, we say that the **radius** of G is 2 and the **diameter** of G is 4. Note also that there are two vertices in G, namely c and f, with least eccentricity (i.e., $e(c) = e(f) = 2$). Each of them is called a **central** vertex, and the set of these two central vertices is called the **center** of G.

Given a connected multigraph G, the **radius** of G, denoted by $\mathrm{rad}(G)$, is defined by

$$\mathrm{rad}(G) = \min\{e(v) \mid v \in V(G)\}$$

and the **diameter** of G, denoted by $\mathrm{diam}(G)$, is defined by

$$\mathrm{diam}(G) = \max\{e(v) \mid v \in V(G)\}.$$

A vertex w in G is called a **central** vertex if $e(w) = \mathrm{rad}(G)$, and the **center** of G, denoted by $C(G)$, is the set of all central vertices of G.

In the graph G of Fig. 1.5.5, the eccentricities of the 12 vertices are shown in parentheses.

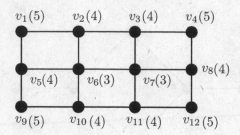

Fig. 1.5.5

Thus, by definition, $\text{rad}(G) = 3$ and $\text{diam}(G) = 5$. The graph G has two central vertices, namely v_6 and v_7. Thus $C(G) = \{v_6, v_7\}$. If G represents the street network of a small town, then, geographically, the junctions v_6 and v_7 are really situated at the 'town center'.

While $\text{diam}(G) = 5$ as pointed out above, we note also that there exist at least two vertices, for instance, v_1 and v_{12}, in G such that $d(v_1, v_{12}) = 5$, and $d(x, y) \leq 5$ for any other two vertices x, y in G. As a matter of fact, we have the following relation (see Problem 1.5.2).

For any connected multigraph G,

$$\text{diam}(G) = \max\{d(x, y) \mid x, y \in V(G)\}.$$

The above notions of radius, diameter and center of a graph are, actually borrowed from those of a circle in plane geometry as shown in Fig. 1.5.6, where r and d denote, respectively, the radius and diameter of the circle with center O. While $d = 2r$ for a circle, is there any relationship between $\text{rad}(G)$ and $\text{diam}(G)$ for a graph G?

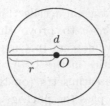

Fig. 1.5.6

It follows immediately by definition that $rad(G) \leq diam(G)$. On the other hand, for the graph G of Fig. 1.5.4, we have $diam(G) = 4 = 2rad(G)$, whereas for the graph G of Fig. 1.5.5, we have $diam(G) = 5 < 6 = 2rad(G)$. Indeed, these are just two instances of the second inequality established below.

Theorem 1.5.5. *Let G be a connected multigraph. Then*

$$rad(G) \leq diam(G) \leq 2rad(G).$$

Proof. We only need to prove that $diam(G) \leq 2rad(G)$. Let u, v be two vertices in G such that $d(u, v) = diam(G)$. Let $w \in C(G)$ (i.e., $e(w) = rad(G)$). We then have

$$
\begin{aligned}
diam(G) = d(u, v) &\leq d(u, w) + d(w, v) \quad \text{(by triangle inequality)} \\
&= d(w, u) + d(w, v) \quad \text{(by symmetry property)} \\
&\leq e(w) + e(w) \quad \text{(by definition)} \\
&= 2e(w) = 2rad(G).
\end{aligned}
$$

$\square$

Remark 1.5.6. Suppose we represent each person on earth by a vertex and put an edge between two vertices if the two persons are acquainted. What will be the diameter of this 'acquaintance graph'? In 1990, John Guare wrote a play called 'Six Degrees of Separation' in which a character claimed to have read that any two persons can be connected by a chain of at most 6 intermediaries. The idea was based on a conjecture by some academics in the mid-20th century that the number of degrees of separation in actual social networks was at most 6, i.e., the diameter of this acquaintance graph is 6 and any two persons can be connected by a chain of at most 5 (wrongly interpreted as 6 in Guare's play) intermediaries. The famous social psychologist, Stanley Milgram conducted an experiment, the result of which was published as an article 'The Small World Problem' in a 1967 issue of *Psychology Today*, to 'prove' this conjecture. In the experiment, Milgram gave some volunteers in Omaha, Nebraska and Wichita, Kansas (both Midwestern cities in the USA) packages with the name of an individual in Boston, Massachusetts (on the east coast of the USA). Each volunteer was asked to pass the package on to an acquaintance who he thought could get the package 'nearer' to the intended recipient, with the instruction to pass it on to another acquaintance, who was to do the same and so on, till it finally reached the intended recipient. The average number of intermediaries for the packages that finally made it to the recipient was 5.5. Certainly, the experiment cannot be used to conclusively prove the 'Six Degrees of Separation' hypothesis, since the packages that did not reach their destination were not counted, and also those which did might not have taken the shortest 'path'. In addition, the experiment was only within the USA and there are many parts of the world that are clearly more isolated than Nebraska or Kansas. Still, the idea of a 'small world' seems plausible and the diameter of the acquaintance graph may indeed be a small number.

Exercise for Section 1.5

1. For each of the graphs G shown in Fig. 1.5.7, find

   (i)   the eccentricity of each vertex in G;
   (ii) rad(G);
   (iii) diam(G);
   (iv) $C(G)$.

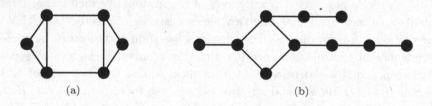

(a)   (b)

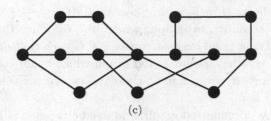

(c)

Fig. 1.5.7

2. Show that $\text{diam}(G) = \max\{d(x,y) \mid x, y \in V(G)\}$ holds for any connected multigraph G.

3. A vertex v in a connected multigraph G is called a **peripheral** vertex if $e(v) = \text{diam}(G)$. Show that G always contains at least two peripheral vertices.

4. For each integer $n \geq 1$, construct a graph G of order n such that $\text{rad}(G) = \text{diam}(G)$.

5. For each integer $n \geq 3$, construct a graph G of order n such that $\text{diam}(G) = 2\,\text{rad}(G)$.

6. Let xy be an edge in a connected multigraph. Show that

$$-1 \leq e(x) - e(y) \leq 1.$$

7. Let G be a graph of order $n \geq 1$ such that $d(x) \geq \frac{n-1}{2}$ for each vertex x in G. Must G be connected? What can be said about $\text{diam}(G)$? (See Problem 1.4.7.)

8. Let G be a connected graph of order $n \geq 3$.

   (i) If $\Delta(G) = n - 1$, find $\text{diam}(G)$.
   (ii) If $\Delta(G) = n - 2$, what can be said about $\text{diam}(G)$?
   (iii) If $\Delta(G) = n - 2$ and $\text{diam}(G) = 2$, show that $e(G) \geq 2(n-2)$.

9. Referring to Example 1.5.4, explain why each vertex other than a has a higher-labeled neighbor in G if the vertices in $V(G) - \{a\}$ are labeled in such a way that

$$d(a, v_{n-1}) \leq d(a, v_{n-2}) \leq \ldots \leq d(a, v_2) \leq d(a, v_1).$$

1.6   Problem Set I

1. Let G be an (n,m)-graph. Show that $\delta(G) \le \frac{2m}{n} \le \Delta(G)$.

2. Let G be an $(n,2n)$-graph, where $n \ge 5$. Show that either G is 4-regular or there exists a vertex z in G such that $d(z) \ge 5$.

3. Let G be a regular connected graph which is not complete. Show that for each vertex x in G, there exists a vertex w in G such that $d(x,w) = 2$.

4. Let G be a connected graph, and u,v be two non-adjacent vertices in G. Show that if $N(u) = N(v)$, then $e(u) = e(v)$.

5. Let G be a connected graph, and u,v be two adjacent vertices in G. Show that if $N(u) - \{v\} = N(v) - \{u\}$, then $e(u) = e(v)$.

6. For any two positive integers r and d with $r \le d \le 2r$, construct a graph G such that $\text{rad}(G) = r$ and $\text{diam}(G) = d$.

7. Let G be a connected graph with $\text{rad}(G) = r$ and $\text{diam}(G) = d$. Show that for each integer k with $r \le k \le d$, there exists a vertex z in G such that $e(z) = k$.

8. Let G be a graph with $\delta(G) \ge k$, where k is a positive integer. Show that

     (i) G contains a path of length k;
     (ii) G contains a cycle C_p, where $p \ge k+1$.

9. [**The n-cube graph**] For $n \ge 1$, let Q_n be the graph with the vertex set $V_n = \{x_1 x_2 \cdots x_n \mid x_i \in \{0,1\}\}$ (thus $V_3 = \{000, 001, 010, 100, 011, 101, 110, 111\}$), and two vertices $a_1 a_2 \cdots a_n$ and $b_1 b_2 \cdots b_n$ are adjacent if and only if $|\{i \mid a_i \ne b_i\}| = 1$ (thus in Q_3, 001 and 101 are adjacent while 001 and 010 are not). We call Q_n the n-cube graph.

     (i) Find $v(Q_n)$.
     (ii) For each $n = 1,2,3,4$, draw the graph Q_n.
     (iii) Show that Q_n is an n-regular connected graph.
     (iv) Find $e(Q_n)$, $\text{rad}(Q_n)$, $\text{diam}(Q_n)$ and $C(Q_n)$.

10. Let G be a graph of order $2r+2$, where $r \ge 1$, such that $\delta(G) \ge r$ and $d(w) > r$ for some vertex w in G. Show that $\text{diam}(G) \le 3$. Construct one such G with $\text{diam}(G) = 3$.

11. Let G be a graph of order $n \ge 3$ and let s denote the sum of the diagonal entries in $A(G)^3$. Find a relation between s and the number of C_3's in G.

12. Let G be a graph of order $n \geq 2$, and $A = \boldsymbol{A}(G)$ the adjacency matrix of G. Let

$$M = A + A^2 + \cdots + A^{n-1}.$$

Show that G is connected if and only if the (i,j)-entry in M is non-zero for all $i, j \in \{1, 2, \cdots, n\}$ with $i \neq j$.

13. Let G be a graph of order $2k + 1$, where $k \geq 1$. Suppose that for any subset A of k vertices, there is a vertex $v \notin A$ such that v is adjacent to each vertex in A. Prove that G contains a vertex z with $d(z) = 2k$.

14. There were n (≥ 2) persons at a party and, as usually happens, some shake hands with others. No one shook hands with the same person more than once. Show that there are at least two persons in the party who had the same number of handshakes.

15. The preceding problem says that in any graph of order $n \geq 2$, there exist two vertices having the same degree. Is the result still valid for multigraphs?

16. Mr. and Mrs. Samy attended an exclusive party where in addition to themselves, there were only 3 other couples. As usually happens, some shake hands with others. No one shook hands with the same person more than once and no one shook hands with his/her spouse. After all the handshakes had been done, Mr. Samy asked each person, including his wife, how many hands he/she had shaken. To everyone's amusement, each one gave a different answer. How many hands did Mrs. Samy shake?

17. In the preceding problem, there were four couples altogether in the party. Solve the general problem where 'four couples' is replaced by 'n (≥ 2) couples'.

18. There are $n \geq 2$ distinct points in the plane such that the distance between any 2 points is at least one. Prove that there are at most $3n$ pairs of these points at distance exactly one.

19. Suppose G is a connected graph with k edges. Prove that it is possible to label the edges $1, 2, ..., k$ in such a way that each vertex belongs to two or more edges (i.e., which is of degree at least two), the greatest common divisor of the integers labeling those edges is 1. (32nd IMO, 1991/4)

20. Nine mathematicians meet at an international conference and discover that among any three of them, at least two speak a common language. If each of the mathematicians can speak at most three

languages, prove that there are at least three of the mathematicians who can speak the same language.

21. Let G be a graph of order $n \geq 2$. Show that its vertex set can be divided into two disjoint subsets such that every vertex v in one subset has at least $\lceil \frac{1}{2} d(v) \rceil$ neighbors in the other subset.

Chapter 2

Graph Isomorphisms, Subgraphs, the Complement of a Graph and Graphic Sequences

2.1 Isomorphic Graphs and Isomorphisms

Suppose now we are asked to draw a graph G which is defined as follows: its vertex set $V(G) = \{w, x, y, z\}$ and edge set $E(G) = \{wx, xy, yz, zw\}$. Some of us may place the four vertices as shown in Fig. 2.1.1(a), others may place them as shown in Fig. 2.1.1(b), (c), etc.

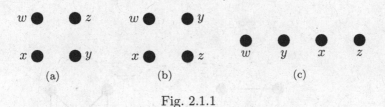

Fig. 2.1.1

By joining some four pairs of vertices with the four edges as given in $E(G)$, we would have their corresponding diagrams as shown in Fig. 2.1.2.

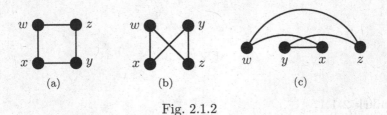

Fig. 2.1.2

Apparently, these three diagrams look very different 'geometrically'.

49

However, in the context of 'graphs', they are absolutely the 'same'.

Intuitively, two graphs G and H are considered the 'same' if it is possible to relocate the vertices of one of the graphs, say G, so that these vertices have the same positions as the vertices in H, the result of which is that the two graphs look identical (imagine that the edges are rubber bands; see Fig. 2.1.3). Mathematically, we use a more fancy term, **isomorphic graphs**, to replace 'same graphs' and define it as follows:

Two graphs G and H are said to be **isomorphic** if there exists a one-one and onto mapping $f : V(G) \to V(H)$ such that two vertices u, v are adjacent in G when and only when their images $f(u)$ and $f(v)$ under f are adjacent in H (i.e., **the adjaceny is preserved under** f).

In this case, we shall write $G \cong H$ and call the mapping f an **isomorphism** from G to H.

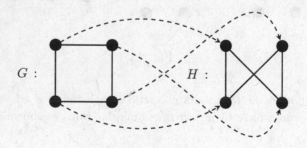

Fig. 2.1.3

Remark 2.1.1.

(1) The phrase 'when and only when' used above means that if u and v are adjacent in G, then $f(u)$ and $f(v)$ **must be** adjacent in H;

and if u and v are **not** adjacent in G, then $f(u)$ and $f(v)$ **must not be** adjacent in H.

(2) The word **isomorphism** is derived from the Greek words *isos* (meaning 'equal') and *morphe* (meaning 'form').

Example 2.1.2.   Consider the graphs G and H as shown in Fig. 2.1.4. We claim that $G \cong H$. Indeed, if we define a mapping $f : V(G) \to V(H)$ by $f(v_i) = u_i$ for each $i = 1, 2, ..., 6$, then it can be checked that f is both one-one and onto, and that the adjacency is preserved under f. Thus G and H are isomorphic under the isomorphism f.

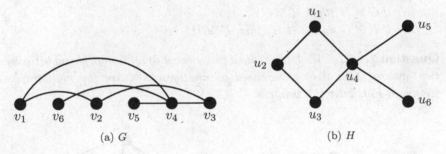

(a) G (b) H

Fig. 2.1.4

Example 2.1.3.   Consider the graphs G and H as shown in Fig. 2.1.5 and define a mapping $f : V(G) \to V(H)$ by $f(w_i) = z_i$ for each $i = 1, 2, ..., 5$. Though f is both one-one and onto, it is clear that f does not preserve the adjacency, and so f is not an isomorphism from G to H. However, it does not mean that G is not isomorphic to H. Indeed, $G \cong H$ and the mapping $g : V(G) \to V(H)$, defined by $g(w_1) = z_4$, $g(w_2) = z_2$, $g(w_3) = z_5$, $g(w_4) = z_3$ and $g(w_5) = z_1$, is an isomorphism from G to H.

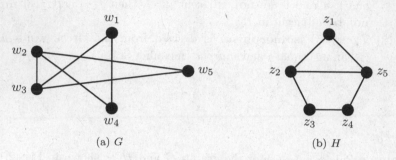

(a) G   (b) H

Fig. 2.1.5

Question 2.1.4.   *Suppose f is an isomorphism from a graph G to a graph H. Let u, v, w be three vertices in G. Assume that u, v, w form a K_3 in G (i.e., any two of them are adjacent in G). Do $f(u), f(v)$ and $f(w)$ also form a K_3 in H?*

Question 2.1.5.   *Let F, G and H be any graphs. Is it true that:*

   (i) $G \cong G$?
   (ii) if $G \cong H$, then $H \cong G$?
   (iii) if $F \cong G$ and $G \cong H$, then $F \cong H$?

Question 2.1.6.   *We have defined the concept of an isomorphism between two graphs. Can it be generalized to multigraphs? Are the multigraphs shown in Fig. 2.1.6 the 'same'?*

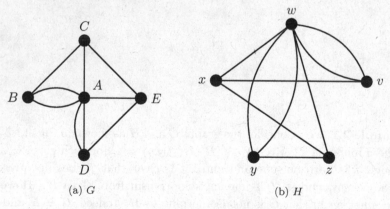

(a) G   (b) H

Fig. 2.1.6

2.2   Testing Isomorphic Graphs

Recall that two graphs are isomorphic if we can find a one-one and onto mapping between their vertex sets which preserves adjacency. It thus follows readily that if $G \cong H$, then G and H must have the same order and size. That is:

Observation 1: If $G \cong H$, then $v(G) = v(H)$ and $e(G) = e(H)$.

We write $G \not\cong H$ if the graphs G and H are not isomorphic. Then, equivalently, Observation 1 says that if $v(G) \neq v(H)$ or $e(G) \neq e(H)$, then $G \not\cong H$.

Question 2.2.1. *Are K_3 and C_3 isomorphic? Are K_4 and C_4 isomorphic?*

Question 2.2.2. *Among the four graphs given in Fig. 2.2.1, are there any two which are isomorphic?*

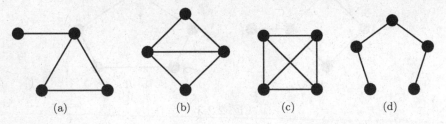

(a)   (b)   (c)   (d)

Fig. 2.2.1

Recall that the degree $d(v)$ of a vertex v in a graph G is the number of edges incident with it. Assume that $V(G) = \{u_1, u_2, ..., u_n\}$. Call $(d(u_1), d(u_2), ..., d(u_n))$ the **degree sequence** of G. We may rename the vertices in G so that $d(u_1) \geq d(u_2) \geq ... \geq d(u_n)$. For instance, in the graph G of Fig. 2.2.2, the five vertices are named as $u_1, ..., u_5$ so that the degree sequence of G is given by $(3, 2, 2, 2, 1)$, which is in non-increasing order.

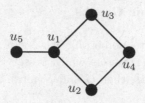

Fig. 2.2.2

Suppose that two graphs G and H are isomorphic under an isomorphism f. As f preserves adjacency, it follows that, for each vertex v in G, $d(v) = d(f(v))$ (see Problem 2.2.7). Thus we have:

Observation 2: If $G \cong H$, then G and H have the same degree sequence, in non-increasing order.

Hence equivalently, if G and H have different degree sequences in non-increasing order, then $G \ncong H$. As an application of this fact, let us consider the following:

Example 2.2.3.   Determine whether the graphs of Fig. 2.2.3 are isomorphic.

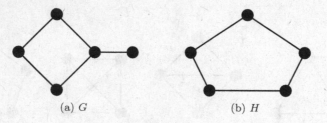

(a) G (b) H

Fig. 2.2.3

Note that although both G and H have the same order and size, the degree sequence of G is $(3, 2, 2, 2, 1)$ while that of H is $(2, 2, 2, 2, 2)$, which is different. Thus $G \ncong H$.

Remark 2.2.4. Actually, in Example 2.2.3, we do not need to use that 'big' notion of degree sequence to conclude that $G \ncong H$. We could arrive at the same by simply pointing out a simple observation that G has an end-vertex (a vertex of degree 1) while H does not.

Given two arbitrary graphs G and H of the same order and same size, is there an 'efficient' procedure which enables us to determine whether $G \cong H$? This problem, known as the **Graph Isomorphism Problem**, is a very difficult problem, and until now, only little progress has been made. There are many practical applications which desire a fast procedure to test graph isomorphism. For example, organic chemists who routinely deal with graphs which represent molecular links would like some system to quickly give each graph a unique name. Thus, many research papers have been published which discuss how to build fast and practical isomorphism testers.

For a good survey on the Graph Isomorphism Problem, the reader may refer to the paper Fortin (1996) or the book Kobler et al. (1993).

Adjacency Matrices and Isomorphism

Recall from Chapter 1 that a graph can be represented by its adjacency matrix. We note first that the adjacency matrix of a graph depends on the ordering of the vertices of the graph. Consider the following graph G:

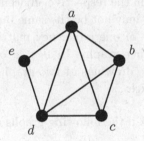

Fig. 2.2.4

If we take the vertices in the ordering a, b, c, d, e, then the adjacency matrix of G is:

$$
\begin{array}{c}
\quad\quad a\ b\ c\ d\ e \\
\begin{array}{c} a \\ b \\ c \\ d \\ e \end{array}
\left(
\begin{array}{ccccc}
0 & 1 & 1 & 1 & 1 \\
1 & 0 & 1 & 1 & 0 \\
1 & 1 & 0 & 1 & 0 \\
1 & 1 & 1 & 0 & 1 \\
1 & 0 & 0 & 1 & 0
\end{array}
\right).
\end{array}
$$

However, if we take the vertices in the ordering a, d, b, c, e, then the adjacency matrix is:

$$
\begin{array}{c}
\quad a\ d\ b\ c\ e \\
\begin{array}{c}
a \\ d \\ b \\ c \\ e
\end{array}
\left(
\begin{array}{ccccc}
0 & 1 & 1 & 1 & 1 \\
1 & 0 & 1 & 1 & 1 \\
1 & 1 & 0 & 1 & 0 \\
1 & 1 & 1 & 0 & 0 \\
1 & 1 & 0 & 0 & 0
\end{array}
\right).
\end{array}
$$

Hence a graph of order n can have up to $n!$ adjacency matrices. On the other hand, if the two graphs G and H possess identical adjacency matrices, we can conclude that G and H are isomorphic. This is true because we can immediately find an isomorphism between $V(G)$ and $V(H)$ by mapping the vertices according to their order in each adjacency matrix. (The reader may like to try to produce a more formal proof for Problem 2.2.12.)

Because it is difficult to find an isomorphism between two (isomorphic) graphs directly from their drawings, it may be easier to show that the graphs are isomorphic from the respective adjacency matrices instead. The initial adjacency matrices may not be the same but subsequent 'intelligent' reordering of the vertices of one adjacency matrix may result in the adjacency matrix of the other graph. In fact, a computer cannot find an isomorphism based on the drawings of two graphs and need inputs in the nature of adjacency matrices.

Example 2.2.5.   Check if the following graphs G and H are isomorphic.

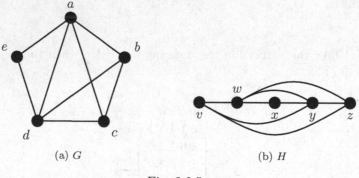

(a) G (b) H

Fig. 2.2.5

The adjacency matrices for G and H in the natural lexicographic ordering of their vertices are as follows:

$$
\boldsymbol{A}(G): \begin{array}{c} \\ a \\ b \\ c \\ d \\ e \end{array}
\begin{array}{c} \cdot a\, b\, c\, d\, e \\ \begin{pmatrix} 0\,1\,1\,1\,1 \\ 1\,0\,1\,1\,0 \\ 1\,1\,0\,1\,0 \\ 1\,1\,1\,0\,1 \\ 1\,0\,0\,1\,0 \end{pmatrix} \end{array}
\qquad
\boldsymbol{A}(H): \begin{array}{c} \\ v \\ w \\ x \\ y \\ z \end{array}
\begin{array}{c} v\, w\, x\, y\, z \\ \begin{pmatrix} 0\,1\,0\,1\,1 \\ 1\,0\,1\,1\,1 \\ 0\,1\,0\,1\,0 \\ 1\,1\,1\,0\,1 \\ 1\,1\,0\,1\,0 \end{pmatrix} \end{array}
$$

If we rearrange the vertices in H as w, v, z, y, x, the resulting adjacency matrix $\boldsymbol{A}'(H)$ is as follows:

$$
\boldsymbol{A}'(H): \begin{array}{c} \\ w \\ v \\ z \\ y \\ x \end{array}
\begin{array}{c} w\, v\, z\, y\, x \\ \begin{pmatrix} 0\,1\,1\,1\,1 \\ 1\,0\,1\,1\,0 \\ 1\,1\,0\,1\,0 \\ 1\,1\,1\,0\,1 \\ 1\,0\,0\,1\,0 \end{pmatrix} \end{array} .
$$

We can now easily see that $\boldsymbol{A}(G) = \boldsymbol{A}'(H)$, and thus conclude that G and H are isomorphic.

Thus, if two graphs are isomorphic, an intelligent system of reordering the vertices of both graphs will finally result in two equal adjacency matrices. On the other hand, if the two graphs are not isomorphic, the person, or more likely the computer, will have to exhaust the large number of reorderings to come to a conclusion that no isomorphism exists.

Question 2.2.6. *Use adjacency matrices to answer Question 2.2.2.*

Exercise for Section 2.2

1. Draw all non-isomorphic graphs of order n with $1 \le n \le 4$.
2. (i) Draw all non-isomorphic $(5, 3)$-graphs.
   (ii) Draw all non-isomorphic $(5, 7)$-graphs.
3. Determine if the following two graphs are isomorphic.

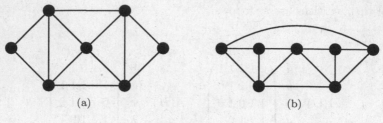

Fig. 2.2.6

4. Determine if the following two graphs are isomorphic.

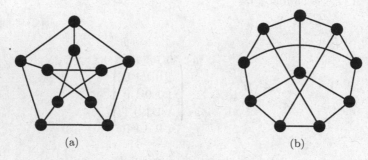

Fig. 2.2.7

5. The following two graphs G and H are isomorphic. List all the isomorphisms from G to H.

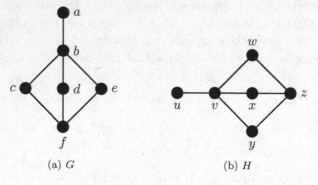

(a) G (b) H

Fig. 2.2.8

6. Prove, by definition of an isomorphism, that the relation '$\cong$' is

reflexive, symmetric and transitive among the family of graphs; that is, the properties listed in Question 2.1.5 hold.

7. Let f be an isomorphism from a graph G to a graph H and w a vertex in G. Show that the degree of w in G is equal to the degree of $f(w)$ in H.

8. A given graph G of order 5 contains at least two vertices of degree 4.

   (i) Assume that not all vertices in G are even. Find all possible degree sequences of G, in non-increasing order; and for each case, construct all such G which are not isomorphic.
   (ii) Assume that all vertices in G are even. Find all possible degree sequences of G, in non-increasing order; and for each case, construct all such G which are not isomorphic.

9. Let H be a graph of order 5 which contains more odd vertices than even. Find all possible degree sequences of H in non-increasing order; and for each case, construct all such H which are not isomorphic.

10. Construct two non-isomorphic 3-regular graphs of order 10.

11. Let G and H be two isomorphic graphs. Show that

    (i) if G is connected, then H is connected;
    (ii) if G is disconnected, then H is disconnected, and $c(G) = c(H)$.

12. Prove that if the adjacency matrices of two graphs G and H are equal, then the graphs G and H are isomorphic.

13. Using adjacency matrices, determine which, if any, of the following three graphs are isomorphic.

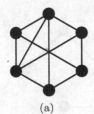
(a)
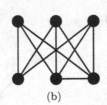
(b)
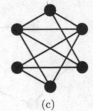
(c)

Fig. 2.2.9

2.3   Subgraphs of a Graph

In studying problems on a graph, quite often, we may wish to consider the 'graphical structures' of **certain portions** of the graph. This leads to the notion of 'subgraphs' of a graph.

> Let G be a graph. A graph H is called a **subgraph** of G if $V(H) \subseteq V(G)$ and $E(H) \subseteq E(G)$.
> By definition, every graph is a subgraph of itself.
> A subgraph H of G is said to be **proper** if $H \not\cong G$.

Example 2.3.1.   Consider the graphs G, H_1, H_2, ..., H_6 as shown in Fig. 2.3.1. We observe that

(i) H_1 is not a subgraph of G as $E(H_1) \nsubseteq E(G)$ even though $V(H_1) \subseteq V(G)$;

(ii) $H_2, ..., H_6$ are subgraphs (indeed, proper subgraphs) of G.

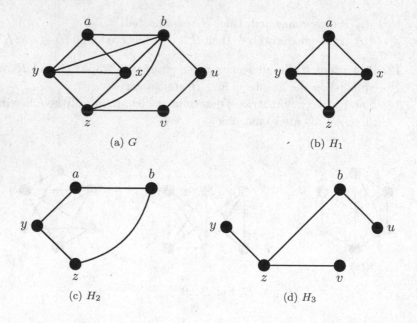

(a) G (b) H_1

(c) H_2 (d) H_3

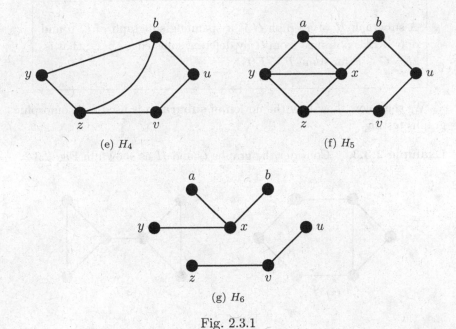

(e) H_4 (f) H_5

(g) H_6

Fig. 2.3.1

Note that $V(H_i) \neq V(G)$ for $i = 2, 3, 4$, but $V(H_5) = V(H_6) = V(G)$.

> A subgraph H of G is said to be **spanning** if $V(H) = V(G)$.

Thus, in Example 2.3.1, the graphs H_5 and H_6 are spanning subgraphs of G, but H_2, H_3 and H_4 are not.

Observe that the spanning subgraph H_5 of G can be obtained by deleting the edges by and bz from G. More generally, let F be a set of edges in G. We denote by $G - F$ the subgraph of G obtained by deleting the edges in F from G. Note that whenever an edge uv is deleted, its two ends (namely, u and v) are still in G. If $F = \{e\}$, consisting of a single edge e, we simply write $G - e$ for $G - \{e\}$.

Question 2.3.2. *Let G be the graph shown in Fig. 2.3.1. Draw the subgraphs $G - yz$, $G - \{uv, vz\}$ and $G - \{yz, bz, ax, uv\}$.*

It can be shown (see Problem 2.3.2) that:

> A subgraph H of a graph G is a spanning subgraph of G if and only if H is obtained from G by deleting some edges in G, that is, $H = G - F$ for some $F \subseteq E(G)$.

We shall now show how the notion of **subgraph** is used for isomorphic graphs testing.

Example 2.3.3.   Consider the graphs G and H as shown in Fig. 2.3.2.

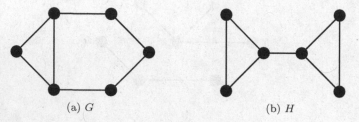

(a) G (b) H

Fig. 2.3.2

Note that G and H have the same degree sequence in non-increasing order, that is, $(3, 3, 2, 2, 2, 2)$, yet $G \not\cong H$ (this shows that the converse of Observation 2 in Section 2.2 is false). How do we argue that $G \not\cong H$? Some ways using the concept of subgraphs are given below:

(i) G contains one K_3 as a subgraph, but H contains two K_3's;
(ii) the two vertices of degree 3 in G are contained in a common K_3, but this is not the case in H;
(iii) G contains a spanning subgraph which is a cycle (i.e., C_6), but H does not have one;
(iv) G contains a C_5, but H does not have one.

Any one of the reasons above would be good enough to justify that $G \not\cong H$.

As we have just seen, to show that $G \not\cong H$, all we need is to find a 'property' that G has but H doesn't have (or vice versa). For this purpose, we state below another useful fact in terms of 'subgraphs'. First of all, we introduce the following notation.

For two graphs G and R, let $n_G(R)$ denote the number of subgraphs of G which are isomorphic to R.

Thus, in Example 2.3.3, $n_G(K_3) = 1$ and $n_H(K_3) = 2$, $n_G(C_5) = n_G(C_6) = 1$ and $n_H(C_5) = n_H(C_6) = 0$.

Observation 3: Let G and H be graphs such that $G \cong H$. Then for any graph R, $n_G(R) = n_H(R)$.

Equivalently, Observation 3 says that if $n_G(R) \neq n_H(R)$ for some graph R, then $G \not\cong H$.

Question 2.3.4. *Consider the graphs G and H as shown in Fig. 2.3.3*

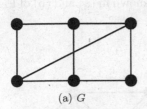
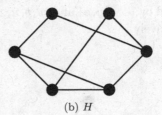

(a) G (b) H

Fig. 2.3.3

  (i) Find $n_G(C_3)$ and $n_H(C_3)$.
  (ii) Find $n_G(C_4)$ and $n_H(C_4)$.
  (iii) Find $n_G(C_5)$ and $n_H(C_5)$.
  (iv) Find $n_G(C_6)$ and $n_H(C_6)$.

Is it true that $G \cong H$?

Question 2.3.5. *Let G be a graph of order $n \geq 2$. What is $n_G(K_1)$? What is $n_G(K_2)$? Is Observation 1 (Section 2.2) a special case of Observation 3?*

Look at the subgraphs H_3 and H_4 of G in Example 2.3.1. By comparing these two subgraphs, we notice that while $V(H_3) = \{b, u, v, y, z\} = V(H_4)$, $E(H_3) \neq E(H_4)$. In H_3, some edges in G which join certain pairs of vertices in H_3 are not present; for instance, yb and uv. On the other hand, **every edge** in G which joins a pair of vertices in H_4 always **remains** in H_4. This feature of H_4 motivates the introduction of the following important type of subgraphs of a graph.

A subgraph H of a graph G is called an **induced subgraph** of G if **any edge** in G that joins a pair of vertices in H is also in H. If H is an induced subgraph of G, we also say that H is the subgraph **induced by its vertex set** $V(H)$ and we write $H = [V(H)]$.

Thus, in Example 2.3.1, among the subgraphs $H_2, ..., H_6$ of G, only H_4 is an induced subgraph of G, and we see that H_4 is induced by $\{b, u, v, y, z\}$ (in notation, $H_4 = [\{b, u, v, y, z\}]$). The subgraphs of G in Example 2.3.1 induced by $\{a, x, y, z\}$ and $\{a, b, u, x, z\}$ are shown in (a) and (b) of Fig. 2.3.4 respectively.

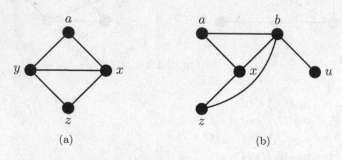

(a)   (b)

Fig. 2.3.4

Question 2.3.6. *Let G be the graph shown in Fig. 2.3.1. Draw the subgraphs of G induced by $\{a, u\}$, $\{a, b, x, y\}$, $\{a, x, y, u, v\}$ and $V(G)$ respectively.*

Question 2.3.7. *Let H be a spanning and induced subgraph of a graph G. What can be said of H?*

Question 2.3.8. *Consider the graph H of Fig. 2.3.5. Note that H is a subgraph of the graph G in Fig. 2.3.1. Is H an induced subgraph of G? Add some more edges in G to H so that the resulting one is $[V(H)]$.*

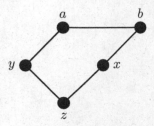

Fig. 2.3.5

Remark 2.3.9. A subgraph H of a graph G is not necessarily an induced subgraph of G. However, it can always be extended to an induced subgraph of G induced by $V(H)$ by adding to H all the missing edges (sharing their two ends in $V(H)$) existing in G.

We have seen that the spanning subgraphs of a graph G are those subgraphs of G that can be obtained from G by deleting some edges in G. In contrast with this, we shall see that induced subgraphs of G can be obtained from G as well, but by deleting some vertices in G as defined below.

Let G be a graph and W a set of vertices in G. We shall denote by $G - W$ the subgraph of G obtained by removing each vertex in W from $V(G)$ together with **all the edges incident with it** from $E(G)$. When W is a singleton, say $W = \{w\}$, we shall write $G - w$ for $G - \{w\}$.

For instance, if G is the graph given in Example 2.3.1, then the subgraphs $G - x$, $G - \{x, y\}$ and $G - \{x, y, z\}$ of G are shown in (a), (b) and (c) of Fig. 2.3.6 respectively. Note that $G - \{x, y, z\} = [\{a, b, u, v\}]$, $G - \{x, y\} = [\{a, b, u, v, z\}]$ and $G - x = [\{a, b, u, v, y, z\}]$.

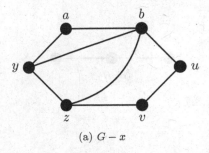

(a) $G - x$

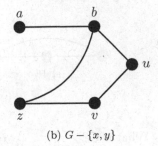

(b) $G - \{x, y\}$

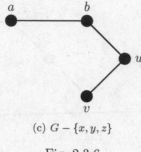

(c) $G - \{x, y, z\}$

Fig. 2.3.6

In general, one can show (see Problem 2.3.4) that:

A subgraph W of a graph G is an induced subgraph of G if and only if $W = G - (V(G) \backslash V(W))$, where $V(G) \backslash V(W)$ consists of those vertices of G which are *not* in W.

Question 2.3.10.   *We have introduced various concepts of subgraphs of a graph. Can they be generalized to multigraphs?*

To end this section, we would like to introduce a very well-known unsolved problem in graph theory, know as the **Reconstruction Conjecture**.

Let us begin with a simple problem. Given a graph G with four vertices v_1, v_2, v_3 and v_4 together with the following information (see Fig. 2.3.7):

(i) $G - v_1 \cong H_1$,
(ii) $G - v_2 \cong H_2$,
(iii) $G - v_3 \cong H_1$,
(iv) $G - v_4 \cong H_2$.

(a) H_1 (b) H_2

Fig. 2.3.7

What is G?

First of all, by (i), G contains the following graph as a subgraph:

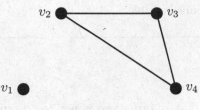

Fig. 2.3.8

Then, by (iii), G must contain the graph of Fig. 2.3.9 as a subgraph.

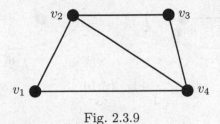

Fig. 2.3.9

Now, it is easily seen that this graph fulfills (ii) and (iv). Furthermore, it can be checked that this graph is the **only** graph that fulfills (i) to (iv). We thus conclude that G is the graph of Fig. 2.3.9.

Let G be a graph with $V(G) = \{u_1, u_2, ..., u_n\}$. We say that G is **reconstructible** if, whenever H is a graph with $V(H) = \{v_1, v_2, ..., v_n\}$ such that $H - v_i \cong G - u_i$ for each $i = 1, 2, ..., n$, then $H \cong G$ (that is, G is uniquely determined by its n subgraphs: $(G - u_1, G - u_2, ..., G - u_n)$).

Thus, the above example shows that the graph in Fig. 2.3.9 is reconstructible. The conjecture can now be stated below.

> **The Reconstruction Conjecture.** Every graph of order at least three is reconstructible.

We note that a graph of order two is not reconstructible. Indeed, take G and H as shown in Fig. 2.3.10:

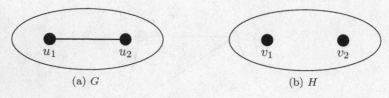

(a) G (b) H

Fig. 2.3.10

It is observed that $G - u_1 \cong H - v_1$ and $G - u_2 \cong H - v_2$, and yet $G \ncong H$.

The above conjecture was first posed by a famous scientist S.M. Ulam (see also Ulam (1960)) and was initially studied by P.J. Kelly in his Ph.D. thesis around 1942. Though the conjecture has been verified to be true for some special families of graphs such as regular graphs and disconnected graphs, it remains unsettled for the general situation. For a very general survey on this conjecture, the reader is referred to the excellent article Bondy (1991).

Exercise for Section 2.3

1. Let G be the graph shown in Fig. 2.3.11:

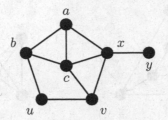

Fig. 2.3.11

(i) Draw the following subgraphs of G:

    (a) $[\{b, v, y\}]$;
    (b) $[\{a, b, c, v, x\}]$; and
    (c) $[\{a, b, u, v, x\}]$.

 (ii) Draw the subgraphs $G - \{ab, cv, xy\}$ and $G - \{b, v\}$ of G.
(iii) Find $E([\{a, b, c, x\}])$.
 (iv) Draw the subgraph $G - E([\{a, b, c, x\}])$.
  (v) Draw a spanning subgraph of G that is connected and that contains a unique C_3 as a subgraph.
 (vi) Draw a spanning subgraph of G that is connected and that contains no cycle as a subgraph.

2. Let H be a subgraph of a graph G. Show that H is a spanning subgraph of G if and only if $H = G - F$, where $F \subseteq E(G)$.
3. Let G be a graph and $X \subseteq V(G)$. Show that $G - X = [V(G)\backslash X]$.
4. Let G be a graph and W a subgraph of G. Show that W is an induced subgraph of G if and only if $W = G - (V(G)\backslash V(W))$.
5. Determine which of the following four graphs are isomorphic and which are not so.

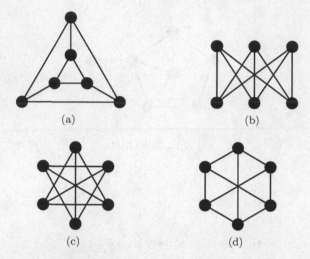

Fig. 2.3.12

6. Let G and H be the two graphs shown in Fig. 2.3.13:

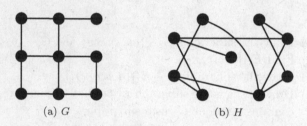

(a) G (b) H

Fig. 2.3.13

Do they have the same degree sequence in non-increasing order? Are they isomorphic?

7. Let G be a graph of order five satisfying the following condition: for any three vertices x, y, z in G, $[\{x, y, z\}]$ is not isomorphic to either O_3 or K_3. What is the graph G? Justify your answer.

8. Draw all non-isomorphic graphs of order 5 which contain a C_5.

9. Let H be a spanning subgraph of a graph G. Which of the following statements is/are true? Why?

   (i) If G is connected, then H is connected.
   (ii) If H is connected, then G is connected.

10. Let G be a disconnected graph with k components. Choose a vertex from each component. What is the subgraph induced by these k vertices?

11. Let H be a spanning subgraph of a graph G. Show that $c(H) \geq c(G)$.

12. Let C be a cycle and S a subset of $V(C)$. Show that $c(C-S) \leq |S|$.

13. Let $G = K_n$. Find

   (i) $n_G(C_3)$, where $n \geq 3$;
   (ii) $n_G(C_4)$, where $n \geq 4$; and
   (iii) $n_G(C_k)$, where $n \geq k \geq 5$.

14. Let G be the Petersen graph. Find $n_G(C_i)$, where $i = 3, 4, 5$. What is the largest cycle in G?

15. Let G be a graph of order 5 which contains at least two vertices of degree 4 and a C_5. Find all possible degree sequences of G, in non-increasing order; and for each case, construct all such G.

16. Let G be a connected graph. An edge e is called a **bridge** if $G - e$ is disconnected.

   (i) Find all bridges in the following graph:

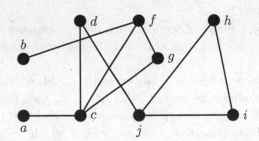

Fig. 2.3.14

   (ii) How many components does $G - e$ have if e is a bridge in G?
   (iii) Show that an edge e in G is a bridge if and only if e is not contained in any cycle in G.

17. Let G be a connected graph in which every vertex is even. Show that G contains no bridges.

18. Let G be a connected graph. A vertex w in G is called a **cut-vertex** if $G - w$ is disconnected.

(i) Find all cut-vertices in the graph shown in Fig. 2.3.14.

(ii) How many components does $G - w$ have if w is a cut-vertex of G?

(iii) Assume that $v(G) \geq 3$. Show that if G contains a bridge, then G contains a cut-vertex.

(iv) Is the converse of (iii) true?

19. Let w be a cut-vertex of a graph G with $d(w) \leq 3$. Show that w is incident with a bridge in G. Is the result still true if $d(w) = 4$?

20. Let G be a connected $(8, 12)$-graph which contains no bridges. Suppose that $\Delta(G) = 4$ and G has exactly two vertices of degree 4.

(i) Find the number of end-vertices in G.

(ii) Find the number of vertices of degree 3 in G.

(iii) Construct three such graphs which are non-isomorphic.

21. Let G be a graph with $\delta(G) = k \geq 2$. Show that there exist at least two vertices in G such that each of them is contained in $\binom{k}{2}$ cycles. (See also Problem I.8(ii).)

22. Let G be a graph of order 9. Assume that $\Delta(G) = 6$ and that G contains at least 4 vertices of degree at least 4. Show that G contains a C_3.

23. Let G be a graph of order n with degree sequence $(d_1, d_2, ..., d_n)$. Construct a graph from G having the degree sequence $(d_1 + 1, d_2 + 1, ..., d_n + 1, n)$.

24. Let G be a connected graph of order n. Show that the vertices in G can always be named as $x_1, x_2, ..., x_n$ such that the induced subgraph $[\{x_1, x_2, ..., x_i\}]$ is connected for each $i = 1, 2, ..., n$.

25. Let G be a connected graph of order 8 which contains two C_4's having no vertex in common.

(i) What is the least possible value of $e(G)$?

(ii) Assume that G contains no cut-vertices. What is the least possible value of $e(G)$?

(iii) Assume that G contains no odd vertices. What is the least possible value of $e(G)$?

(iv) Assume that G contains no even vertices. What is the least possible value of $e(G)$?

For each of the above cases, construct a corresponding G which has its $e(G)$ attaining your least possible value.

26. Let G be a graph with $V(G) = \{x_1, x_2, x_3, x_4\}$ such that

   (i) $G - x_1 \cong H_1$,
   (ii) $G - x_2 \cong H_2$,
   (iii) $G - x_3 \cong H_3$,
   (iv) $G - x_4 \cong H_3$,

   where H_1, H_2 and H_3 are shown in Fig. 2.3.15.

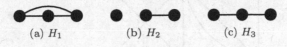

(a) H_1 (b) H_2 (c) H_3

Fig. 2.3.15

Determine G and justify your answer.

27. Let G be a graph with $V(G) = \{y_1, y_2, y_3, y_4, y_5\}$ such that

   (i) $G - y_1 \cong H_1$,
   (ii) $G - y_2 \cong H_2$,
   (iii) $G - y_3 \cong H_3$,
   (iv) $G - y_4 \cong H_4$,
   (v) $G - y_5 \cong H_3$,

   where H_1, H_2, H_3 and H_4 are shown in Fig. 2.3.16.

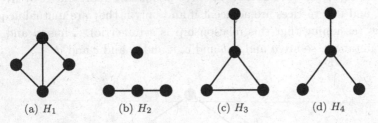

(a) H_1 (b) H_2 (c) H_3 (d) H_4

Fig. 2.3.16

Determine G and justify your answer.

28. Let G be a graph with $V(G) = \{u_1, u_2, ..., u_n\}$, where $n \geq 3$. Let $m = e(G)$, $m_i = e(G - u_i)$, $i = 1, 2, ..., n$. Show that

   (i) the degree of u_i in G is equal to $m - m_i$, $i = 1, 2, ..., n$;
   (ii) $m = (m_1 + m_2 + ... + m_n)/(n - 2)$.

29. Let G be a connected multigraph of order at least two and A be

a subset of $V(G)$. Denote by $e(A, V(G)\backslash A)$ the number of edges having one end in A and the other in $V(G)\backslash A$.

(i) Let H be the multigraph shown in Fig. 2.3.17 and $A = \{u, v, z\}$. Find $e(A, V(H)\backslash A)$.

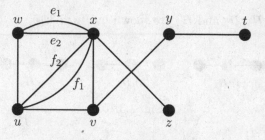

Fig. 2.3.17

(ii) Show that $e(A, V(G)\backslash A)$ is even if and only if A contains an even number of odd vertices in G.

2.4   The Complement of a Graph

The graph of Fig. 2.4.1 shows the 'acquaintance' relationship among a group of five people: a, b, c, d and e. The five people are represented by five vertices, and two vertices are adjacent if and only if they are mutual acquaintances (assuming that this relationship is symmetric). Thus, a and b are acquaintances, so are a and e, b and e, e and c, and c and d.

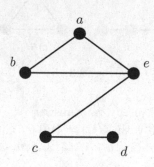

Fig. 2.4.1

Based on the graph of Fig. 2.4.1, we construct a new graph on the same vertex set in Fig. 2.4.2 which now shows the 'stranger' relationship among the five.

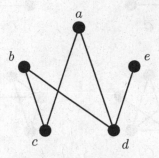

Fig. 2.4.2

What is the relationship between these two graphs?

Firstly, they have the **same vertex set**. Secondly, if two vertices are **adjacent** in the first graph (such as a and b), then they are **not adjacent** in the second; and if two vertices are **not adjacent** in the first (such as a and c), then they are **adjacent** in the second. We call the second graph the **complement** of the first, and vice versa.

For a given graph G, the **complement** of G, denoted by $\overline{G}$, is the graph with $V(\overline{G}) = V(G)$ such that two vertices are adjacent in $\overline{G}$ if and only if they are not adjacent in G.

Thus $v(G) = v(\overline{G})$. If $v(G) = 1$, then note that $\overline{G} \cong G \cong K_1$.

Example 2.4.1. Figure 2.4.3 displays six graphs G of order 5. The complements of the first three are shown. You are invited to construct the complements of the remaining three.

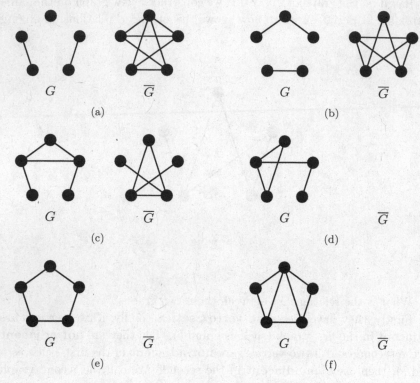

Fig. 2.4.3

Question 2.4.2.

(i) What is $\overline{O_n}$?

(ii) What is $\overline{K_n}$?

(iii) Define $\overline{\overline{G}}$ as $\overline{(\overline{G})}$. What is $\overline{\overline{G}}$?

(iv) Suppose G and H are two graphs such that $G \cong H$. Is it true that $\overline{G} \cong \overline{H}$? (See Problem 2.4.3.)

(v) If G is a graph with degree sequence $(4, 4, 3, 3, 2, 2)$, what is the degree sequence of $\overline{G}$, in non-increasing order?

(vi) Let G be a 3-regular graph of order 10. Is $\overline{G}$ also regular? If 'yes', what is the degree of each vertex in $\overline{G}$?

(vii) In Example 2.4.1, for each pair $\{G, \overline{G}\}$, if we superimpose $\overline{G}$ onto G so that the same vertices are identified, what is the resulting graph? Is it a K_5?

(viii) *What is the sum $e(G) + e(\overline{G})$ for each of the six graphs G in Example 2.4.1?*

In general, given a graph G of order n, by superimposing $\overline{G}$ onto G so that the same vertices are identified, we would obtain the complete graph K_n. Thus, we have:

Observation 4: For any graph G of order n, $e(G) + e(\overline{G}) = e(K_n) = \binom{n}{2}$.

Question 2.4.3. *In Fig. 2.4.3, we notice that the first graph is disconnected but its complement becomes connected. Consider the disconnected graph H in Fig. 2.4.4.*

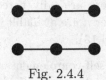

Fig. 2.4.4

Construct $\overline{H}$. Is $\overline{H}$ connected?

Indeed, we have the following general result.

Theorem 2.4.4. *Let G be a graph. If G is disconnected, then $\overline{G}$ is connected.*

Proof. Assume that G is disconnected. To show that $\overline{G}$ is connected, we show that every two vertices in $\overline{G}$ are joined by a path in $\overline{G}$. Thus, let $u, v \in V(\overline{G})(= V(G))$. If u, v are in different components in G, then u and v are joined by an edge in $\overline{G}$. If u and v are in the same component of G, let w be any vertex in another component of G, then uwv is a $u - v$ path in $\overline{G}$. This completes the proof.

$\square$

Question 2.4.5. *For a disconnected graph G, the above proof not just shows that $\overline{G}$ is connected, but further reveals that the distance between any two vertices is very small in $\overline{G}$. How small is it? (See Problem 2.4.4.)*

It is interesting to note that there are graphs G such that $\overline{G} \cong G$.

Question 2.4.6. *Which of the graphs G in Fig. 2.4.3 are such that $\overline{G} \cong G$?*

Graphs of this type have attracted some researchers' attention.

A graph G is said to be **self-complementary** if $\overline{G} \cong G$.

Thus, the graphs (c) and (e) in Fig. 2.4.3 are self-complementary.

Question 2.4.7. *Construct a self-complementary graph of order n, where $2 \leq n \leq 4$.*

Self-complementary graphs are quite rare. Indeed, among all graphs of order n, $2 \leq n \leq 7$, there are only three such graphs. Also, self-complementary graphs of order n are available only for some values of n as stated below (see Problem 2.4.13).

Observation 5: Let G be a self-complementary graph. Then

   (i) G is connected and
 • (ii) $v(G) = 4k$ or $v(G) = 4k + 1$ for some integer k.

Exercise for Section 2.4

1. Consider Problem 2.2.2. Is there any relation between the family of graphs found in (i) and the family of graphs in (ii)?
2.   (i) Draw all non-isomorphic $(6,3)$-graphs.
     (ii) Find the number of non-isomorphic $(6,12)$-graphs.
3. Let G and H be two graphs. Show that $G \cong H$ if and only if $\overline{G} \cong \overline{H}$.
4. Let G be a disconnected graph. Show that $\text{diam}(\overline{G}) \leq 2$.
5. Let G be a k-regular graph of order n, where $n > k \geq 1$. Is $\overline{G}$ also regular? If 'yes', what is the degree of each vertex in $\overline{G}$?
6. Let G be a graph of order $n \geq 2$ with degree sequence $(d_1, d_2, ..., d_n)$ in non-increasing order. Find the degree sequence of $\overline{G}$ in non-increasing order.
7. Draw all non-isomorphic 4-regular graphs of order 7.
8. How many non-isomorphic graphs are there with degree sequence $(5,5,4,4,4,4)$? Construct one such graph.
9. How many non-isomorphic graphs are there with degree sequence $(5,5,5,4,4,3)$? Construct one such graph.
10. What is the value of each diagonal entry in the matrix $\boldsymbol{A}(G)\boldsymbol{A}(\overline{G})$?

11. For each of the graphs in Fig. 2.4.5,

    (i) construct its complement and

    (ii) determine if it is self-complementary.

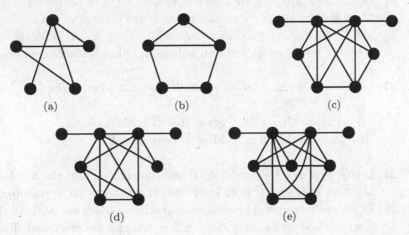

(a)   (b)   (c)

(d)   (e)

Fig. 2.4.5

12. Show that every self-complementary graph is connected.

13. Let G be a self-complementary graph of order $n \geq 2$. Show that

    (i) $e(G) = \frac{1}{4}n(n-1)$ and

    (ii) $n = 4k$ or $n = 4k + 1$ for some positive integer k.

14. Determine the values of n, where $n \geq 3$, for which the cycle C_n is self-complementary.

15. Let G be a self-complementary graph of order n. Show that if G is regular, then $n = 4k + 1$ for some positive integer k.

16. Construct a regular self-complementary graph of order 9.

17. (i) Let G be a self-complementary graph of order 9. Show that G contains at least one vertex of degree 4.

    (ii) Generalize the result in (i).

18. Let G be a graph and x be a vertex in G.

    (i) Is it true that $\overline{G - x} \cong \overline{G} - x$?

    (ii) If x is a cut-vertex of G, is $\overline{G} - x$ connected?

    Justify your answers.

19. (i) Show that at a gathering of any six persons, some three of them are either mutual acquaintances or complete strangers

to one another.

(ii) Does the result in (i) still hold for 'five' persons?

20. Let G be a graph of order 6. If G does not contain O_3 as an induced subgraph, what is the least possible value for $n_G(C_3)$?

21. Let G be a graph with $\Delta(G) \geq r$, where r is a positive integer. Show that either G contains a triangle or $\overline{G}$ contains a K_r.

22. Let G be a graph of odd order and $\delta(G) \geq 5$. Assume that G contains no O_3 as an induced subgraph. Show that G contains a K_4.

23. Let G be a graph of order n which contains no triangles.

    (i) Assume that $n = 9$. Show that $\overline{G}$ contains a K_4.
    (ii) Assume that $n = 8$. Must $\overline{G}$ contain a K_4?

24. Let G be a graph of order 5. Prove that the sum of the numbers of cycles in G and $\overline{G}$ is at least two. Show that 'two' is possible.

25. 17 people correspond by mail with another — each one with all the rest. In their letters only three different topics are discussed. Each pair of correspondents deals with only one of these topics. Prove that there are at least three people who write to each other about the same topic. (IMO 1964/4)

2.5   Graphic Sequences

Any graph G of order n with $V(G) = \{v_1, v_2, ..., v_n\}$ has its **degree sequence** defined to be $(d(v_1), d(v_2), ..., d(v_n))$. We usually assume that the sequence is non-increasing, that is, $d(v_1) \geq d(v_2) \geq ... \geq d(v_n)$.

A non-increasing sequence $\boldsymbol{d} = (d_1, d_2, ..., d_n)$ of non-negative integers is called a **graphic sequence** if $\boldsymbol{d}$ is the degree sequence of some graph. If $\boldsymbol{d}$ is the degree sequence of a graph G, we say that $\boldsymbol{d}$ is **representable** by G or G **represents** $\boldsymbol{d}$.

Remark 2.5.1. All graphs considered for graphic sequences need not be connected, but are always assumed to be simple.

Example 2.5.2.   Which of the following sequences are graphic?

(i) $(0,0,0,0)$ (ii) $(1,0,0,0)$

(iii) $(1,1,0,0)$ (iv) $(1,1,1,0)$

(v) $(1,1,1,1)$ (vi) $(2,1,1,0)$

(vii) $(2,2,0,0)$ (viii) $(2,1,1,1,0)$

(ix) $(2,1,1,1,1)$ (x) $(3,2,2,1,1)$

(xi) $(4,3,2,1,0)$ (xii) $(4,3,2,2,1)$

It is clear that the sequences (ii), (iv), (viii) and (x) are non-graphic. The reason is that the number of odd vertices in any graph must be even.

The sequence (vii) is not graphic. For if it were representable by, say G, then G would have two isolated vertices, and thus have its remaining two vertices joined by '2' parallel edges, which is not allowed for 'simple' graphs.

Consider the sequence (xi). Though the number of odd numbers (namely, '3' and '1') in (xi) is two, which is even, the sequence is non-graphic. Why? (See Problem 1.6.15.)

The remaining sequences are graphic, and the graphs representing them are shown in Fig. 2.5.1.

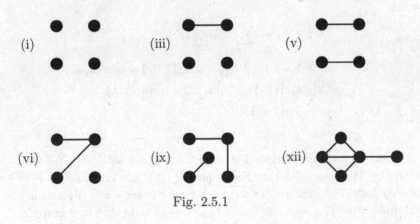

Fig. 2.5.1

Two very natural questions arise: Which sequence d is graphic? How do we determine whether a given d is graphic?

Erdös and Gallai (1960) gave the following first set of necessary and sufficient conditions for a sequence to be graphic.

Theorem 2.5.3. *Let $d = (d_1, d_2, ..., d_n)$ be a non-increasing sequence of non-negative integers. Then d is graphic if and only if* *(i)* $\sum_{1 \leq i \leq n} d_i$ *is even and*

(ii) $\sum_{1 \leq i \leq k} d_i \leq k(k-1) + \sum_{k+1 \leq i \leq n} \min\{k, d_i\}$ *for each* $k = 1, 2, ..., n$.

$\square$

Example 2.5.4.   Determine if the following sequences are graphic by Theorem 2.5.3.

(i) $(4, 3, 3, 1, 1)$,
(ii) $(4, 3, 2, 2, 1)$.

Consider the sequence $(4, 3, 3, 1, 1)$ where $n = 5$. Clearly, the condition (i) in Theorem 2.5.3 is satisfied. For condition (ii), by letting $k = 2$, it is checked that

$$\sum_{1 \leq i \leq 2} d_i = d_1 + d_2 = 4 + 3 = 7$$

while

$$k(k-1) + \sum_{3 \leq i \leq 5} \min\{k, d_i\}$$
$$= 2 \times 1 + \min\{2, d_3\} + \min\{2, d_4\} + \min\{2, d_5\}$$
$$= 2 + \min\{2, 3\} + \min\{2, 1\} + \min\{2, 1\}$$
$$= 2 + 2 + 1 + 1$$
$$= 6.$$

That is, condition (ii) in Theorem 2.5.3 is not satisfied. We thus conclude by Theorem 2.5.3 that the sequence (i) is not graphic.

We leave it to the reader to check that the sequence (ii) satisfies both (i) and (ii) in Theorem 2.5.3, and thus to conclude that it is graphic.

The second necessary and sufficient condition for a sequence to be graphic was found by Havel (1955) and Hakimi (1962) independently. The result is recursive in nature. Before establishing it, we first introduce a sequence $\boldsymbol{d}^*$ with $n - 1$ terms from a given sequence $\boldsymbol{d}$ with n terms.

Thus, given $\boldsymbol{d} = (d_1, d_2, ..., d_n)$, define

$$\boldsymbol{d}^* = (d_2 - 1, d_3 - 1, ..., d_{d_1+1} - 1, d_{d_1+2}, ..., d_n).$$

For instance, if $d = (3,3,3,2,2,1)$, then $d_1 + 1 = 4$, and so

$$d^* = (d_2 - 1, d_3 - 1, d_4 - 1, d_5, d_6)$$
$$= (3-1, 3-1, 2-1, 2, 1)$$
$$= (2,2,1,2,1).$$

In general, to obtain d^* from d, we first delete the first term 'd_1', then subtract '1' from each of the next 'd_1' terms, and leave the remaining terms unchanged. Note that d^* may not be non-increasing. With this explanation, it should be easier to understand the following result due to Havel and Hakimi.

Theorem 2.5.5.   *Let $d = (d_1, d_2, ..., d_n)$ be a non-increasing sequence of non-negative integers. Then d is graphic if and only if d^* is a permutation of a graphic sequence.*

Proof. ($\Leftarrow$) Let H be a graph with $n-1$ vertices $v_2, v_3, ..., v_n$ whose degree sequence is

$$d^* = (d_2 - 1, d_3 - 1, ..., d_{d_1+1} - 1, d_{d_1+2}, ..., d_n).$$

Form a new graph G from H by adding a new vertex v_1 adjacent to each of $v_2, v_3, ..., v_{d_1+1}$. Then d is a degree sequence of G.

($\Rightarrow$) Let G be a graph with vertices $v_1, v_2, ..., v_n$ such that $d(v_i) = d_i$ for $i = 1, 2, ..., n$. We claim that G can be chosen such that v_1 is adjacent to each vertex in $A = \{v_2, v_3, ..., v_{d_1+1}\}$. Suppose this is not the case. Let G be chosen among all graphs with $d(v_i) = d_i$ such that v_1 is adjacent to as many vertices in A as possible. By assumption, v_1 is not adjacent to some vertex v_j in A. Thus v_1 is adjacent to a vertex v_k not in A. Since $d_j \geq d_k$, there exists a vertex v_r such that

$$v_j v_r \in E(G) \text{ but } v_k v_r \notin E(G).$$

Form a new graph G' from G by deleting the edges $v_1 v_k$ and $v_j v_r$ but adding two new edges $v_1 v_j$ and $v_k v_r$. (See Fig. 2.5.2. Note that it is not important whether v_r belongs to A or not. In the diagram, we assume $v_r \notin A$.)

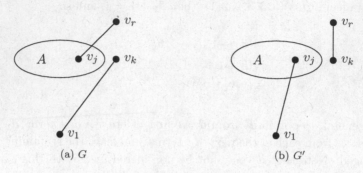

(a) G (b) G'

Fig. 2.5.2

It is evident that d is also a degree sequence of G'. Observe that in G', v_1 is adjacent to more vertices in A as compared to those in G. This however, contradicts the choice of G. We thus conclude that G can be chosen so that v_1 is adjacent to all vertices in A. Let $H = G - v_1$. It is now clear that d^* is a permutation of the degree sequence of H.

$\square$

Example 2.5.6.   Apply Theorem 2.5.5 to determine whether the two sequences in Example 2.5.4 are graphic.

  (i) Let $d = (4, 3, 3, 1, 1)$. Then $d^* = (2, 2, 0, 0)$. As d^* is not graphic (see (vii) in Example 2.5.2), by Theorem 2.5.5, d is not graphic.
 (ii) Let $d = (4, 3, 2, 2, 1)$. Then $d^* = (2, 1, 1, 0)$. As d^* is graphic (see (vi) in Example 2.5.2), by Theorem 2.5.5, d is also graphic.

Comparing the discussions given in Example 2.5.4 and Example 2.5.6, it appears to us that Theorem 2.5.5 is more practical than Theorem 2.5.3 as far as determination of graphic sequences is concerned. Indeed, based on Theorem 2.5.5, an efficient algorithm for determining if a given sequence is graphic can naturally be derived as shown in Algorithm 2.5.7.

Algorithm 2.5.7.   Let $d = (d_1, d_2, ..., d_n)$ be a sequence of n integers.

Step 1: If $d_i \geq n$ for some $i = 1, 2, ..., n$, then d is not graphic.

Step 2: If each term in the current sequence is '0', then d is graphic.

Step 3: If there is a 'negative' term in the current sequence, then d is not graphic.

Step 4: Arrange the current sequence in non-increasing order.

Step 5: Let 'r' be the first term of the current sequence. Form a new sequence by deleting 'r' and subtracting '1' from each of the next r terms. Go to Step 2.

Example 2.5.8. Determine if the sequence $(5, 4, 4, 3, 1, 1)$ is graphic.
By applying Algorithm 2.5.7, we have:

$$
\begin{array}{cccccc}
5 & 4 & 4 & 3 & 1 & 1 \\
  & 3 & 3 & 2 & 0 & 0 \\
  &   & 2 & 1 & -1 & 0 \\
\end{array}
$$

As the last sequence contains a negative term (namely, -1), the given sequence is not graphic.

Example 2.5.9. Determine if the sequence $(4, 4, 3, 2, 2, 1)$ is graphic.
By applying Algorithm 2.5.7, we have:

$$
\begin{array}{cccccc}
4 & 4 & 3 & 2 & 2 & 1 \\
  & 3 & 2 & 1 & 1 & 1 \\
  &   & 1 & 0 & 0 & 1 \\
  &   & 1 & 1 & 0 & 0 \\
  &   &   & 0 & 0 & 0 \\
\end{array}
$$

As the last sequence consists of '0' only, the given sequence is graphic.

Remark 2.5.10.

(1) When applying Algorithm 2.5.7 manually, the procedure may terminate at the place where you know if the current sequence is graphic. For instance, in Example 2.5.9, we already have the answer when the current sequence $(1, 1, 0, 0)$ (see (iii) in Example 2.5.2) is obtained.

(2) When applying Algorithm 2.5.7, remember in Step 4 that the current sequence should be sorted in non-increasing order first before performing Step 5.

Knowing that a given sequence d is graphic (say, by Algorithm 2.5.7), one may naturally wish to find a way to construct a graph which represents

d. How can this be done?

Indeed, the proof of the implication ($\Leftarrow$) in Theorem 2.5.5 shows us a way to construct a graph G representing $\boldsymbol{d}$ from a graph G^* representing $\boldsymbol{d}^*$.

That is, if G^* is a graph representing

$$\boldsymbol{d}^* = (d_2 - 1, d_3 - 1, ..., d_{d_1+1} - 1, d_{d_1+2}, ..., d_n),$$

let G be the graph obtained from G^* by adding a new vertex w and d_1 new edges joining w to all the d_1 vertices of degrees $d_2 - 1, d_3 - 1, ..., d_{d_1+1} - 1$. Then G represents $\boldsymbol{d} = (d_1, d_2, ..., d_n)$.

With this idea in mind, one may reverse the procedure of Algorithm 2.5.7 repeatedly to obtain a graph representing $\boldsymbol{d}$ from a graph representing a shorter current sequence.

Example 2.5.11. By Example 2.5.9, the sequence $(4, 4, 3, 2, 2, 1)$ is graphic. Construct a graph representing it.

The following is what we have obtained by applying Algorithm 2.5.7:

$$
\begin{array}{cccccc}
4 & 4 & 3 & 2 & 2 & 1 \\
  & 3 & 2 & 1 & 1 & 1 \\
  &   & 1 & 0 & 0 & 1 \\
  &   & 1 & 1 & 0 & 0 \\
\end{array}
$$

It is clear that $(1, 1, 0, 0)$ is representable by the graph G_1 shown in Fig. 2.5.3.

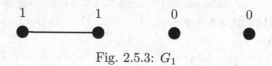

Fig. 2.5.3: G_1

A graph representing $(3, 2, 1, 1, 1)$, shown in Fig. 2.5.4, can then be obtained using the method mentioned above.

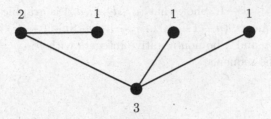

Fig. 2.5.4: G_2

Repeating this procedure, we finally obtain the following graph G representing the sequence $(4, 4, 3, 2, 2, 1)$

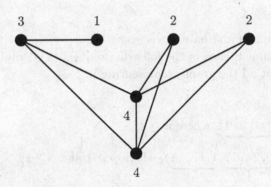

Fig. 2.5.5: G

Exercise for Section 2.5

1. Determine if each of the following sequences is graphic, and if it is so, construct a graph representing it.

    (i) $(5, 4, 3, 3, 2, 2, 1)$
   (ii) $(5, 4, 4, 4, 3, 2, 2)$
  (iii) $(5, 5, 4, 4, 3, 2, 1)$
   (iv) $(6, 5, 4, 3, 3, 3, 3, 1)$
    (v) $(4, 4, 3, 3, 3, 3, 2, 2, 2)$
   (vi) $(9, 7, 6, 4, 3, 3, 3, 1, 1, 1)$
  (vii) $(8, 6, 6, 5, 5, 5, 4, 3, 3, 3, 2)$

2. Let $(d_1, d_2, ..., d_n)$ be a sequence of non-increasing integers, where

$0 \leq d_i \leq n - 1$. Show that $(d_1, d_2, ..., d_n)$ is graphic if and only if the sequence $(n - 1 - d_n, ..., n - 1 - d_2, n - 1 - d_1)$ is graphic.

3. Let n, p and q be non-negative integers with $p + q = n \geq 3$. Show that the sequence

$$(\underbrace{n - 2, n - 2, ..., n - 2}_{p}, \underbrace{n - 3, n - 3, ..., n - 3}_{q})$$

is graphic if and only if p is even.

4. Determine if each of the following sequences is graphic. If it is so, find out all the graphs representing it.

   (i) $(\underbrace{1, 1, ..., 1}_{n})$, where $n \geq 2$;

   (ii) $(\underbrace{2, 2, ..., 2}_{p}, \underbrace{1, 1, ..., 1}_{q})$, where $p \geq 1$ and $q \geq 1$;

   (iii) $(\underbrace{2, 2, ..., 2}_{n})$, where $n \geq 3$;

   (iv) $(3, \underbrace{2, 2, ..., 2}_{k}, 1)$, where $k \geq 2$;

   (v) $(3, 3, \underbrace{2, 2, ..., 2}_{k})$, where $k \geq 2$;

5. (i) Given that the sequence $(6, 5, 4, 3, 2, 1, 0, d)$ is a permutation of a graphic sequence, find the value of d.

   (ii) Given that the sequence $(2p, 2p-1, 2p-2, ..., 2, 1, 0, d)$ is a permutation of a graphic sequence, where p is a positive integer, find the value of d.

6. Let G be a graph of order p and H, a graph of order q. The **join** of G and H, denoted by $G + H$, is the graph of order $p + q$ obtained from the disjoint union of G and H by joining each vertex in G to each vertex in H. An example is shown in Fig. 2.5.6:

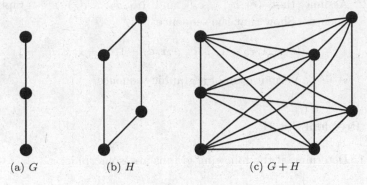

(a) G (b) H (c) $G + H$

Fig. 2.5.6

Let $(c_1, c_2, ..., c_p)$ be the degree sequence of G and $(d_1, d_2, ..., d_q)$, the degree sequence of H. Find the degree sequence of $G + H$, expressed in terms of c_i's and d_j's.

7. Let $(d_1, d_2, ..., d_n)$ be a graphic sequence. Show that the sequence
$$(2d_1, d_2, d_2, d_3, d_3, ..., d_n, d_n)$$
is also graphic.

8. Let $(d_1, d_2, ..., d_n)$ be a graphic sequence. Show that the sequence
$$(d_1 + 1, d_1 + 1, d_2 + 1, d_2 + 1, ..., d_n + 1, d_n + 1)$$
is also graphic.

9. Let $(d_1, d_2, ..., d_n)$ be a graphic sequence. Show that, for each $k \geq 1$, the sequence
$$(d_1 + k, d_2 + k, ..., d_n + k, \underbrace{n, n, ..., n}_{k})$$
is also a permutation of a graphic sequence.

10. Let $(d_1, d_2, ..., d_n)$ be a graphic sequence. Show that, for each $k \geq 1$, the sequence
$$(d_1 + k, d_2 + k, ..., d_n + k, \underbrace{n + k - 1, n + k - 1, ..., n + k - 1}_{k})$$
is also a permutation of a graphic sequence.

11. For any integer $p \geq 3$, construct a connected graph of order $2p$ with degree sequence
$$(\underbrace{p, p, ..., p}_{p}, \underbrace{3, 3, ..., 3}_{p}).$$

12. Assume that $(c_1, c_2, ..., c_n)$ and $(d_1, d_2, ..., d_n)$ are graphic sequences. Show that the sequence

$$(c_1 + 1, c_2 + 1, ..., c_n + 1, d_1 + 1, d_2 + 1, ..., d_n + 1)$$

is also a permutation of a graphic sequence.

2.6   Problem Set II

1. Determine if the following graphs are isomorphic.

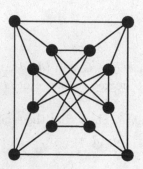 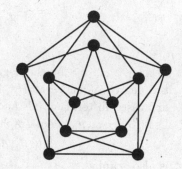

Fig. 2.6.1

2. Recall that $c(G)$ denotes the number of components of a graph G.

   (i) Show that $c(G) \le c(G - e) \le c(G) + 1$ for each edge e in G.
   (ii) Show that $c(G) \le c(G - v)$ for each vertex v in G.
   (iii) For each postitive integer k, construct a graph G such that $c(G - v) = c(G) + k$ for some vertex v in G.

3. Let G be a graph in which every vertex is a central vertex. Show that G contains no cut-vertex.

4. Let H be a graph. Show that there exists a graph G such that $H \cong C(G)$, where $C(G)$ is the center of G.

5. Let G be a non-trivial graph in which any two vertices are contained in a common cycle. Show that

   (a) G contains no cut-vertex;
   (b) $e(G) \ge 2\mathrm{diam}(G)$.

   What can be said about the structure of G when $e(G) = 2\mathrm{diam}(G)$?

6. [**The girth of a graph**] The **girth** of a graph, denoted by $g(G)$, is defined as the order (or length) of a shortest cycle in G if G contains a cycle; and $g(G) = \infty$, otherwise.

   (a) Let G be a k-regular graph with $g(G) = 5$. Show that $v(G) \geq k^2 + 1$.

   (b) For $k = 2, 3$, construct a k-regular graph G with $g(G) = 5$ and $v(G) = k^2 + 1$.

   (c) Let G be a k-regular graph with $g(G) = 5$ and $\text{diam}(G) = 2$. Show that $v(G) = k^2 + 1$.

7. (a) Let G be a graph with $g(G) \geq 4$. Show that $v(G) \geq \Delta(G) + \delta(G)$.

   (b) Is the inequality in (a) still valid if $g(G) = 3$?

   (c) For each $n \geq 4$, construct a graph H of order n with $g(H) = 4$ such that $\Delta(H) + \delta(H) = n$.

8. [**Chord**] Let G be a graph of order n and C_k a cycle of order k in G, where $n \geq k \geq 4$. A **chord** of C_k is an edge in G which joins two non-consecutive vertices along C_k.

   (a) Show that if $e(G) \geq 2n - 3$, then G contains a cycle with a chord.

   (b) Does the conclusion in (a) still hold if $e(G) = 2n - 4$?

9. [**Powers of graphs**] Let G be a connected graph and k, a positive integer. The kth **power** of G, denoted by G^k, is the graph defined by

$$V(G^k) = V(G) \quad \text{and} \quad E(G^k) = \{uv \mid d_G(u, v) \leq k\}.$$

We call G^2 the **square** of G, and G^3, the **cube** of G.

   (a) What can you say about the graph G^k, where $k = \text{diam}(G)$?

   (b) Show that

   (i) for any integers p, q with $q \geq p \geq 1$, G^p is a spanning subgraph of G^q;

   (ii) if H is a spanning subgraph of G; then H^p is a spanning subgraph of G^p.

10. [**Strongly regular graphs**] A graph G of order n is said to be **strongly regular** if there exist non-negative integers k, λ, μ such that the following two conditions are satisfied:

   (1) G is k-regular and

(2) for any two distinct vertices u and v in G,

$$|N(u) \cap N(v)| = \begin{cases} \lambda \text{ if } u \text{ and } v \text{ are adjacent,} \\ \mu \text{ otherwise.} \end{cases}$$

The numbers n, k, λ and μ are called the **parameters** of G.

(a) Determine if the following graphs are strongly regular:
    (i) C_4,   (ii) C_5,   (iii) the *Petersen* graph,
    (iv) O_n,   (v) K_n.
    Find also their parameters if they are strongly regular.

(b) Construct a strongly regular graph with parameters $(n, k, \lambda, \mu) = (6, 4, 2, 4)$.

(c) Let G be a strongly regular graph with parameters (n, k, λ, μ).

    (i) Prove that $\overline{G}$ is also strongly regular. Find also the parameters of $\overline{G}$ in terms of n, k, λ and μ.

    (ii) Show that $k(k - \lambda - 1) = (n - k - 1)\mu$, where $\mu \neq 0$.

(Bose (1963))

11. **[The line graph of a graph]** Let G be a graph with no isolated vertices. The **line graph** of G, denoted by $L(G)$, is the graph with vertex set $V(L(G)) = E(G)$ such that two vertices e and f in $L(G)$ are adjacent if and only if e and f are adjacent edges in G. An example is shown in Fig. 2.6.2:

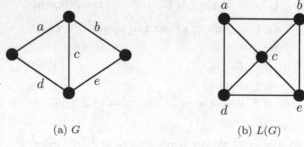

(a) G (b) $L(G)$

Fig. 2.6.2

(a) Construct the line graph of each of the following graphs:

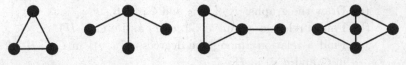

Fig. 2.6.3

(b) Construct two graphs G and H such that $G \not\cong H$ but $L(G) \cong L(H)$.

(c) Let G be an (n, m)-graph. Show that $L(G)$ is of order m and size

$$\frac{1}{2}\left(\sum_{v \in V(G)} d(v)^2\right) - m.$$

(d) Let $e = uv$ be an edge in G. Show that the degree of e in $L(G)$ is given by $d(u) + d(v) - 2$.

(e) Show that if G is regular, then so is $L(G)$. Is the converse true?

(f) Show that G is connected if and only if $L(G)$ is connected.

12. [**The Cartesian product of graphs**] Let G be a graph of order p and H, a graph of order q. The **Cartesian product** of G and H, denoted by $G \,\square\, H$, is the graph of order $p \times q$ with vertex set

$$V(G) \times V(H) = \{(u, v) \mid u \in V(G) \text{ and } v \in V(H)\}$$

such that two vertices (u, v), (x, y) are adjacent if and only if *either* $u = x$ in G and vy in $E(H)$ or ux in $E(G)$ and $v = y$ in H. An example is shown in Fig. 2.6.4:

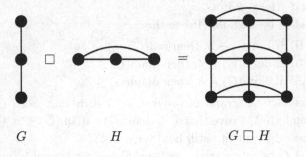

G H $G \,\square\, H$

Fig. 2.6.4

(a) Draw the graphs: $K_2 \,\square\, C_5$ and $C_4 \,\square\, K_4$.

(b) Find a relation among $e(G)$, $e(H)$ and $e(G \,\square\, H)$.

(c) Find a relation among the degrees $d((u,v))$ in $G \,\square\, H$, $d(u)$ in G and $d(v)$ in H.

(d) Find a relation among $\delta(G)$, $\delta(H)$ and $\delta(G \,\square\, H)$.

(e) Find a relation among $\Delta(G)$, $\Delta(H)$ and $\Delta(G \,\square\, H)$.

(f) Show that if G and H are regular, then so is $G \,\square\, H$.

(g) Is the converse of the result in (f) true?

13. Let G be a graph of order p with degree sequence $(c_1, c_2, \cdots, c_p)$ and H, a graph of order q with degree sequence $(d_1, d_2, \cdots, d_q)$. Express the degree of each vertex of $G \,\square\, H$ in terms of c_i's and d_j's.

14. Show that the Cartesian product of the p-cube graph Q_p and the q-cube graph Q_q, where $p \geq q \geq 1$, is the $(p+q)$-cube graph. Deduce that for $n \geq 1$, $Q_n \cong K_2^n$, where

$$K_2^n = \overbrace{((\cdots(K_2 \,\square\, K_2) \,\square\, \cdots) \,\square\, K_2)}^{n}.$$

15. Let G be a graph of order n, where $n = 2k \geq 4$. Show that if $\delta(G) \geq k$, then G contains a C_4 as a subgraph.

16. Let G be a graph of order $n \geq 4$ in which the subgraph induced by any four vertices contains a C_3.

(i) Show that $e(G) \geq \binom{n-1}{2}$.

(ii) Characterize such G for which the equality in (i) holds.

17. Let w be a vertex in a graph G. Show that $e(w) \geq 3$ in G if and only if w is adjacent to a vertex z in $\overline{G}$ such that $N(w) \cup N(z) = V(\overline{G})$ and $e(w) \leq 2$ in $\overline{G}$.

18. Let G be a graph. Prove that

(i) if $\mathrm{rad}(G) \geq 3$, then $\mathrm{rad}(\overline{G}) \leq 2$;

(ii) if $\mathrm{diam}(G) \geq 4$, then $\mathrm{diam}(\overline{G}) \leq 2$;

(iii) if $\mathrm{diam}(G) \geq 3$, then $\mathrm{diam}(\overline{G}) \leq 3$.

19. Let G be a graph of order $n \geq 5$ such that both G and $\overline{G}$ are connected. Prove that $4 \leq \mathrm{diam}(G) + \mathrm{diam}(\overline{G}) \leq n + 1$. Does the second inequality still hold for $n = 4$?

20. Let G be a non-trivial self-complementary graph. Show that

(i) $\mathrm{rad}(G) = 2$;

(ii) $2 \leq \mathrm{diam}(G) \leq 3$;

(iii) $\mathrm{diam}(G) = 3$ if and only if there is an edge xy in G such that $N(x) \cup N(y) = V(G)$.

Note: In what follows, for $n \geq 3$, let $P_n = C_n - e$ where e is an edge in the cycle C_n.

21. Let G be a graph of order $n \geq 5$. Show that either G contains a P_3 as a subgraph or $\overline{G}$ contains a C_3 as a subgraph. Does the result still hold for $n = 4$?

22. Let G be a graph of order $n \geq 5$. Show that either G contains a P_4 as a subgraph or $\overline{G}$ contains a C_4 as a subgraph. Does the result still hold for $n = 4$?

23. Let G be a graph of order $n \geq 6$. Assume that every two edges in G have an end in common. Show that $\overline{G}$ contains a K_4 as a subgraph. Is the result still true for $n = 5$?

24. A graph G is said to be P_4-free if it does not contain P_4 as an induced subgraph. Show that G is P_4-free if and only if $\overline{G}$ is so.

25. Let G be a P_4-free graph (see the problem above). Show that either G or $\overline{G}$ is disconnected.

(Seinische (1974))

26. Let G be a non-trivial graph such that both G and $\overline{G}$ have the same degree sequence.

   (i) Is it true that G is self-complementary? Justify your answer.
   (ii) Show that $2 \leq \mathrm{diam}(G) \leq 4$.

(Xu (1993))

27. Let G be a connected graph of order $n \geq 2$ with vertex set V. A subset R of V is called a **resolving set** for G if for every two distinct vertices x, y in V, there exists a vertex z in R such that $d(z, x) \neq d(z, y)$. The **metric dimension** of G, denoted by $\dim(G)$, is the minimum cardinality of a resolving set for G.

   (a) Let H be the following graph:

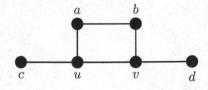

Fig. 2.6.5

     (i) Verify that the set $\{a, c\}$ is a resolving set for H.

     (ii) Show that $\dim(H) = 2$.

(b) Verify that

     (i) $\dim(C_n) = 2$, where $n \geq 3$;

     (ii) $\dim(G) = 3$, where G is the Petersen graph.

(c) Prove that $\dim(G) = 1$ if and only if $G \cong P_n$.

(d) Prove that $\dim(G) = n - 1$ if and only if $G \cong K_n$.

(e) Prove that $\dim(G) \leq n- \operatorname{diam}(G)$. Give an example to show that the bound is sharp.

Note: The notions of resolving set and $\dim(G)$ were first introducted by Slater (1975) and Harary and Melter (1976) independently. For the above results (c) - (e), see Chartrand et al. (2000).

Chapter 3

Bipartite Graphs and Trees

3.1 Bipartite Graphs

In Example 1.2.3, the following graph which models a job-application situation is shown.

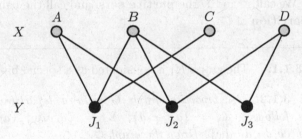

Fig. 3.1.1

This is a $(7, 8)$-graph, and the above diagram is drawn in such a way that the seven vertices are divided into two sets, $X = \{A, B, C, D\}$ and $Y = \{J_1, J_2, J_3\}$, such that each of the eight edges has an end in X and the other in Y.

Another example of this type of graphs is shown in Fig. 3.1.2:

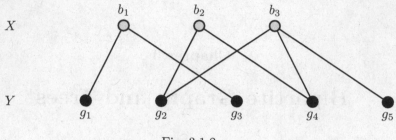

Fig. 3.1.2

Here, X is a group of 3 boys and Y is a group of 5 girls in a party, and that 'x' and 'y' are adjacent means that the boy 'x' dances with the girl 'y' in the party. Graphs of this type are very important and useful. We call them bipartite graphs.

> A graph G of order at least 2 is said to be **bipartite** if its vertex set $V(G)$ can be partitioned into two non-empty subsets X and Y such that each of the edges in G joins a vertex in X to a vertex in Y. We call X and Y the **partite sets**, and call the pair (X, Y) a **bipartition** of G.

Remark 3.1.1.   The graph K_1 is considered as a special bipartite graph.

Question 3.1.2.   *A bipartite graph G with a bipartition (X, Y) is defined as follows: $X = \{a, b, c, d\}$, $Y = \{u, v, w\}$, and $E(G) = \{aw, bu, bw, cv, du, dv, dw\}$. Draw the graph G.*

Question 3.1.3.   *A graph G is defined as follows: $V(G) = X \cup Y$, where*

$$X = \{2, 3, 5\} \quad and \quad Y = \{6, 7, 8, 9, 10\},$$

and $E(G) = \{xy \mid x \in X, y \in Y \text{ and } y \text{ is divisible by } x\}$.

     (i) Is ' 2 ' adjacent to ' 6 '? Is ' 3 ' adjacent to ' 8 '? Is ' 5 ' adjacent to ' 10 '?

     (ii) Draw the graph G.

     (iii) Is G bipartite?

Consider the graph H of Fig. 3.1.3. Is it a bipartite graph?

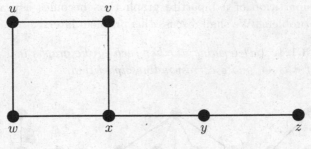

Fig. 3.1.3

To answer this question, by definition, we ask ourselves: Does H have a bipartition (X, Y)?

If we take $A = \{u, v\}$ and $B = \{w, x, y, z\}$, is (A, B) a bipartition of H? The answer is 'no'. Why? Since there is an edge joining two vertices in A. Though, in this case, (A, B) is not a bipartition of H, it does not mean that H is not bipartite. Indeed, if we take $X = \{u, x, z\}$ and $Y = \{w, v, y\}$, then we observe that

(i) $V(H) = X \cup Y$ and
(ii) each edge in H joins a vertex in X to a vertex in Y.

It follows by definition that (X, Y) is a bipartition of H, and thus H is a bipartite graph.

Let us re-draw the graph H in a 'natural' way as shown in Fig. 3.1.4. It is now clear that H is a bipartite graph.

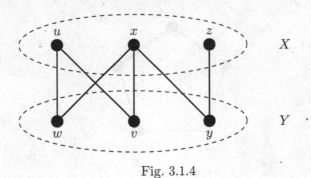

Fig. 3.1.4

Quite often, bipartite graphs are not drawn in this 'natural' form. How to find a bipartition of a bipartite graph thus becomes a practical and interesting problem. We shall discuss this problem later.

Question 3.1.4.   *Determine whether each of the graphs in Fig. 3.1.5 is bipartite. If it is so, find a corresponding bipartition.*

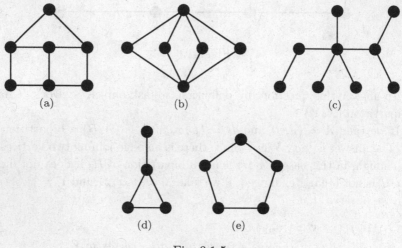

Fig. 3.1.5

In Fig. 3.1.5, while the graphs (a), (b) and (c) are bipartite, the graphs (d) and (e) are not. Let us examine, for instance, the graph (e), which is C_5. For convenience, we name its five vertices as shown in Fig. 3.1.6.

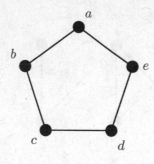

Fig. 3.1.6

Suppose on the contrary that it is bipartite and has its bipartition

(X, Y). We may assume that $a \in X$. As no two adjacent vertices can be in the same part, $b \in Y$. This, in turn, implies (anti-clockwise) that $c \in X$, $d \in Y$ and $e \in X$. We thus arrive at the situation that both a and e are in X and they are adjacent, which, however, is not allowed. This shows that C_5 has no bipartition, and is therefore not bipartite.

A cycle C_k is said to be **odd** if k is odd, and **even** if k is even.

From the discussion above, it is not hard to see that the argument can be similarly carried out to lead to a contradiction if the graph contains an odd cycle. Thus, we conclude that

if G contains an odd cycle, then G is not bipartite;

or equivalently,

if G is bipartite, then G contains no odd cycles.

Is the converse of this result true? That is, if G contains no odd cycles, must G be bipartite?

Question 3.1.5. *The graphs (a), (b) and (c) in Fig. 3.1.5 contain no odd cycles and they are bipartite. Draw the 3-cube graph Q_3 (see Problem I.9).*

(i) Does Q_3 contain any odd cycle?

(ii) Is Q_3 bipartite?

Yes! If a graph contains no odd cycles, then it must be bipartite! This result was found by Hungarian combinatorist, Denes König (1884–1944), in 1916. König wrote the first book on graph theory in 1936.

Theorem 3.1.6. *A graph is bipartite if and only if it contains no odd cycles.*

The proof of the 'necessity', namely, 'if G is bipartite, then G contains no odd cycles', can be carried out as how we did above, and is left to the reader (see Problem 3.1.3).

We shall now prove the 'sufficiency'; namely, 'if G contains no odd cycles, then G is bipartite'. For this purpose, we first prove the following simple observation.

Lemma 3.1.7. *Every closed walk of odd length in a graph always contains an odd cycle.*

Proof. Let W be a closed walk of odd length $p \geq 3$. We shall prove the statement by induction on p.

For $p = 3$, we have $W = w_1 w_2 w_3 w_1$, which obviously forms a C_3. Assume that it is true for all closed walks of odd length less than p, where $p \geq 5$.

Now consider a closed walk of odd length $p : W = w_1 w_2 ... w_p w_1$. Our aim is to show that W contains an odd cycle. If W itself forms a C_p, we are through.; otherwise, some vertices are repeated and there exist i and j with $1 \leq i < j < p$ such that $w_i = w_j$, as shown in Fig. 3.1.7.

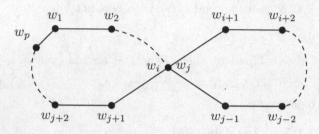

Fig. 3.1.7

Consider the two closed walks of smaller length:

$$W_1 = w_1 w_2 ... w_i w_{j+1} ... w_p w_1 \quad \text{and}$$
$$W_2 = w_i w_{i+1} ... w_{j-1} w_j \quad (\text{note that } w_i = w_j).$$

One of them must be of odd length (why?), say W_1. As the (odd) length of W_1 is less than p, by the induction hypothesis, W_1 contains an odd cycle. Clearly, this odd cycle is contained in W. The proof is thus complete.

$\square$

With the help of Lemma 3.1.7, we are now ready to prove the sufficiency of Theorem 3.1.6.

Proof of Theorem 3.1.6. Let G be a graph that does not contain any odd cycles. We aim to show that G is bipartite by providing a bipartition of G. We may assume that G is connected (why?). Let w be a fixed vertex in G and let

$$X = \{v \in V(G) \mid d(w, v) \text{ is even}\} \quad \text{and} \quad Y = \{v \in V(G) \mid d(w, v) \text{ is odd}\}.$$

We now claim that (X, Y) **is a bipartition** of G.

It is obvious that X and Y are disjoint, and as G is connected, $V(G) = X \cup Y$ (note that $w \in X$). It remains to show that each edge in G joins a vertex in X to a vertex in Y. Suppose this is not the case. Then there exist $u, v \in X$ or $u, v \in Y$, say the former, such that $uv \in E(G)$. As $u, v \in X$, by definition, $d(w, u)$ and $d(w, v)$ are even, and there exist a $w - u$ path P of even length and a $v - w$ path Q of even length in G.

Consider the walk W which begins at w, follows P to reach u, passes 'uv' to reach v, and finally follows Q to return to w. This walk W is closed and of odd length (why?). By Lemma 3.1.7, W contains an odd cycle. But then this implies that G contains an odd cycle, a contradiction.

The proof is thus complete.

$\square$

Remark 3.1.8. The argument given in the proof above suggests a way to find a bipartition of a connected graph G if G contains no odd cycles. The procedure is as follows:

> Begin with an (arbitrary) vertex and label it '1'. Suppose a vertex has been labeled '1', label all its neighbors '2'; and if a vertex has been labeled '2', label all its neighbors '1'. Repeat this until all vertices have been labeled. Then the set of all vertices with label '1' and the set of vertices with label '2' form a bipartition of G.

Question 3.1.9. *Apply the above procedure to find a bipartition for each of the graphs in Fig. 3.1.5(a), (b) and (c), and Q_3.*

Question 3.1.10. *The graph of Fig. 3.1.8 is not bipartite. Apply the above procedure to identify an odd cycle.*

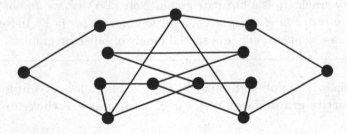

Fig. 3.1.8

Example 3.1.11. Problem 2.3.7 is restated below:

Let G be a graph of order five satisfying the following condition: for any three vertices x, y, z in G, $[\{x, y, z\}]$ is not isomorphic to either O_3 or K_3. What is the graph G?

There are different methods to solve this problem. Here, we solve it by applying Theorem 3.1.6.

Suppose G is bipartite with bipartition (X, Y). As $v(G) = 5$, either $|X| \geq 3$ or $|Y| \geq 3$. In either case, G contains an O_3 as an induced subgraph, a contradiction.

Thus G is not bipartite. Now, by Theorem 3.1.6, G contains an odd cycle C_k. By assumption, $k \neq 3$. As $v(G) = 5$, $k = 5$ and G contains C_5 as a spanning subgraph. If $G \not\cong C_5$, then G would contain a K_3, a contradiction. We therefore conclude that $G \cong C_5$.

The Handshaking Lemma (Theorem 1.3.6) states that for any multi-graph G,

$$\sum_{v \in V(G)} d(v) = 2e(G).$$

Suppose now that G is a bipartite graph with bipartition (X, Y). Then $\sum_{v \in V(G)} d(v)$ can be split naturally into two parts, namely $\sum_{x \in X} d(x)$ and $\sum_{y \in Y} d(y)$. What can we say about these two sums? Indeed, it is easily seen that each of them counts $e(G)$. This simple but useful relation is stated below.

Observation 1: Let G be a bipartite graph with bipartition (X, Y). Then

$$\sum_{x \in X} d(x) = e(G) = \sum_{y \in Y} d(y).$$

By definition, a bipartite graph possesses a bipartition (X, Y) such that each edge in the graph joins a vertex in X to a vertex in Y. Note that the empty graph O_n is a bipartite graph. Note also that we *do not require* that *every vertex in X must be adjacent to every vertex in Y*. Indeed, this extreme case happens only in a special family of bipartite graphs.

A bipartite graph with bipartition (X, Y) is called a **complete bipartite graph** if each vertex in X is adjacent to each vertex in Y.

Question 3.1.12. *Which of the following bipartite graphs are complete bipartite graphs?*

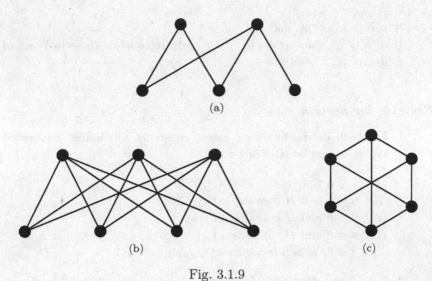

(a)

(b)   (c)

Fig. 3.1.9

There is one and only one (up to isomorphism) *complete* bipartite graph with a given bipartition (X, Y). If $|X| = p$ and $|Y| = q$, we shall denote this complete bipartite graph by $K(p, q)$ or $K_{p,q}$. In particular, we call $K(1, q)$ or $K(p, 1)$ a **star** (see Fig. 3.1.10).

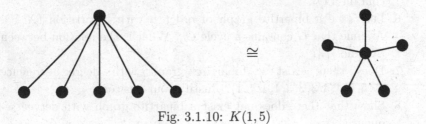

Fig. 3.1.10: $K(1, 5)$

Question 3.1.13.

 (i) Draw $K(3, 5)$ and $K(5, 3)$.
 (ii) Is $K(3, 5)$ isomorphic to $K(5, 3)$?
 (iii) Find $e(K(3, 5))$.
 (iv) Find the degree of each vertex in $K(3, 5)$.

Remark 3.1.14.   For all positive integers p and q, we have:

(1) $K(p,q) \cong K(q,p)$;
(2) $e(K(p,q)) = pq$; and
(3) in $K(p,q)$, the vertices in X are of degree q while those in Y are of degree p.

Exercise for Section 3.1

1. For each of the following cases, construct all desired connected bipartite graphs H of order n:

    (i) $2 \leq n \leq 4$;
    (ii) $n = 5$ and H contains no cycles;
    (iii) $n = 5$ and H contains a cycle;
    (iv) $n = 6$ and H contains a C_6;
    (v) $n = 8$, H is 3-regular and contains a C_8.

2. Let G be a connected bipartite graph. Then G has a bipartition (X,Y). Is $\{X,Y\}$ always unique? What if G is disconnected?

3. Show that if G is bipartite, then G contains no odd cycles.

4. Let G be a bipartite graph with bipartition (X,Y). Show that if G is k-regular, where $k \geq 1$, then $|X| = |Y|$.

5. Construct all non-isomorphic $(8,10)$-graphs that are bipartite and contain a C_8.

6. Let G be a bipartite graph of order n with bipartition (X,Y). Assume that G contains a cycle C_n. What is the relation between $|X|$ and $|Y|$?

7. Does there exist a bipartite graph with degree sequence $(5,5,5,4,4,3,3,3,1,1,1,1)$? Justify your answer.

8. Show that there does not exist a bipartite graph with degree sequence

$$(6,6,...,6,5,3,3,...,3).$$

9. At a party, assume that no boy dances with every girl but each girl dances with at least one boy. Prove that there are two couples b, g and $b^{'}, g^{'}$ which dance, whereas b does not dance with $g^{'}$ nor does g dance with $b^{'}$. (Putnam Exam (1965))

10. Let G be a bipartite graph with bipartition (X, Y). Assume that $e(G) = v(G)$ and that $d(x) \leq 5$ for each $x \in X$. Show that $|Y| \leq 4|X|$.

11. Let H be a bipartite graph with bipartition (X, Y). Assume that $e(H) \leq 2v(H)$ and that $d(x) \geq 3$ for each $x \in X$. Show that $|X| \leq 2|Y|$.

12. Let G be a bipartite graph of order $2k$, where $k \geq 1$. What is the maximum size of G? Find all such bipartite graphs with maximum size.

13. Let G be a bipartite graph of order $2k + 1$, where $k \geq 1$. What is the maximum size of G? Find all such bipartite graphs with maximum size.

14. Find, in terms of p and q, the number of C_4's in $K(p, q)$, where $2 \leq p \leq q$.

15. Find, in terms of p and q, the number of C_6's in $K(p, q)$, where $3 \leq p \leq q$.

16. Let H be a graph obtained from $K(p, q)$, $2 \leq p \leq q$, by adding a new edge joining two non-adjacent vertices in $K(p, q)$.

    (i) Is H bipartite?
    (ii) What is the largest number of triangles that H could contain?
    (iii) What is the largest number of C_5's that H could contain?

17. What is the largest cycle in $K(p, q)$, where $2 \leq p \leq q$?

18. What can be said about the complement of $K(p, q)$?

19. Let G be a bipartite graph.

    (i) Is $\overline{G}$ also bipartite?
    (ii) Is $\overline{G}$ always connected?
    (iii) What conditions should be imposed on G so that $\overline{G}$ is connected?

20. A connected graph G has the following property:

    For each pair of distinct vertices u and v, either all $u - v$ paths are of even length or all $u - v$ paths are of odd length.

    What can be said about G? Justify your answer.

21. Let G be a graph. A cycle C in G is said to be **induced** if C is induced by $V(C)$.

    (i) Consider the graph H shown in Fig. 3.1.11 . Which cycles in H are induced cycles?

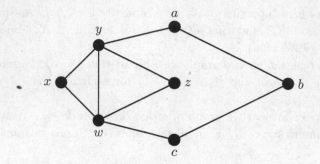

Fig. 3.1.11

    (ii) Show that G is bipartite if and only if G contains no *induced* cycles of odd order.

22. A graph H has the property that each edge in H is incident with an even vertex and an odd vertex. What can be said about H? Construct one such H.

23. Let G be a bipartite graph of order 7 such that every vertex in G is contained in a cycle.

    (i) Construct one such G.
    (ii) Must G be connected?
    (iii) What is the least possible value of $e(G)$?
    (iv) Construct all possible non-isomorphic bipartite graphs G which have their $e(G)$ attaining the least possible value obtained in (iii).

24. Let G be a connected bipartite $(p + q, pq)$-graph, where $1 \leq p \leq q$. Is it true that $G \cong K(p, q)$?

25. Let G be a bipartite $(p + q, pq)$-graph, where $2 \leq p \leq q$, and $\delta(G) \geq 1$. Show that $G \cong K(p, q)$ if and only if every two edges in G are contained in a common C_4.

3.2   Trees

In Fig. 3.1.5(c), we show a special bipartite graph, which is *connected* and *contains no cycles*. It is one of a very important family of graphs that we shall study in this section.

A graph is called a **tree** if it is connected and contains no cycles.

Question 3.2.1. *Which of the graphs in Fig. 3.2.1 is a tree? Why?*

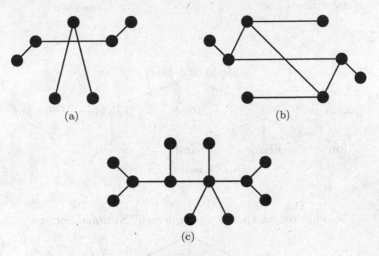

Fig. 3.2.1

Example 3.2.2.

(i) Every star $K(1,q)$, where $q \geq 1$, is a tree. It is noted that $d(u,v) \leq 2$ for any two vertices u, v in $K(1,q)$.

(ii) The **path** of order $n \geq 1$, denoted by P_n (see the Note preceding Problem II.21), is defined as $P_1 = K_1$, $P_2 = K_2$ and, for $n \geq 3$, $P_n = C_n - e$ where e is an edge in the cycle C_n. Clearly, P_n is a tree with exactly two end-vertices, say u and v, and $d(u,v) = n-1$. Recall that in Section 1.4, a 'path' is introduced as a special walk in a multigraph, but now a path is regarded as a graph. What a 'path' really means should be clear from the context when it is mentioned.

(iii) It should be noted that, among all trees of order $n \geq 3$, while the star $K(1, n-1)$ has the smallest diameter (equals 2), the path P_n has the largest diameter (equals $n-1$).

(iv) Some other examples of trees are shown in Fig. 3.2.2.

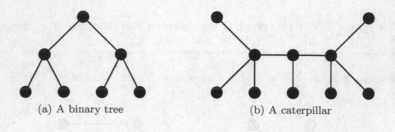

(a) A binary tree   (b) A caterpillar

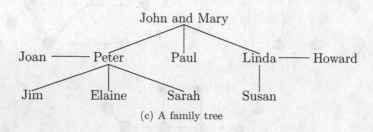

(c) A family tree

To what extent should we implement National Service?

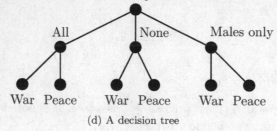

(d) A decision tree

Fig. 3.2.2

A tree is a very special graph, and so it must have certain properties that other graphs do not have. In what follows, we shall find out some of them.

Question 3.2.3. *Consider the two connected graphs in Fig. 3.2.3. Observe that graph G is a tree while graph H is not.*

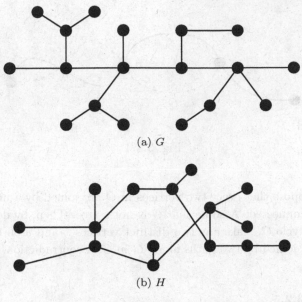

(a) G

(b) H

Fig. 3.2.3

(i) *Is it true that every two vertices in G are joined by exactly one path?*

(ii) *Is it true that every two vertices in H are joined by exactly one path?*

A tree is connected, and thus it is not surprising that every two of its vertices are joined by a path. An important feature of trees that other connected graphs do not have is the *uniqueness* of such a path joining any two vertices.

Theorem 3.2.4. *Let G be a graph. Then G is a tree if and only if every two vertices in G are joined by a unique path.*

Proof. ($\Rightarrow$) Suppose that G is a tree and assume on the contrary that there exist pairs of vertices in G which are joined by two different paths. Choose x and y be such a pair of distinct vertices which are joined by two different $x - y$ paths, say P and Q such that the total length of P and Q is the minimum. Then P and Q form a cycle in G (see Fig. 3.2.4), a contradiction.

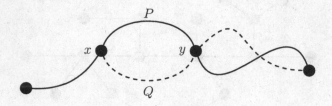

Fig. 3.2.4

($\Leftarrow$) Suppose that every two vertices in G are joined by a unique path. Then G is connected. Assume that G is not a tree. Then, by definition, G contains a cycle C. Take any two distinct vertices x and y on C. Clearly, x and y are joined by two paths along C in G, a contradiction.

$\square$

Question 3.2.5.

    (i) For each of the trees in Fig. 3.1.11, Fig. 3.2.1(c), Fig. 3.2.2, Fig. 3.2.3(a), find its order and size. Do you notice any relationship between the order and the size?

    (ii) For each connected graph shown in Fig. 3.2.1(b) and Fig. 3.2.3(b), find its order and size. Do they have the same relationship as the one you have found in (i)?

Theorem 3.2.6. *Let G be a connected (n, m)-graph. Then G is a tree if and only if $m = n - 1$.*

Proof. ($\Rightarrow$) We prove it by induction on n. The result is trivial if $n = 1$. Assume that the result is true for all trees of order less than n, where $n \geq 2$, and let G be a tree of order n. Choose an edge, say xy, in G. By Theorem 3.2.4, xy is the unique $x - y$ path in G. Thus, $G - xy$ is a disconnected graph having two components, say G_1 and G_2, both of which are also trees. Let n_i and m_i be, respectively, the order and size of G_i, $i = 1, 2$. By the induction hypothesis, we have $m_i = n_i - 1$ for each $i = 1, 2$. Thus,

$$m = m_1 + m_2 + 1 = (n_1 - 1) + (n_2 - 1) + 1 = n_1 + n_2 - 1 = n - 1,$$

as was to be shown.

($\Leftarrow$) Assume that G is connected and $m = n - 1$. We shall show that G is a tree. We first establish the following result:

Claim: G contains an end-vertex if $n \geq 2$.

If not, then $d(v) \geq 2$ for each vertex v in G, and we have, by applying Euler's Handshaking Lemma,

$$2m = \sum_{v \in V(G)} d(v) \geq 2n,$$

which implies that $m \geq n$, a contradiction and thus the claim is proved.

We shall now prove that G is a tree by induction on n. The result is trivial if $n = 1$. Assume that $n \geq 2$. By the above claim, G contains an end-vertex, say w. Clearly, $G - w$ is connected, and $e(G - w) = v(G - w) - 1$. By the induction hypothesis, $G - w$ is a tree. It follows that G is a tree, as desired.

$\square$

The proof of the 'sufficiency' (that is, the ($\Leftarrow$) implication) in Theorem 3.2.6 shows that every tree of order at least two has at least one end-vertex. We have also pointed out that every path P_n, $n \geq 2$, has exactly two end-vertices. Indeed, we find more end-vertices in other trees shown above.

Question 3.2.7. *Look at the tree T of Fig. 3.2.5.*

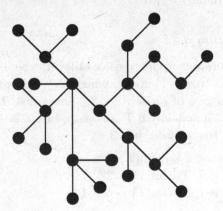

Fig. 3.2.5

For convenience, let us denote by n_i the number of vertices of degree i in T, where $i = 1, 2, \ldots$. Thus n_1 counts the number of end-vertices in T.

(i) *Verify that* $\Delta(T) = 5$.

(ii) *Count* n_i, $i = 1, 2, 3, 4, 5$.

(iii) *Evaluate the sum:* $2 + n_3 + 2n_4 + 3n_5$.

(iv) *Is your sum obtained in (iii) equal to* n_1?

Indeed, we have in general the following interesting result which expresses the number of end-vertices of a tree T in terms of the numbers of vertices of degrees $3, 4$, and so on, up to $\Delta(T)$. The reader is strongly encouraged to prove it (see Problem 3.2.14).

Theorem 3.2.8. *Let T be a tree having n_i vertices of degree i, where $i = 1, 2, ..., k$ with $k = \Delta(T)$. Then*

$$n_1 = 2 + n_3 + 2n_4 + 3n_5 + ... + (k - 2)n_k.$$

$\square$

Remark 3.2.9.

(1) It follows from Theorem 3.2.8 that every tree of order $n \geq 2$ contains at least two end-vertices.

(2) The number n_2 is not involved in the above expression. That is, the number of end-vertices in a tree is independent of the number of vertices of degree two.

Exercise for Section 3.2

1. Draw all non-isomorphic trees of order n, where $2 \leq n \leq 6$.

2. Let G be an $(n, n-1)$-graph, where $n \geq 4$. Must G be a tree?

3. Let T be a tree of order $n \geq 2$. Show that T has exactly two end-vertices if and only if T is a path, i.e., $T \cong P_n$.

4. Let T be a tree of order $n \geq 3$.

   (i) Show that $2 \leq \text{diam}(T) \leq n - 1$.

   (ii) What is T if $\text{diam}(T) = 2$?

   (iii) What is T if $\text{diam}(T) = n - 1$?

5. Find all trees T of order $n \geq 2$ such that $\overline{T}$ is a tree. Is there any tree of order $n \geq 2$ which is self-complementary?

6. A connected graph is said to be **unicyclic** if it contains one and only one cycle as a subgraph.

   (i) Is every cycle unicyclic?

(ii) Construct two unicyclic graphs of order 8 which are not C_8.

(iii) How many edges are there in each of your graphs in (ii)?

7. Let G be a unicyclic graph.

   (i) What is the relation between $e(G)$ and $v(G)$? Justify your answer.

   (ii) Show that there exist at least three edges e in G such that $G - e$ is a tree.

8. Let G be a unicyclic graph and let n_1 denote the number of end-vertices in G. Find an expression for n_1 similar to that in Theorem 3.2.8.

9. Let G be a connected graph. Show that G is a tree if and only if every edge in G is a bridge. (See Problem 2.3.16.)

10. Let T be a tree of order k. Show that if G is a graph with $\delta(G) \geq k - 1$, then T is isomorphic to some subgraph of G.

11. Let T be a tree of order 15 such that $1 \leq d(v) \leq 4$ for each vertex v in T. Suppose that T contains exactly 9 end-vertices and exactly 3 vertices of degree 4. How many vertices of degree 3 does T have? Construct one such tree T.

12. The degrees of the vertices of a tree T of order 18 are $1, 2$ and 5. If T has exactly 4 vertices of degree 2, how many end-vertices does T have?

13. Let T be a tree and let n_i be the number of vertices of degree i in T. Which of the following statements is/are true?

   (i) If T is not a path, then $n_1 \geq n_2$.

   (ii) If $n_2 = 0$, then T has more end-vertices than other vertices.

14. Let T be a tree having n_i vertices of degree i, where $i = 1, 2, ..., k$ with $k = \Delta(T)$. Show that
$$n_1 = 2 + n_3 + 2n_4 + 3n_5 + ... + (k - 2)n_k.$$

15. Let G be a graph of order n with degree sequence $(d_1, d_2, ..., d_n)$. Show that if G is a tree, then
$$\sum_{i=1}^{n} d_i = 2(n - 1).$$
Is the converse true?

16. Show that every non-increasing sequence $(d_1, d_2, ..., d_n)$ of positive integers with
$$\sum_{i=1}^{n} d_i = 2(n - 1)$$

is a degree sequence of a tree.

17. A **forest** is a graph which contains no cycle as a subgraph. An example is given in Fig. 3.2.6:

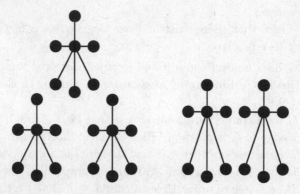

Fig. 3.2.6

(i) Is it true that every tree is a forest?
(ii) Is it true that every forest is a tree?
(iii) Is it true that every component of a forest is a tree?
(iv) Let F be a forest. Find a relationship linking $v(F), e(F)$ and $c(F)$, and prove your result.

18. Let G be an $(n, n-1)$-graph. Prove that G is connected if and only if G contains no cycles.

3.3   Spanning Trees of a Graph

Consider the graph G of Fig. 3.3.1(a). It is connected. Indeed, it remains connected even if some edges are removed. For instance, as shown in Fig. 3.3.1(b), $T = G - \{uv, xy\}$ is still connected. However, we now cannot afford to miss any edge from T to maintain the connectedness. Observe that T is both (i) a spanning subgraph of G and (ii) a tree. It is called a spanning tree of G.

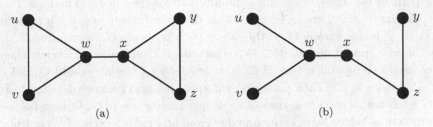

Fig. 3.3.1

In general:

A graph H is called a **spanning tree** of a graph G if H is both (i) a spanning subgraph of G and (ii) a tree.

Two more spanning trees of G in Fig. 3.3.1(a) other than T are shown in Fig. 3.3.2.

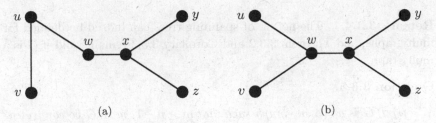

Fig. 3.3.2

Question 3.3.1. *Find all spanning trees of the graph G in Fig. 3.3.1(a). How many are there?*

The existence of a spanning tree of a graph G is directly linked to the connectedness of G as shown in Theorem 3.3.2.

Theorem 3.3.2. *Let G be a graph. Then G is connected if and only if G contains a spanning tree.*

Proof. ($\Leftarrow$) Suppose that G contains a spanning tree, say T. We shall show that G is connected by showing that every two vertices in G are joined by a path in G. Thus, let x and y be any two vertices in G. Then, as T is spanning, x and y are in T. Since T is connected, there is a $x - y$ path in T. As T is a subgraph of G, this $x - y$ path is also in G, as required.

($\Rightarrow$) Suppose now that G is connected. If G contains no cycles, then G itself is a spanning tree of G. Otherwise, let C be a cycle of G and e be an edge in C. Then $G - e$ is still spanning and connected (why?). If $G - e$ contains no cycles, then $G - e$ is a spanning tree of G. Otherwise, we proceed as before by deleting an edge from an existing cycle. We continue this procedure repeatedly until a spanning tree of G is eventually found after a finite number of steps.

$\square$

If G is a connected graph, then, by Theorem 3.3.2, G contains a spanning tree T as a subgraph. Thus, by Theorem 3.2.6, we have:

$$e(G) \geq e(T) = v(T) - 1 = v(G) - 1.$$

Corollary 3.3.3.   *If G is a connected (n, m)-graph, then*

$$m \geq n - 1.$$

$\square$

Remark 3.3.4.   The notion of spanning trees can indeed be defined for multigraphs, and Theorem 3.3.2 and Corollary 3.3.3 remain valid if G is a multigraph.

Question 3.3.5.

(i) *If G is an (n, m)-graph such that $m \geq n - 1$, must G be connected?*
(ii) *If H is a $(100, 98)$-graph, can H be connected?*

We shall now give an example to show an application of Corollary 3.3.3.

Example 3.3.6.   Let G be a connected bipartite graph with bipartition (X, Y). Assume now that $d(x) \leq 5$ for each x in X. As G is connected and each vertex in X has degree at most 5, it is clear that $|Y|$ cannot be too large as compared to $|X|$. It is thus reasonable to ask: what is the best upper bound for $|Y|$ in terms of $|X|$?

Let us consider $e(G)$. By Observation 1 in Section 3.1,

$$e(G) = \sum_{x \in X} d(x).$$

As $d(x) \leq 5$ for each x in X,

$$e(G) = \sum_{x \in X} d(x) \leq 5|X|.$$

On the other hand, as G is connected, by Corollary 3.3.3,

$$e(G) \geq v(G) - 1 = |X| + |Y| - 1.$$

Now combining the above two inequalities through $e(G)$, we have:

$$|X| + |Y| - 1 \leq e(G) \leq 5|X|$$

and so

$$|Y| \leq 4|X| + 1.$$

That is, Y can have at most $4|X| + 1$ vertices.

To show that the above inequality is sharp (i.e., the equality can be attained), the reader is invited to construct such a bipartite graph with $|Y| = 4|X| + 1$ and $|X| = 1, 2, \ldots$

Graphs, in particular trees, can be used to model systems as diverse as oil pipelines and internet search programs. The following describes one application of trees in mechanical engineering.

An application. Figure 3.3.3 shows a 3-D folded structure and one of its flat layouts.

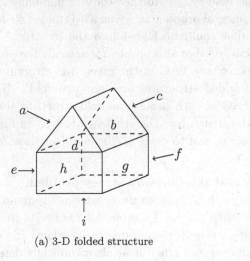

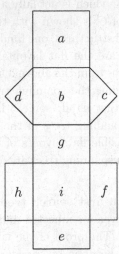

(a) 3-D folded structure   (b) Flat layout

Fig. 3.3.3

Let us introduce a graph to study the relationship between the 3-D folded structure and its flat layout. The 3-D folded structure has nine faces as indicated. Construct a graph G with $V(G) = \{a, b, c, d, e, f, g, h, i\}$, where each vertex represents a face, such that two vertices are adjacent in G if and only if the faces they represent have one edge in common. This graph G, called the **face adjacency graph** (FAG) of the 3-D folded structure, is shown in Fig. 3.3.4.

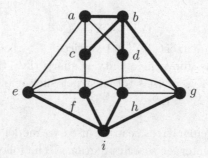

Fig. 3.3.4

If we construct the FAG of the flat layout of Fig. 3.3.3, we obtain a graph, which is actually a spanning tree of G. In Fig. 3.3.4, this spanning tree of G is shown with the bold edges. In the study of unfolding of 3-D folded structures, one fundamental problem is: given a 3-D folded structure, what are the flat layouts that could be folded into the structure? It is obvious from the above discussion that this problem is actually the problem of finding the spanning trees of the FAG of the given 3-D structure. On the other hand, some 3-D folded structures are 'non-manifold'. Not all the spanning trees of the FAG of such a non-manifold structure give rise to feasible flat layouts of the structure. The challenge here is to find out which spanning trees are feasible and to develop efficient algorithms for the search.

Depth-first search trees and the one-way street problem

Given a connected multigraph G, how do we construct a spanning tree of G? The proof of the necessity part of Theorem 3.3.2 suggests that this could be done by 'breaking a cycle' (deleting one of its edges) in G step by step. This method is, however, not efficient as algorithmically detecting the existence of a cycle in G is a difficult task. In what follows, we shall introduce an efficient algorithm, called the **depth-first search algorithm**, to construct a spanning tree of G.

Algorithm 3.3.7. (Finding a spanning tree in a connected multigraph G of order n.)

Step 1: Pick a vertex and label it '1'. Set $i := 1$, $j := 1$, $X := \{1\}$ and $E := \emptyset$.

Step 2(a): If $i = n$, output the n labeled vertices together with E as a spanning tree of G. Otherwise, go to Step 2(b).

Step 2(b): If there are unlabeled neighbors of j, pick one and label it '$i+1$', set $E := E \cup \{j(i+1)\}$, $X := X \cup \{i+1\}$, $j := i+1$, and $i := i+1$, and return to Step 2(a). Otherwise, go to Step 2(c).

Step 2(c): If there are no unlabeled neighbors of j, set $X := X \setminus \{j\}$ and $j := \max\{k \mid k \in X\}$, and return to Step 2(b).

Remark 3.3.8. The spanning tree of G obtained from the above algorithm is called a **depth-first search spanning tree** of G and the labeling '1', '2',..., 'n' of the vertices in G is called a **depth-first search labeling** of G.

Example 3.3.9. Let G be the graph shown in Fig. 3.3.5.

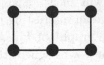

Fig. 3.3.5

We apply Algorithm 3.3.7 to obtain a depth-first search labeling as shown in Fig. 3.3.6.

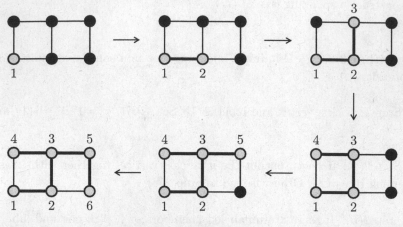

Fig. 3.3.6

A depth-first seach tree of G with the associated depth-first search labeling is shown in Fig. 3.3.7.

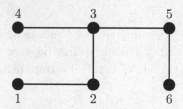

Fig. 3.3.7

To end this section, we will show an application of the notion of a depth-first search tree to solve an interesting problem, the so-called **One-way Street Problem**. The graphs of Fig. 3.3.8 model certain sections of the street systems of some towns where the roads are two-way. The edges represent the roads and the vertices represent the road junctions. To speed up traffic flow when the number of vehicles increases or during certain occasions such as when there is a large sports event at the nearby stadium, it may be better to make the roads one-way. Thus, the question is: Can we convert these two-way systems into one-way systems where every two vertices are still mutually reachable in the system?

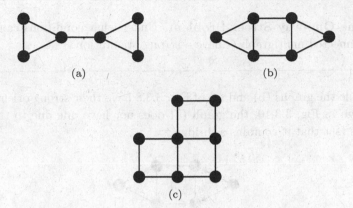

(a)

(b)

(c)

Fig. 3.3.8

An **orientation** of a connected multigraph G is a one-way system
obtained from G by assigning a direction to each edge of G. An
orientation of G is said to be **strong** if every two vertices of G are
mutually reachable (via directed paths) in it.

Example 3.3.10. Figure 3.3.9(a) shows a graph G while Fig. 3.3.9(b)
(resp. (c)) shows a strong (resp. non-strong) orientation of G.

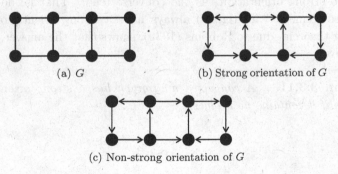

(a) G

(b) Strong orientation of G

(c) Non-strong orientation of G

Fig. 3.3.9

The One-way Street Problem may now be formulated using graph terminology as follows:

> **The One-way Street Problem.** Under what conditions can a connected multigraph G have a strong orientation?

While the graphs (b) and (c) of Fig. 3.3.8 have their strong orientations as shown in Fig. 3.3.10, the graph (a) does not have one due to the very obvious fact that it contains a 'bridge'.

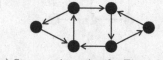

(a) Strong orientation for Fig. 3.3.8(b)

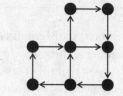

(b) Strong orientation for Fig. 3.3.8(c)

Fig. 3.3.10

It is trivial to see that if a connected multigraph contains a bridge, then it has no strong orientation. Is the converse true? That is, does every bridgeless connected multigraph always have a strong orientation? The following theorem, due to Robbins (1939), shows that the answer is in the affirmative.

Theorem 3.3.11.   *A connected multigraph has a strong orientation if and only if it contains no bridges.*

$\square$

The proof given by Robbins on the existence of a strong orientation for a bridgeless connected multigraph is basically by induction. No explicit procedure for obtaining such a strong orientation is spelt out therein. In what follows, we shall see how Roberts (1978) made use of a depth-first search tree to obtain such a strong orientation.

Algorithm 3.3.12. (Designing a strong orientation of a bridgeless connected multigraph G.)

Step 1: Construct a depth-first search spanning tree T of G with a depth-first search labeling of G.

Step 2: For each edge in T, orient it from the smaller label to the larger label; for each of the remaining edges in G, orient it from the larger label to the smaller label.

Example 3.3.13. Consider the graph G in Fig. 3.3.11. Three of its depth-first search trees and their corresponding depth-first search labelings are shown in Fig. 3.3.12(a), (b) and (c). By Algorithm 3.3.12, we design three respective strong orientations of G, as shown in Fig. 3.3.12(d), (e) and (f).

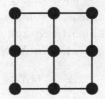

Fig. 3.3.11

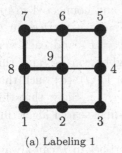

(a) Labeling 1

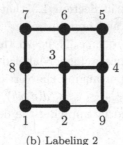

(b) Labeling 2

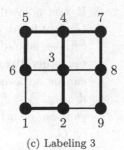

(c) Labeling 3

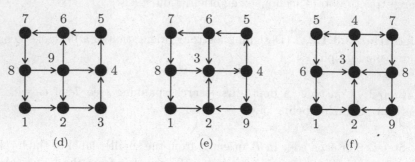

Fig. 3.3.12

Exercise for Section 3.3

1. Let G be a connected graph and e a bridge in G. Must e be contained in any spanning tree of G? Why?

2. Let H be a $(12, 10)$-graph. Find the least and largest possible values of $c(H)$.

3. Let n and r be integers such that $1 \leq r \leq n$. Prove that every $(n, n - r)$-graph has at least r components.

4. Let G be a connected bipartite graph with bipartition (X, Y). Assume that $d(x) \leq 7$ for each x in X. Show that

$$|Y| \leq 6|X| + 1.$$

For each $|X| = 1, 2, ...$, construct one such bipartite graph G with $|Y| = 6|X| + 1$.

5. Let G be a connected bipartite graph with bipartition (X, Y). Assume that G is not a tree and $d(x) \leq 4$ for each x in X. Find the best upper bound for $|Y|$, in terms of $|X|$.

6. It was pointed out in Section 1.5 that for any connected graph G of order $n \geq 2$ and any vertex 'a' in G, the n vertices in G can be labeled $v_1, v_2, ..., v_{n-1}, v_n$ so that $a = v_n$ and each vertex other than a has a higher-labeled neighbor in G; and this can be achieved by arranging the vertices in $V(G) \backslash \{a\}$ such that $d(a, v_{n-1}) \leq d(a, v_{n-2}) \leq ... \leq d(a, v_2) \leq d(a, v_1)$. Show that this can also be achieved by applying the depth-first search algorithm on G.

7. Let T be a tree of order n. Show that the vertices in T can always be named as $x_1, x_2, ..., x_n$ so that every x_i has one and only one

neighbor in $\{x_1, x_2, ..., x_{i-1}\}$, for $i = 2, 3, ..., n$.

8. Design a strong orientation of each of the following graphs by applying Algorithm 3.3.12.

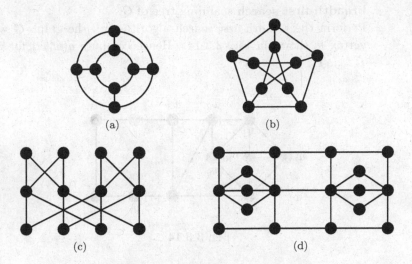

(a)   (b)

(c)   (d)

Fig. 3.3.13

9. The following algorithm is known as the **breadth-first search algorithm**, which when applied to a connected graph G with a starting vertex w, results in a spanning tree of G and allows us to compute $d(w, v)$ for all vertices v in G.

   (**Breadth-first search**) Let G be a connected graph and w be a vertex in G.

   Step 1: Set $i := 0$, $L_0 := \{w\}$ and $E_0 := \emptyset$.

   Step 2(a): Set $X := V(G) - \bigcup_{k=0}^{i} L_k$. If $X = \emptyset$, then $T = (\bigcup_{k=0}^{i} L_k, \bigcup_{k=0}^{i} E_k)$ is a spanning tree of G. Otherwise, set $Y := \emptyset$ and $Z := \emptyset$ and go to Step 2(b).

   Step 2(b): If there exists some $x \in X$ that is adjacent to some vertex $y \in L_i$, set $Z := Z \cup \{xy\}$. If no such x exists, go to Step 3.

   Step 2(c): Set $X := X - \{x\}$, $Y := Y \cup \{x\}$ and return to Step 2(b).

Step 3: Set $L_{i+1} := Y$, $E_{i+1} := Z$. Increase i by 1 and return to Step 2(a).

The spanning tree of G obtained from this algorithm is called a **breadth-first search** spanning tree of G.

Perform the breadth-first search algorithm on the graph G with vertex w shown in Fig. 3.3.14. Hence compute $d(w, v_i)$ for $i = 1, 2, ..., 11$.

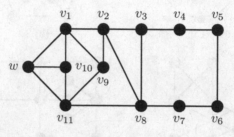

Fig. 3.3.14

3.4   The Number of Spanning Trees

Given a connected multigraph G, we know (Theorem 3.3.2) that G contains a spanning tree. We may proceed to ask: how many spanning trees could G contain? We shall study this problem in this and the next section.

In what follows, unless otherwise stated, G is a connected multigraph of order n. For convenience, we shall denote by $\tau(G)$, our key quantity, namely, the number of different spanning trees of G.

$$\tau(G) = \text{the number of different spanning trees of } G.$$

However, before we proceed, we need to clarify what it means by saying that two spanning trees of G are different. Let T_1 and T_2 be two spanning trees of G. We say that T_1 is *different* from T_2 if there exists an edge of G

which appears in T_1 but not in T_2 (or vice versa).

Example 3.4.1. Let G be the multigraph shown in Fig. 3.4.1.

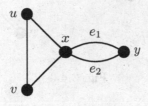

Fig. 3.4.1

The following two spanning trees T_1 and T_2 of G, though isomorphic, are regarded as different, since the edge e_1 of G appears in T_1 but not in T_2.

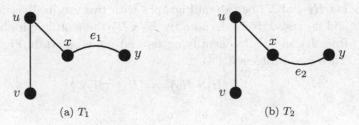

(a) T_1 (b) T_2

Fig. 3.4.2

Indeed, there are exactly six different spanning trees of G (besides T_1 and T_2, the other four are shown in Fig. 3.4.3). Thus, $\tau(G) = 6$.

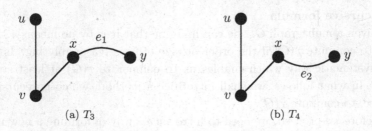

(a) T_3 (b) T_4

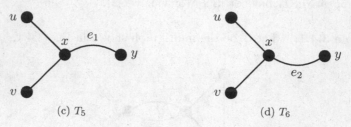

(c) T_5 (d) T_6

Fig. 3.4.3

Some simple observations on $\tau(G)$ are given below.

(1) If H is a disconnected multigraph, then $\tau(H) = 0$. Thus, for any multigraph H, $\tau(H) \geq 1$ if and only if H is connected.

(2) If H is a tree, then H itself is its only spanning tree, and so $\tau(H) = 1$. Indeed, let H be a multigraph. Then $\tau(H) = 1$ if and only if H is a tree. (See Problem 3.4.1.)

(3) For the cycle C_n of order $n \geq 2$, $\tau(C_n) = n$.

(4) Let H_1 and H_2 be two multigraphs with two specified vertices w_1 and w_2 respectively. Denote by $H_1 \bullet H_2$ the multigraph obtained from H_1 and H_2 by identifying w_1 and w_2 as shown in Fig. 3.4.4. Then (see Problem 3.4.3)

$$\tau(H_1 \bullet H_2) = \tau(H_1)\tau(H_2).$$

$$w_1 = w_2$$

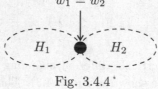

Fig. 3.4.4

A recursive formula

Given a multigraph G, one can imagine that it is by no means a simple task to compute $\tau(G)$ if the order or size of G is reasonably big. Is there any systematic way which enables us to enumerate $\tau(G)$ at least step by step? In what follows, we shall introduce a method, which is recursive in nature, to compute $\tau(G)$.

Before we proceed, we need to introduce a way of forming a new multigraph from the given G with a specific edge. Thus suppose that $e = uv$ is a specific edge in G. Let us denote by $G \circ e$ the multigraph obtained from G by first deleting all edges joining u and v, and then identifying u and v. We note that the order of $G \circ e$ is one less than that of G and the size of $G \circ e$ is always less than that of G.

Example 3.4.2. Figure 3.4.5 shows a way to form $G \circ e$ given an edge e in a multigraph G.

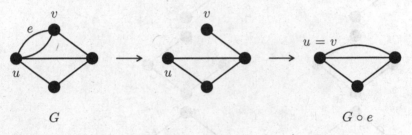

$$G \qquad\qquad\qquad\qquad G \circ e$$

Fig. 3.4.5

Recall that $G - e$ is the multigraph obtained by deleting e from G. Thus, if G is the multigraph with a specific edge e as shown in Fig. 3.4.6(a), then $G - e$ and $G \circ e$ are shown in Fig. 3.4.6(b) and (c) respectively.

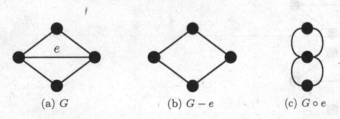

(a) G (b) $G - e$ (c) $G \circ e$

Fig. 3.4.6

Let us list all the spanning trees of G, $G - e$ and $G \circ e$ respectively in the three columns of the following table.

G	$G - e$	$G \circ e$

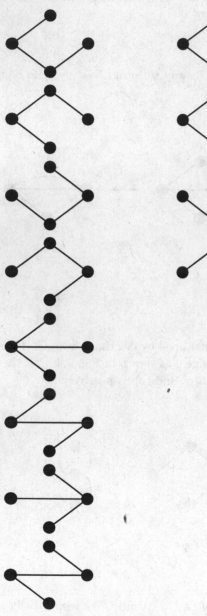 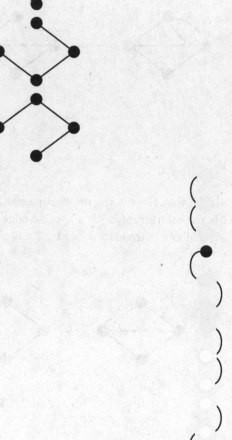

We note that

(1) $\tau(G) = \tau(G - e) + \tau(G \circ e)$;
(2) the four spanning trees of $G - e$ are the same as the first four spanning trees of G;
(3) there is a one-one correspondence between the four spanning trees of $G \circ e$ and the last four spanning trees of G.

With this in mind, we are now in a position to establish the following result.

Theorem 3.4.3. *Let G be any multigraph and e an edge in G. Then*

$$\tau(G) = \tau(G - e) + \tau(G \circ e).$$

Proof. Let $\mathcal{A}$ be the set of spanning trees of G that contain the given edge e and let $\mathcal{B}$ be the set of spanning trees of G that do not contain e. Then there is a one-one correspondence between $\mathcal{A}$ and the set of spanning trees of $G \circ e$ (see Problem 3.4.5) and, evidently, $\mathcal{B}$ is the same as the set of spanning trees of $G - e$. Thus

$$\tau(G \circ e) = |\mathcal{A}| \quad \text{and} \quad \tau(G - e) = |\mathcal{B}|,$$

and so $\tau(G) = |\mathcal{A}| + |\mathcal{B}| = \tau(G \circ e) + \tau(G - e)$, as was to be shown. $\square$

Given G, both $G - e$ and $G \circ e$ are either of smaller order or of smaller size, as compared to G. Thus, by applying Theorem 3.4.3 successively, the enumeration of $\tau(G)$ is reduced to the computations of $\tau(H)$'s, where H's are multigraphs of much smaller order or size, which are certainly much easier. An illustration is given below.

Example 3.4.4. Consider the following multigraph G.

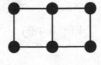

Fig. 3.4.7

By applying Theorem 3.4.3 repeatedly, we have:

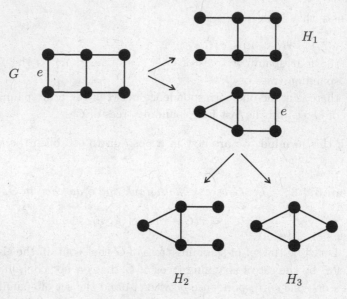

Fig. 3.4.8

Note that $\tau(H_1) = 4$ and $\tau(H_2) = 3$ (why?) while $\tau(H_3) = 8$ (see the table following Fig. 3.4.6). Thus, $\tau(G) = \tau(H_1) + \tau(H_2) + \tau(H_3) = 4 + 3 + 8 = 15$.

Computing $\tau(G)$ using another sequence of edges, we have:

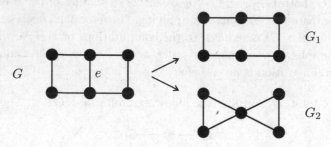

Fig. 3.4.9

Thus in this case, we have $\tau(G) = \tau(G_1) + \tau(G_2) = 6 + 9 = 15$, as obtained previously.

Note: The value of $\tau(G)$ by applying Theorem 3.4.3 is independent of the choice of the sequence of edges.

Exercise for Section 3.4

1. Let H be a multigraph. Show that $\tau(H) = 1$ if and only if H is a tree.
2. Let G be a unicyclic graph which contains C_k as a subgraph, where $k \geq 3$. Evaluate $\tau(G)$.
3. Let H_1 and H_2 be two multigraphs. Show that $\tau(H_1 \bullet H_2) = \tau(H_1)\tau(H_2)$, where $H_1 \bullet H_2$ is defined in Section 3.4. (See Fig. 3.4.4.)
4. Let G be a connected multigraph. Show that

   (i) if H is a spanning submultigraph of G, then $\tau(H) \leq \tau(G)$; and more generally,

   (ii) if H is a submultigraph of G, then $\tau(H) \leq \tau(G)$.

5. Let G be a connected multigraph and e an edge in G. Establish a one-one correspondence between the set of spanning trees of G that contain e and the set of spanning trees of $G \circ e$.
6. Evaluate $\tau(G)$ for each of the following multigraphs G:

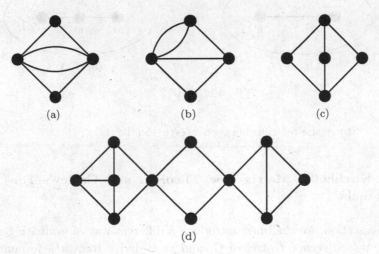

(a)   (b)   (c)

(d)

Fig. 3.4.10

7. Let G be an (n, n)-multigraph, where $n \geq 3$. Which G has the largest $\tau(G)$?

8. Consider the graph H shown in Fig. 3.4.11, obtained from two cycles C_p and C_q by identifying an edge in C_p with an edge in C_q, where $p, q \geq 3$. Evaluate $\tau(H)$ in terms of p and q.

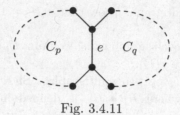

Fig. 3.4.11

9. Let G be an $(n, n+1)$-multigraph, where $n \geq 3$. Which G has the largest $\tau(G)$?

10. Let G be a multigraph and e an edge in G. Denote by $G(e)$ the multigraph obtained from G by inserting a new vertex of degree 2 into e as shown in Fig. 3.4.12:

(a) G (b) $G(e)$

Fig. 3.4.12

Study the relation between $\tau(G(e))$ and $\tau(G)$.

3.5 Kirchhoff's Matrix-Tree Theorem and Cayley's Formula

In this section, we shall first introduce a different way of evaluating $\tau(G)$ using the adjacency matrix of G, and then derive from it a formula for $\tau(K_n)$.

Let G be a multigraph with $V(G) = \{v_1, v_2, ..., v_n\}$. Recall that (see Section 1.2) the adjacency matrix of G is the $n \times n$ matrix $\boldsymbol{A}(G) = (a_{ij})$, where a_{ij} is the number of edges joining vertices v_i and v_j for $1 \leq i, j \leq n$.

Thus, if G is the multigraph in Fig. 3.5.1,

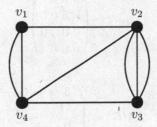

Fig. 3.5.1

then we have

$$A(G) = \begin{pmatrix} 0 & 1 & 0 & 2 \\ 1 & 0 & 3 & 1 \\ 0 & 3 & 0 & 1 \\ 2 & 1 & 1 & 0 \end{pmatrix}.$$

Notice that the matrix $A(G)$ is symmetric and all the diagonal entries in $A(G)$ are zero.

Let us define another matrix associated with G, called its degree matrix.

Let G be a multigraph of order n with
$$V(G) = \{v_1, v_2, ..., v_n\}.$$
The **degree matrix** of G is the $n \times n$ matrix
$$D(G) = (d_{ij})_{n \times n},$$
where
$$d_{ij} = \begin{cases} \text{the degree of } v_i, \text{ if } j = i, \\ 0, \qquad\qquad\qquad \text{otherwise.} \end{cases}$$

Thus, $D(G)$ is a diagonal matrix in which the diagonal entries are the degrees of the corresponding vertices in G. For instance, for the multigraph

G of Fig. 3.5.1,

$$D(G) = \begin{pmatrix} 3\,0\,0\,0 \\ 0\,5\,0\,0 \\ 0\,0\,4\,0 \\ 0\,0\,0\,4 \end{pmatrix}.$$

Observe that

$$D(G) - A(G) = \begin{pmatrix} 3 & -1 & 0 & -2 \\ -1 & 5 & -3 & -1 \\ 0 & -3 & 4 & -1 \\ -2 & -1 & -1 & 4 \end{pmatrix}.$$

We shall see that this resulting matrix $D(G) - A(G)$ plays a prominent role in evaluating $\tau(G)$.

Let M be an $n \times n$ matrix. For $i, j = 1, 2, ..., n$, the *cofactor* of the (i, j)-entry in M is defined as $(-1)^{i+j}$ times the determinant of the $(n-1) \times (n-1)$ matrix obtained from M by deleting the ith row and the jth column in M.

For instance, if M is the 4×4 matrix $D(G) - A(G)$ shown above, then the cofactor of the $(1, 1)$-entry in M is given by

$$(-1)^{1+1} \begin{vmatrix} 5 & -3 & -1 \\ -3 & 4 & -1 \\ -1 & -1 & 4 \end{vmatrix} = 29$$

and the cofactor of the $(3, 2)$-entry in M is given by

$$(-1)^{3+2} \begin{vmatrix} 3 & 0 & -2 \\ -1 & -3 & -1 \\ -2 & -1 & 4 \end{vmatrix} = 29.$$

Note that both cofactors are equal in this case (see Remark 3.5.3 below).

We are now ready to state, without proof, the following beautiful and surprising result, known as the Matrix-Tree Theorem, which is implicit in the work of Gustav Kirchhoff (1824–1887).

Theorem 3.5.1. *For any multigraph G, $\tau(G)$ is equal to the cofactor of any entry in $D(G) - A(G)$.*

$\square$

Example 3.5.2. Let G be the graph shown in Fig. 3.5.2. Verify that $\tau(G) = 8$ by the Matrix-Tree Theorem.

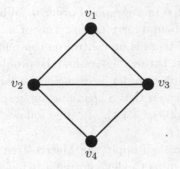

Fig. 3.5.2

We first compute the matrix $D(G) - A(G)$.

$$D(G) - A(G) = \begin{pmatrix} 2 & -1 & -1 & 0 \\ -1 & 3 & -1 & -1 \\ -1 & -1 & 3 & -1 \\ 0 & -1 & -1 & 2 \end{pmatrix}.$$

Evaluating the cofactor of the $(2,3)$-entry in $D(G) - A(G)$ gives, by Theorem 3.5.1, the following:

$$\tau(G) = \begin{vmatrix} 2 & -1 & 0 \\ -1 & -1 & -1 \\ 0 & -1 & 2 \end{vmatrix} = 8.$$

Remark 3.5.3.

(1) In general, given a square matrix M, the cofactors of different entries in M need not be the same. However, due to the special feature of $D(G) - A(G)$ (note that the sum of the entries in each row or column is zero), the cofactors of any two different entries in $D(G) - A(G)$ are the same. Thus, as highlighted in Theorem 3.5.1, the value of $\tau(G)$ is independent of the choice of the entry in $D(G) - A(G)$.

(2) As far as computing $\tau(G)$ is concerned, it seems that Theorem 3.4.3 is more user friendly than Theorem 3.5.1. In fact, if the order and size of G are large, Theorem 3.4.3 is impractical. On the other hand, there are efficient algorithms available to compute the determinant of a square matrix. Thus, one can compute $\tau(G)$ efficiently by Theorem 3.5.1.

Let G be a connected multigraph of order n. While $\tau(G) \geq 1$, we now look for a good upper bound for $\tau(G)$ in terms of n. Clearly, this problem does not make sense if there is no control on the number of parallel edges joining pairs of vertices. Let us thus confine the problem to (simple) graphs and ask: if G is a graph of order n, what is the largest value that $\tau(G)$ can attain? We note that if G is a spanning subgraph of a graph H, then $\tau(G) \leq \tau(H)$ (see Problem 3.4.4). It thus follows that $\tau(G) \leq \tau(K_n)$. What is the value of $\tau(K_n)$?

In what follows, we shall apply the Matrix-Tree Theorem to derive a celebrated result, known as Cayley's formula, which counts $\tau(K_n)$ for all $n \geq 1$.

Cayley's formula

More than 120 years ago, Arthur Cayley, a famous British mathematician, published his paper Cayley (1889) in which he established the following beautiful result.

Theorem 3.5.4.   *For each integer $n \geq 2$, $\tau(K_n) = n^{n-2}$.*

$\square$

For example, when $n = 4$, there are $4^{4-2} = 16$ different spanning trees of K_4, as shown in Fig. 3.5.3(b).

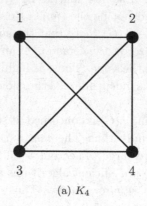

(a) K_4

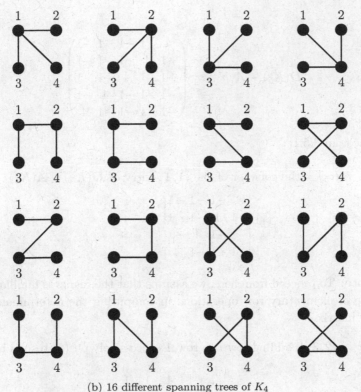

(b) 16 different spanning trees of K_4

Fig. 3.5.3

We shall now apply Kirchhoff's Matrix-Tree Theorem to derive Cayley's formula for the case when $n = 5$. The reader will easily find that the argument can be naturally extended to the general case.

Thus our objective now is to show that $\tau(K_5) = 5^3$ by apply Theorem 3.5.1. Observe that

$$D(K_5) = \begin{pmatrix} 4 & 0 & 0 & 0 & 0 \\ 0 & 4 & 0 & 0 & 0 \\ 0 & 0 & 4 & 0 & 0 \\ 0 & 0 & 0 & 4 & 0 \\ 0 & 0 & 0 & 0 & 4 \end{pmatrix} \quad \text{and} \quad A(K_5) = \begin{pmatrix} 0 & 1 & 1 & 1 & 1 \\ 1 & 0 & 1 & 1 & 1 \\ 1 & 1 & 0 & 1 & 1 \\ 1 & 1 & 1 & 0 & 1 \\ 1 & 1 & 1 & 1 & 0 \end{pmatrix}.$$

Thus,

$$
D(K_5) - A(K_5) = \begin{pmatrix}
4 & -1 & -1 & -1 & -1 \\
-1 & 4 & -1 & -1 & -1 \\
-1 & -1 & 4 & -1 & -1 \\
-1 & -1 & -1 & 4 & -1 \\
-1 & -1 & -1 & -1 & 4
\end{pmatrix}.
$$

By Theorem 3.5.1,

$\tau(K_5) =$ the cofactor of the $(1,1)$-entry of $D(K_5) - A(K_5)$

$$
= (-1)^{1+1} \begin{vmatrix}
4 & -1 & -1 & -1 \\
-1 & 4 & -1 & -1 \\
-1 & -1 & 4 & -1 \\
-1 & -1 & -1 & 4
\end{vmatrix}. \tag{*}
$$

Note: To proceed from here, we assume that the reader is familiar with the use of elementary row operations in computing the determinant of a square matrix.

For $i = 2, 3, 4$, adding row i to row 1 successively on (*) results in

$$
\tau(K_5) = \begin{vmatrix}
1 & 1 & 1 & 1 \\
-1 & 4 & -1 & -1 \\
-1 & -1 & 4 & -1 \\
-1 & -1 & -1 & 4
\end{vmatrix}. \tag{**}
$$

Now, for $i = 2, 3, 4$, adding row 1 to row i successively on (**) yields

$$
\tau(K_5) = \begin{vmatrix}
1 & 1 & 1 & 1 \\
0 & 5 & 0 & 0 \\
0 & 0 & 5 & 0 \\
0 & 0 & 0 & 5
\end{vmatrix} = 1 \times 5 \times 5 \times 5.
$$

We thus arrive at $\tau(K_5) = 5^3$, as asserted.

Exercise for Section 3.5

1. For each of the following multigraphs G, compute $\tau(G)$ by using Theorem 3.5.1.

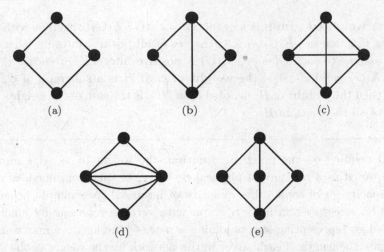

(a)   (b)   (c)

(d)   (e)

Fig. 3.5.4

2. Prove Cayley's formula for general K_n using the Matrix-Tree Theorem.

3. Consider the special complete bipartite graph $K(2, r)$, where $r \geq 1$. Show that $\tau(K(2, r)) = r2^{r-1}$ by (i) applying the Matrix-Tree Theorem; and (ii) any other method.

4. Let $G = K_n - e$, where e is an edge in K_n. Show that

$$\tau(G) = (n - 2)n^{n-3}.$$

3.6 Two Problems on Weighted Graphs

Very often, using graphs as mathematical models may not be good enough, as we may wish to include more information than simply the presence (or absence) of a relationship between pairs of elements in a set. One way of including more information is to assign a *weight* to each edge in a graph and this gives rise to the following definition.

A **weighted graph** is a graph $G = (V(G), E(G))$ together with a function $w : E(G) \to \mathbb{Z}^+$ (the set of all positive integers) that assigns to each edge $e \in E(G)$ a positive integer. For each $e \in E(G)$, $w(e)$ is called the **weight** of e. If H is a subgraph of G, then the **weight** of H, denoted by $w(H)$, is the sum of the weights of all the edges in H.

Depending on the practical situation, the weight of an edge may be interpreted as a measure of physical distance, of time consumed, of cost, of capacity, or of some other quantity of interest. For example, when we model a street system using a graph with vertices representing junctions and edges representing streets joining a pair of junctions, we may want to indicate the length of each street in the network by the weight of the edge.

Example 3.6.1.   Let G be the following weighted graph, with the weight of each edge indicated next to the edge.

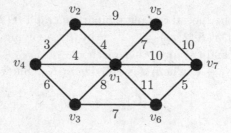

Fig. 3.6.1

Suppose each vertex in G represents a remote village and an edge between two vertices represents the possibility to build a transport route between the two villages. The **cost** of building such a route between the villages is the weight on each edge. There are no existing routes currently and we would like to construct some routes between the villages such that

(a) every pair of villages are connected either directly or indirectly (passing through some other villages);
(b) the total cost of building such routes is minimized.

We know that to satisfy (a), we need to have a spanning connected subgraph H of G where the edges in H represent those routes that are selected to be constructed. In order to minimize the total cost, we need to minimize the number of routes constructed (that is, the edges in H) as well as the sum of the weights of the edges in H (that is, the weight of H). It is now clear that we need to find a **spanning tree** of G with **minimum weight**.

> **The Minimum Spanning Tree Problem.** Let G be a connected weighted graph. A spanning tree T of G is said to be minimum if $w(T) \leq w(T')$ for any other spanning tree T' of G. Find a minimum spanning tree of G.

We will next present two different algorithms to solve the Minimum Spanning Tree Problem. The descriptions of the algorithms involve the notion of a subgraph induced by a set of edges. We will first introduce this notion.

Suppose G is a graph and F is a subset of $E(G)$. The subgraph of G **induced by** F, denoted by $[F]$, is the graph with edge set F and vertex set

$$\{v \in V(G) \mid v \text{ is incident with an edge in } F\}.$$

Example 3.6.2. Consider the graph G in Fig. 3.6.2(a). If $F = \{e_1, e_4, e_6\}$, then $[F]$ is shown in Fig. 3.6.2(b).

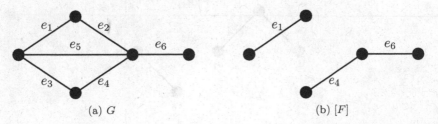

(a) G (b) $[F]$

Fig. 3.6.2

The first algorithm is due to Kruskal (1956).

Algorithm 3.6.3. (Kruskal's algorithm) Let G be a connected weighted graph of order n.

Step 1: Set $i := 0$, $E_0 := \emptyset$.

Step 2: Set $X := E(G) - E_i$. Choose an edge $e \in X$ such that (i) $[E_i \cup \{e\}]$ does not contain any cycle in G and (ii) $w(e) \leq w(e')$ for all those edges e' satisfying condition (i).

Step 3: Set $E_{i+1} := E_i \cup \{e\}$. If $|E_{i+1}| = n-1$, then $[E_{i+1}]$ is a minimum spanning tree of G. Otherwise, increase i by 1 and return to Step 2.

Example 3.6.4.   Applying Algorithm 3.6.3 on the weighted graph in Example 3.6.1, we have the following sequence of subgraphs of G.

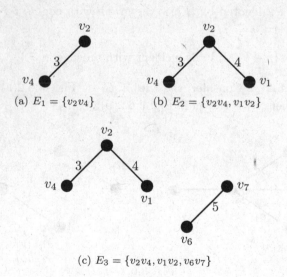

(a) $E_1 = \{v_2v_4\}$

(b) $E_2 = \{v_2v_4, v_1v_2\}$

(c) $E_3 = \{v_2v_4, v_1v_2, v_6v_7\}$

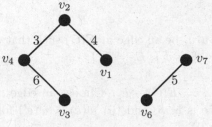

(d) $E_4 = \{v_2v_4, v_1v_2, v_6v_7, v_3v_4\}$

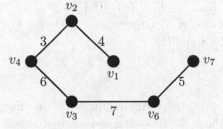

(e) $E_5 = \{v_2v_4, v_1v_2, v_6v_7, v_3v_4, v_3v_6\}$

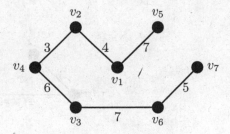

(f) $E_6 = \{v_2v_4, v_1v_2, v_6v_7, v_3v_4, v_3v_6, v_1v_5\}$

Fig. 3.6.3

The tree T induced by E_6 is a minimum spanning tree of G. The weight of T is $3 + 4 + 5 + 6 + 7 + 7 = 32$.

The next algorithm, widely attributed to Prim (1957), also gives us a minimum spanning tree of a connected graph G.

Algorithm 3.6.5. (Prim's algorithm) Let G be a connected weighted graph of order n.

Step 1: Set : $i = 1$.

Step 2: Let $e = u_1v_1$ be an edge in $E(G)$ such that $w(e) \leq w(e')$ for all $e' \in E(G)$. Set $S_i := \{u_1, v_1\}$, $E_i := \{e\}$.

Step 3(a): Set $X := E(G) - E_i$. Choose an edge $e \in X$ such that (i) exactly one end of e is in S_i and (ii) $w(e) \leq w(e')$ for all those edges e' satisfying condition (i).

Step 3(b): Set $E_{i+1} := E_i \cup \{e\}$. If $|E_{i+1}| = n - 1$, then $[E_{i+1}]$ is a minimum spanning tree of G. Otherwise, update $S_{i+1} := S_i \cup \{v\}$, increase i by 1 and return to Step 3(a).

Example 3.6.6.   Applying Algorithm 3.6.5 on the weighted graph in Example 3.6.1, we have the following sequence of subgraphs of G.

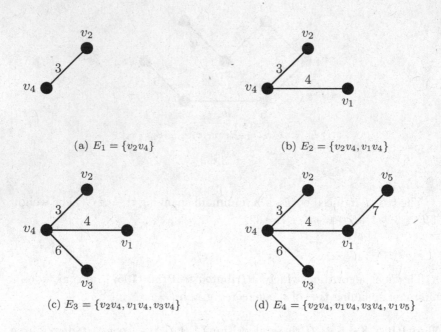

(a) $E_1 = \{v_2v_4\}$

(b) $E_2 = \{v_2v_4, v_1v_4\}$

(c) $E_3 = \{v_2v_4, v_1v_4, v_3v_4\}$

(d) $E_4 = \{v_2v_4, v_1v_4, v_3v_4, v_1v_5\}$

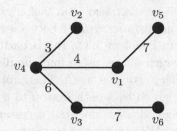

(e) $E_5 = \{v_2v_4, v_1v_4, v_3v_4, v_1v_5, v_3v_6\}$

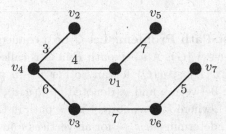

(f) $E_6 = \{v_2v_4, v_1v_4, v_3v_4, v_1v_5, v_3v_6, v_6v_7\}$

Fig. 3.6.4

The tree T induced by E_6 is a minimum spanning tree of G. The weight of T is $3 + 4 + 5 + 6 + 7 + 7 = 32$.

Remark 3.6.7.

(1) Note that using Algorithm 3.6.5, the subgraph induced by E_i for each i is always connected whereas this is generally not true if we use Algorithm 3.6.3.

(2) One often discusses the **complexity** of an algorithm when investigating how **efficient** is the algorithm in solving that problem in question. The discussion of complexities of various graph algorithms is omitted in this book. Interested readers may wish to see Chartrand and Oellermann (1993) for such discussion.

(3) Given any weighted connected graph G, the *weight* of its minimum spanning tree is **unique** but not necessarily the spanning tree that has this unique weight. For example, the above minimum spanning trees obtained by Kruskal's or Prim's algorithm are different but the weights are equal.

The second problem to be discussed is in fact a generalization of the question that computes the distance $d(u, v)$ between any two vertices in an unweighted graph G. In Problem 3.3.9, the **breadth-first search** algorithm was introduced which allows us to find the distance $d(u, v)$ from a fixed vertex u to all other vertices v in G. This problem becomes more difficult if the graph is weighted as a path from u to v traversing through a minimum number of weighted edges does not necessarily provide a $u - v$ path with minimum weight. We now present the **Shortest Path Problem** for weighted graphs formally.

The Shortest Path Problem. Let G be a connected weighted graph, and $u, v \in V(G)$. A $u - v$ path P in G is called a **shortest** $u - v$ path if $w(P) \leq w(Q)$ for any other $u - v$ path Q in G. The **distance** between u and v, denoted by $d(u, v)$, is defined by $d(u, v) = w(P)$, where P is a shortest $u - v$ path in G. For a fixed vertex w in G, determine $d(w, v)$ for all vertices v in $V(G) - \{w\}$.

Remark 3.6.8. Any (unweighted) connected graph G can be considered as a weighted graph where all the edges in G have weights equal to 1.

Let G be a connected weighted graph with $V(G) = \{u_1, u_2, ..., u_n\}$. The following algorithm due to Dijkstra (1959) computes, for a fixed vertex, say u_1, the distance $d(u_1, u_i)$ to **all other** vertices $u_i \in V(G)$. In this algorithm, each vertex is assigned a series of **temporary** labels $(j, L(i))$ terminating in a **permanent** label $[j, L(i)]$. The value $L(i)$ in $[j, L(i)]$ is the weight of a shortest $u_1 - u_i$ path and the j in $[j, L(i)]$ means that vertex u_j is the vertex on this shortest $u_1 - u_i$ path just preceding u_i.

Algorithm 3.6.9. (Dijkstra's algorithm) Let G be a connected weighted graph with $V(G) = \{u_1, u_2, ..., u_n\}$.

Step 1: Assign vertex u_1 the permanent label $[-, 0]$. All other vertices in $V(G)$ are assigned temporary labels $(-, *)$. Set $z := 1$.

Step 2(a): Set

$$A_z := \{u_k \in V(G) \mid u_k \in N(u_z), \text{ and } u_k\text{'s label is temporary}\}.$$

Step 2(b): For each $u_k \in A_z$, compare $L(u_k)$ and $L(u_z) + w(u_z u_k)$. If $L(u_z) + w(u_z u_k) \geq L(u_k)$, then the label for u_k is unchanged. If $L(u_z) + w(u_z u_k) < L(u_k)$, then update u_k's label to be the temporary label $(u_z, L(u_z) + w(u_z u_k))$.

Step 3: Set $T := \{u_j \in V(G) \mid$ label at u_j is temporary$\}$. Choose $u_r \in T$ such that $L(u_r) = \min\{L(u_j) \mid u_j \in T\}$. Change the label for u_r to permanent. If all vertices in G now have permanent labels, the process ends. Otherwise, set $z := r$ and return to Step 2(a).

Example 3.6.10. Apply Algorithm 3.6.9 to compute $d(u_1, u_j)$ for $j = 2, ..., 8$ in the weighted graph shown in Fig. 3.6.5.

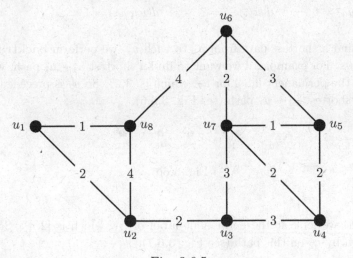

Fig. 3.6.5

The following table shows the series of labels assigned to each vertex u_i. Each row represents one iteration of Algorithm 3.6.9 before returning to Step 2(a) and the last column indicates the vertex whose label is changed from temporary to permanent. Once the label of a vertex is permanent, it is no longer shown in subsequent iterations (that is, rows).

u_1	u_2	u_3	u_4	u_5	u_6	u_7	u_8	Change
$[-,0]$	$(-,*)$	$(-,*)$	$(-,*)$	$(-,*)$	$(-,*)$	$(-,*)$	$(-,*)$	u_1
	$(1,2)$	$(-,*)$	$(-,*)$	$(-,*)$	$(-,*)$	$(-,*)$	$(1,1)$	u_8
		$(1,2)$	$(-,*)$	$(-,*)$	$(8,5)$	$(-,*)$		u_2
			$(2,4)$	$(-,*)$	$(8,5)$	$(-,*)$		u_3
			$(3,7)$	$(-,*)$	$(8,5)$	$(3,7)$		u_6
			$(3,7)$	$(6,8)$		$(3,7)$		u_4
				$(6,8)$		$(3,7)$		u_7
				$(6,8)$				u_5
								End

So we have found the distances between u_1 and u_j, $j = 2, ..., 8$ as follows:

$$d(u_1, u_2) = 2 \qquad d(u_1, u_3) = 4 \qquad d(u_1, u_4) = 7 \qquad d(u_1, u_5) = 8$$
$$d(u_1, u_6) = 5 \qquad d(u_1, u_7) = 7 \qquad d(u_1, u_8) = 1$$

To find a shortest path from u_1 to each u_j, we perform **backtracking** as follows. For example, if we want to find a shortest $u_1 - u_7$ path, we first look at the permanent label for u_7, which is $[3, 7]$. So u_7 is preceded by u_3 on this shortest $u_1 - u_7$ path (see Fig. 3.6.6).

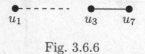

Fig. 3.6.6

Next, we look at the permanent label of u_3, which is $[2, 4]$. So u_3 is preceded by u_2 on this path (see Fig. 3.6.7).

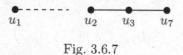

Fig. 3.6.7

Looking at the permanent label for u_2, which is $[1, 2]$, we see that u_2 is preceded by u_1 on this path. Thus $u_1 u_2 u_3 u_7$ is a shortest $u_1 - u_7$ path in G, whose weight is $2 + 2 + 3 = 7$.

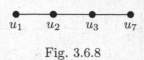

Fig. 3.6.8

You may wish to find all the other shortest $u_1 - u_j$ paths.

Remark 3.6.11. If we only want to compute the distance between u_1 and a particular vertex u_j, then the algorithm can terminate once the label for u_j changes to permanent.

Exercise for Section 3.6

1. Applying Algorithm 3.6.3 (Kruskal's algorithm), construct a minimum spanning tree of the weighted graph of order 7 whose weight matrix is given below. (Note: The (i, j)-entry gives the weight of the edge ij.) Repeat the problem by applying Algorithm 3.6.5 (Prim's algorithm) instead.

	1	2	3	4	5	6	7
1	0	3	8	5	2	6	5
2	3	0	8	5	3	5	6
3	8	8	0	9	7	9	8
4	5	5	9	0	6	4	4
5	2	3	7	6	0	4	6
6	6	5	9	4	4	0	4
7	5	6	8	4	6	4	0

2. Applying Algorithm 3.6.3 (Kruskal's algorithm), construct a minimum spanning tree of the weighted graph of order 9 whose weight matrix is given below. (Note: A '-' in the (i, j)-entry indicates there is no edge between vertices i and j.) Repeat the problem by applying Algorithm 3.6.5 (Prim's algorithm) instead.

	1	2	3	4	5	6	7	8	9
1	0	6	5	4	4	2	6	8	5
2	6	0	5	3	8	7	3	6	3
3	5	5	0	9	9	-	-	4	7
4	4	3	9	0	6	-	-	4	6
5	4	8	9	6	0	8	3	3	-
6	2	7	-	-	8	0	2	2	4
7	6	3	-	-	3	2	0	1	1
8	8	6	4	4	3	2	1	0	4
9	5	3	7	6	-	4	1	4	0

3. For each of the weighted graphs shown in Fig. 3.6.9, construct a minimum spanning tree by applying Algorithm 3.6.3 (Kruskal's algorithm). Repeat the problem by applying Algorithm 3.6.5 (Prim's algorithm) instead.

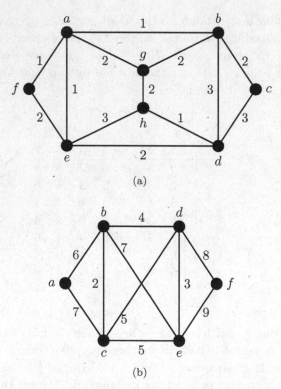

(a)

(b)

Fig. 3.6.9

4. Apply Dijkstra's algorithm to find the shortest paths (and the respective weights) from u_1 to all other vertices in the following weighted graph.

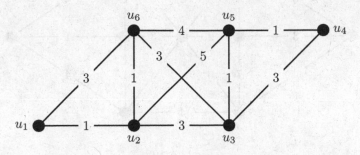

Fig. 3.6.10

5. Apply Dijkstra's algorithm to find the shortest paths (and the respective weights) from u_1 to all other vertices in the following weighted graph.

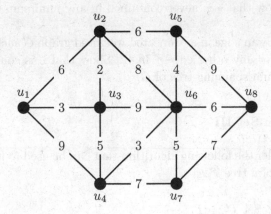

Fig. 3.6.11

6. Apply Dijkstra's algorithm to find the shortest paths (and the respective weights) from a to all other vertices in the following weighted graph.

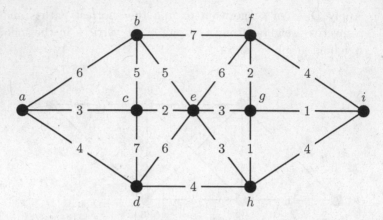

Fig. 3.6.12

7. Let G be a connected weighted graph in which the weights of the edges are all distinct. Show that G has one and only one minimum spanning tree.

8. Let G be a connected weighted graph and C, a cycle in G. Suppose e is an edge on C such that $w(e) > w(f)$ for any other edge f on C. Show that e is never contained in any minimum spanning tree of G.

9. Let e be an edge in a connected weighted graph G such that $w(e) < w(f)$ for any other edge f in G. Show that e is contained in any minimum spanning tree of G.

3.7   Problem Set III

1. Consider the following algorithm that can be used to find the center $C(T)$ of a tree T.

   Step 1: Set $T' := T$.

   Step 2: If $T' = K_1$ or K_2, then $C(T) = V(T')$. Otherwise, go to Step 3.

   Step 3: Remove all end vertices in T'. Let the resulting graph be T' and return to Step 2.

For each of the trees T shown in Fig. 3.7.1, apply the above algorithm to find $C(T)$.

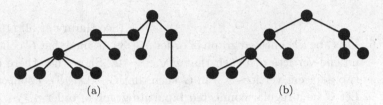

(a)   (b)

Fig. 3.7.1

2. Prove that the center $C(T)$ of a tree T consists of either a single vertex or a pair of adjacent vertices.
3. Let G be the graph shown in Fig. 3.7.2.

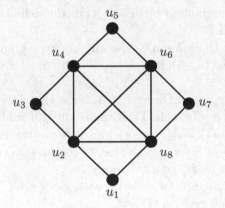

Fig. 3.7.2

(a) Find a bipartite subgraph H of G with largest size $e(H)$.
(b) Find a spanning tree of G that contains the least number of cut-vertices.
(c) Find a spanning tree T of G such that $T \cong G - E(T)$.

4. Show that $\tau(K(p,q)) = p^{q-1}q^{p-1}$ for any two positive integers p and q.

5. Show that

$$\tau(Q_n) = 2^{2^n - n - 1} \prod_{k=2}^{n} k^{\binom{n}{k}}.$$

(See Harary et al. (1988))

6. Let G be a connected graph of order $n \geq 3$. Assume that G contains no end-vertices u, v such that $d(u, v) = 2$. Show that there exist two adjacent vertices x, y in G such that $G - \{x, y\}$ is connected.

7. Let G be a regular connected bipartite graph of order $n \geq 4$. Can G contain a bridge?

8. Let G be a k-regular graph of order $2k + p$, where $k \geq 4$ and $1 \leq p < \frac{k}{2}$. Show that if G is non-bipartite, then G contains a C_3.

9. Let G be a graph with $\delta(G) \geq 3$. Show that G contains at least three cycles and at least one of them is an even cycle.

10. Let G and H be graphs. Show that the Cartesian product $G \,\square\, H$ is bipartite if and only if both G and H are bipartite. Deduce that the n-cube graph is bipartite. (For the definition of $G \,\square\, H$, see Problem II.12.)

11. Show that there doesn't exist any bipartite graph G whose degree sequence is $(10, \cdots, 10, 7, 5, \cdots, 5)$ with at least one 10 and at least one 5.

12. Let G be a graph of order $n \geq 6$. Show that either G or $\overline{G}$ contains $K(1, 3)$ as a subgraph. Does the result still hold if $n = 5$?

13. Let $n = (p-1)(q-1)$, where p and q are positive integers. Construct a graph G of order n such that G contains no tree of order p and $\overline{G}$ contains no K_q as a subgraph.

14. Let G be a graph of order n with girth $g(G) \geq 4$. Let C be a shortest odd cycle in G.

    (a) Show that each vertex in $G - V(C)$ is adjacent to at most two vertices in C.

    (b) Deduce that $\sum_{v \in V(C)} d(v) \leq 2n$.

    (c) Does the inequality in (b) still hold if $g(G) = 3$?

15. Let G be a bipartite graph with bipartition (X, Y) such that $|X| > |Y|$. Assume that $c(G - B) \leq |B|$ for all $B \subset Y$ with $|B| = |Y| - 1$. Show that

$$e(G) \geq |Y|(|X| - |Y| + 2).$$

16. Let G be an (n, m)-graph with $g(G) \geq 4$.

(a) Show that $mn \geq \sum_{v \in V} d(v)^2 \geq \frac{4m^2}{n}$.

(b) Deduce that $m \leq \frac{n^2}{4}$.

(c) Determine all such G for which the equality in (b) holds.

17. Prove that if a graph G contains no bridges, then neither does $L(G)$. (For the definition of $L(G)$, see Problem II.11.)

18. Let G be a graph and $n \geq 4$. Show that $L(G) \cong K_n$ if and only if $G \cong K(1, n)$.

19. Let G be a connected graph of order $n \geq 3$. Show that $L(G) \cong G$ if and only if $G \cong C_n$.

20. Let G and H be connected graphs. Show that $L(G) \cong L(H)$ if and only if $G \cong H$ or $\{G, H\} = \{K_3, K(1, 3)\}$.

21. Show that if a graph G contains a vertex of degree 3, then $L(L(G))$ contains an even vertex.

22. Let H be a graph and $e = uv$ be an edge in H. Call e an even (resp. odd) edge if $d(u)$ and $d(v)$ have the same (resp. different) parity in H. Show that

    (a) a vertex e in $L(H)$ is even if and only if e is an even edge in H.

    (b) the number of even (resp. odd) vertices in $L(H)$ is equal to the number of even (resp. odd) edges in H.

    (c) if a graph G contains a vertex of degree $k \geq 3$, then the number of even vertices in $L(L(G))$ is at least $\binom{\lceil \frac{k}{2} \rceil}{2} + \binom{\lfloor \frac{k}{2} \rfloor}{2}$.

    (Koh (2009, preprint))

23. Let G be a graph of order $n \geq 4$ satisfying the following condition: for any set $A \subseteq V$ with $|A| = 4$, $[A]$ contains $K(1, 3)$ as a subgraph.

    (a) Show that for any set $B \subseteq V$ with $|B| = 4$, B contains a vertex of degree $n - 1$ in G.

    (b) Let n^* denote the number of vertices of degree $n-1$ in G. Show that $n^* \geq n - 3$, and construct one such G with $n^* = n - 3$.

24. **[Multi-partite graphs]** A graph G is said to be r-**partite**, where $r \geq 2$, if $V(G)$ can be partitioned into r subsets $V_1, V_2, \cdots, V_r$ (called **partite sets**) such that each edge joins a vertex of V_i to a vertex of V_j for some i, j in $\{1, 2, \cdots, r\}$ with $i \neq j$. (Thus, G is bipartite when $r = 2$.) A 3-partite graph is shown in Fig. 3.7.3:

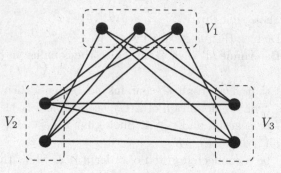

Fig. 3.7.3

An r-partite graph G with partite sets $V_1, V_2, \cdots, V_r$ is said to be **complete** if every vertex in V_i is adjacent to every vertex in V_j for all i, j in $\{1, 2, \cdots, r\}$ with $i \neq j$. If $|V_i| = p_i$ for each $i = 1, 2, \cdots, r$, such a complete r-partite graph G is unique (up to isomorphism), and is denoted by $K(p_1, p_2, \cdots, p_r)$. The complete 3-partite graph $K(3, 2, 2)$ is shown in Fig. 3.7.4:

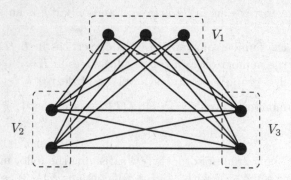

Fig. 3.7.4: $K(3, 2, 2)$

Let G be a non-empty graph. Show that G is a complete r-partite graph for some positive integer r if and only if the following implication holds for all vertices x, y, z in G:

$(x$ not adjacent to $y)$ and $(y$ not adjacent to $z)$

$\Rightarrow (x$ not adjacent to $z)$.

25. Let $G = K(p_1, p_2, \cdots, p_r)$, where $r \geq 3$ and $p_1 \geq p_2 \geq \cdots \geq p_r$. Evaluate in G

(a) $e(G)$;

(b) the number of C_3's;

(c) the number of K_4's;

(d) the degree of each vertex;

(e) $\delta(G)$;

(f) $\Delta(G)$.

26. Let G be a 3-partite graph of order $n \geq 3$. Show that $e(G) \leq \lfloor \frac{n^2}{3} \rfloor$. Determine all such G for which the equality holds.

27. Let G be a 3-partite graph with partite sets X, Y and Z such that $|X| = |Y| = |Z| = p$ and $d(v) = p+1$ for each vertex v in G. Show that G contains a C_3 as a subgraph.

28. There are 99 members in a club. Every one of them wishes to play bridge but only with his/her friends. Assume that each member has at least 67 friends in the club. Show that any member in the club can always have three of his/her friends to play bridge together. Is the conclusion still true if each member has at most 66 friends?

29. Let G be a graph. Show that both G and $\overline{G}$ are connected if and only if neither G nor $\overline{G}$ contains a complete bipartite graph as a spanning subgraph.

30. Let p, q be integers such that $p \geq 2$, $q \geq 2$ and $q - 1$ is a multiple of $p - 1$.

    (a) Construct a graph G of order $p+q-2$ such that G contains no tree of order p and $\overline{G}$ contains no star $K(1, q)$ as a subgraph.

    (b) Let G be a graph of order at least $p + q - 1$. Assume that $\overline{G}$ contains no star $K(1, q)$ as a subgraph. Show that G contains a tree of order p as a subgraph. [You may wish to apply the result of Problem 3.2.10.]

31. Let G be a graph of order at least three with $\delta(G) \geq 1$. Assume that $[\{x, y, z\}]$ is isomorphic to O_3 or P_3, for any 3 distinct vertices x, y, z in G.

    (a) Show that G contains no odd cycles.

    (b) Is G a complete bipartite graph? Justify your answer.

32. **[Balanced Incomplete Block Designs]** A **balanced incomplete block design** (BIBD) consists of a set X of v (≥ 2) objects, called **varieties**, together with a collection of b (> 0) subsets of X, called **blocks**, satisfying the following conditions:

    (i) each block contains exactly k (> 0) varieties,

   (ii) each variety appears in exactly r of the b blocks,

   (iii) every 2 varieties appear simultaneously in exactly λ of the b blocks, and

   (iv) $k < v$.

Such a BIBD is called a (b, v, r, k, λ)-design. We may represent a BIBD, $\boldsymbol{B}$, by a bipartite graph $G(\boldsymbol{B})$ with bipartitiion (X, Y), where X is the set of v varieties and Y is the set of b blocks, such that there is an edge joining a variety 'x' in X and a block 'A' in Y if and only if $x \in A$.

  (a) A $(7, 7, 3, 3, 1)$-design $\boldsymbol{B}$ is given below. Construct $G(\boldsymbol{B})$.

$$X = \{1, 2, \cdots, 7\} \quad \text{and}$$
$$Y = \{\{1, 2, 4\}, \{1, 3, 7\}, \{1, 5, 6\}, \{2, 3, 5\}, \{2, 6, 7\}, \{3, 4, 6\}, \\ \{4, 5, 7\}\}.$$

  (b) Let $\boldsymbol{B}$ be a (b, v, r, k, λ)-design.

    (i) Show, by counting the size of $G(\boldsymbol{B})$, that $bk = vr$.

    (ii) Fix a vertex 'x' in X. Show, by counting the number of paths xyz of length 2 in $G(\boldsymbol{B})$, that $r(k - 1) = \lambda(v - 1)$.

33. **[Graceful graphs]** Let $G = (V, E)$ be an (n, m)-graph, where $m \geq n - 1 \geq 1$. A **graceful valuation** of G is a $1 - 1$ mapping $\theta : V \to \{0, 1, \cdots, m\}$ such that the induced mapping π, defined by $\pi(uv) = |\theta(u) - \theta(v)|$ for each edge uv in G, is a bijection between E and $\{1, 2, \cdots, m\}$. We call G a **graceful** graph if it admits a graceful valuation. A graceful graph with graceful valuation is shown in Fig. 3.7.5:

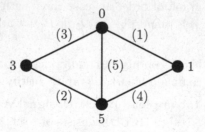

Fig. 3.7.5

(a) Let θ be a graceful valuation of G. Show that the mapping $\theta^* : V \to \{0, 1, \cdots, m\}$ defined by

$$\theta^*(v) = m - \theta(v)$$

is also a graceful valuation of G.

(b) Show that the complete graph K_n is graceful if and only if $2 \leq n \leq 4$.

(c) Show that the path P_n, $n \geq 2$, is graceful.

(d) A tree T is called a **caterpillar** if $T - A$ is a path, where A is the set of end-vertices in T (see Fig. 3.2.2(b)). Show that every caterpillar is graceful.

(e) Show that the complete bipartite graph $K(p, q)$ is graceful for all $p \geq q \geq 1$.

Note: Graceful graphs were introduced by Rosa (1966) and Golomb (1972) independently. A famous conjecture in this area is that **every tree is graceful**, which has not been settled yet. For a general survey on this topic, see the excellent article Gallian (2013).

34. Let $G + H$ denote the join of graphs G and H (for definition, see Problem 2.5.6). The **fan** of order $n \geq 4$, denoted by F_n, is defined as the join $P_{n-1} + O_1$. The **wheel** of order $n \geq 4$, denoted by W_n, is defined as the join $C_{n-1} + O_1$. Figure 3.7.6 shows a graceful fan and a graceful wheel.

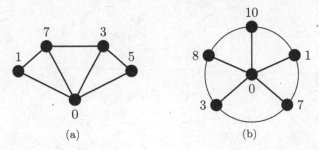

Fig. 3.7.6

(a) Show that every fan F_n, where $n \geq 4$, is graceful.

(b) Show that every wheel W_n, where $n \geq 4$, is graceful.

(Frucht (1979) and Hoede and Kuiper (1987))

35. Show that, for any graceful tree T, the following joins of graphs are also graceful:

    (a) $T + K(1, p)$;
    (b) $T + O_p$, where $p \geq 1$.

                                        (Koh, Rogers and Lim (1979))

36. Show that the following Cartesian products of graphs are graceful:

    (a) $P_n \,\square\, P_2$;
    (b) $K(1, 2p) \,\square\, P_2$;

                                                (Maheo (1980))

    (c) $C_n \,\square\, P_2$.

                                        (Frucht and Gallian (1988))

Chapter 4

Eulerian Multigraphs and The Chinese Postman Problem

4.1 Euler Circuits and Eulerian Multigraphs

We introduced the Königsberg bridge problem at the beginning of this book and, in Section 1.3, we mentioned how Euler generalized it to the following more general problem:

Let G be a multigraph. Assume that, starting with an arbitrary vertex in G, we can have a walk which passes through each edge once and only once, and then be able to terminate at the starting vertex. What could be said about G?

As promised, we shall discuss Euler's solution in this chapter. Much more than that, its further development and applications will be presented.

Let us begin by introducing some basic terms and note that in this chapter we shall not confine ourselves to 'simple graphs'.

Let G be a connected multigraph. Recall that a *circuit* in G is a closed walk in which no edge is repeated. A circuit W in G is called an **Euler circuit** if W contains all the edges in G. The multigraph G is called an **Eulerian multigraph** if G possesses an Euler circuit.

Example 4.1.1. The multigraph of Fig. 4.1.1(a) is Eulerian as it has an Euler circuit

$$W: \quad xe_1we_4ye_5we_3xe_7ze_8ye_9ze_6we_2x$$

as shown in Fig. 4.1.1(b).

165

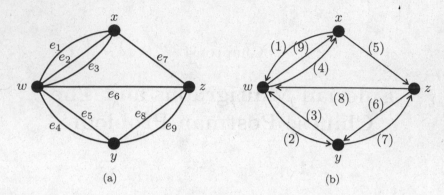

Fig. 4.1.1

Question 4.1.2. *Show that each of the following multigraphs is Eulerian by exhibiting an Euler circuit. Is there an odd vertex in any of the multigraphs?*

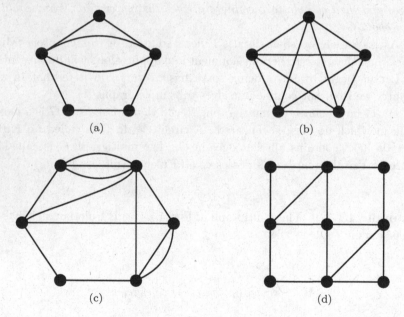

Fig. 4.1.2

Question 4.1.3.   *Are the following multigraphs Eulerian? Are there any odd vertices in each multigraph?*

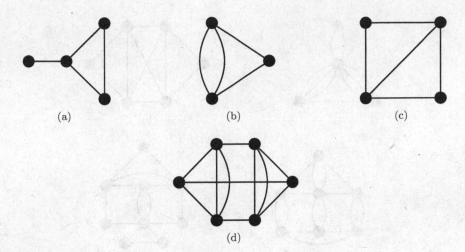

(a)   (b)   (c)

(d)

Fig. 4.1.3

Exercise for Section 4.1

1. Consider the following multigraph G of order 5.

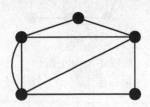

Fig. 4.1.4

  (i) Find $e(G)$.
 (ii) Find in G a circuit with 2 edges; with 3 edges; with 4 edges.
(iii) Find in G a circuit with 5 edges that is not a cycle.
 (iv) Find in G a circuit with 6 edges.
  (v) If W is an Euler circuit in G, exactly how many edges are contained in W?
 (vi) Does G contain an Euler circuit? Show one if there is.

2. Five multigraphs are shown in Fig. 4.1.5. Show that each of them
   is Eulerian by exhibiting an Euler circuit.

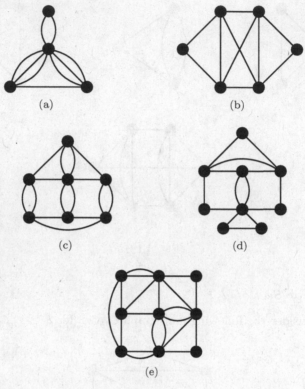

Fig. 4.1.5

4.2   Characterizations of Eulerian Multigraphs

Let us recall some observations found in the discussion of Question 4.1.2
and Question 4.1.3. We see that all the multigraphs in Question 4.1.2
are Eulerian and at the same time they contain no odd vertices. On the
other hand, all multigraphs in Question 4.1.3 are not Eulerian and they do
contain some odd vertices.

A natural question is: *Does whether a given multigraph G is Eulerian
or not depend on the non-existence of odd vertices in G?*

Euler (1736) studied this question. Assume that G is a connected

Eulerian multigraph. Then G possesses an Euler circuit, say W. He claimed that every vertex in G is even.

Let v be a vertex in G.

Case (1). v is not the initial vertex in W (and hence not the terminal vertex in W). Then each time we traverse along W to visit v, there must be two edges in W incident with v such that one is for us to enter v and the other to leave v. Since all edges incident with v are contained in W, the number of edges incident with v is even, that is, $d(v)$ is even.

Case (2). v is the initial vertex in W (and so is the terminal vertex in W). For the first move, there is an edge in W for us to leave v, and for the last move, there is another edge in W for us to return to v. Besides these, if there is an opportunity to visit v, then just like in Case (1), there are two edges in W such that one is for entering v and the other for leaving v. Thus, again, $d(v)$ is even.

We now have the following conclusion:

(*) If G is a connected Eulerian multigraph, then every vertex in G is even.

Remark 4.2.1.

(1) We now revisit the Königsberg bridge problem. The multigraph that models the situation is re-depicted below:

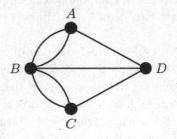

Fig. 4.2.1

Note that no vertex in it is even. It follows immediately from (*) that the answer to the Königsberg bridge problem is 'no'. Likewise, all the multigraphs in Question 4.1.3 contain some odd vertices, and accordingly, they all are not Eulerian.

(2) Is the converse of (*) true? That is, if G is a connected multigraph in which every vertex is even, must G be Eulerian? Euler thought that the answer is 'yes', but did not provide a proof. Unaware of Euler's work, Carl Hierholzer (1840–1871), a young German mathematician, gave not only a proof of (*), but also a proof of its converse.

(3) Veblen (1912) found another characterization of Eulerian multigraphs G in terms of the cycle-partition of $E(G)$. This result and the previous results are now combined together as shown in Theorem 4.2.2.

Theorem 4.2.2. *Let G be a connected multigraph. The following statements are equivalent:*

(1) G is Eulerian;

(2) every vertex in G is even;

(3) the edge set $E(G)$ can be partitioned into cycles.

Proof. For (1)$\Rightarrow$(2), see the above proof of (*).

(2)$\Rightarrow$(3): We prove this by induction on $e(G)$. Thus, let G be a connected multigraph with no odd vertices. If $e(G) = 0$, then (3) is obvious since G is a singleton in this case. Assume that (3) is true for all such G with $e(G) < k$, where $k > 0$; and now, consider the case when $e(G) = k$. As $\delta(G) \geq 2$, G contains a cycle C. Let $G' = G - E(C)$. Note that every vertex in each component of G' is again even. Thus, by the induction hypothesis, the edge set of each component can be partitioned into cycles. Accordingly, the edge set $E(G)$ can be partitioned into cycles, in which C is a member.

(3)$\Rightarrow$(1): Let S_1 be a cycle in this partition of $E(G)$. If S_1 is the only cycle in this partition, then $G = S_1$ and G is Eulerian. Otherwise, as G is connected, there is another cycle, say S_2, in the partition which has a vertex, say v, in common with S_1. We now form a circuit W_1 as follows: Begin with v, traverse S_1, reach v, traverse S_2, terminate at v. If S_1 and S_2 are the two cycles in the partition, then G is Eulerian. Otherwise, by applying a similar argument, W_1 can be extended to a new circuit W_2 of G. Continuing in this manner, an Eulerian circuit will eventually be obtained, and so G is Eulerian.

$\square$

Example 4.2.3. The following multigraph G is Eulerian since all vertices are even.

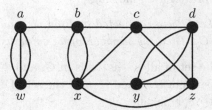

Fig. 4.2.2

Also, a partition of $E(G)$ into cycles is shown in Fig. 4.2.3:

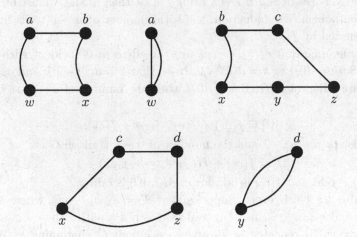

Fig. 4.2.3

Can you find an Euler circuit for G?

Question 4.2.4.

    (i) Is every cycle Eulerian?

    (ii) Can any tree be Eulerian?

    (iii) Which K_n, $n \geq 2$, are Eulerian?

    (iv) Which $K(p, q)$, $p \geq q \geq 2$, are Eulerian?

Toida (1973) found an interesting property of Eulerian multigraphs. To get to this, he first proved the following:

Lemma 4.2.5.   *Let G be a connected multigraph, and u and v be the only two odd vertices in G. Then the number of $u - v$ paths in G is odd.*

Proof. Let Ω be the set of all $u - v$ trails in G, and Φ the set of all $u - v$ trails that are not paths in G. Then $\Omega \backslash \Phi$ is the set of all $u - v$ paths in G. We shall prove that $|\Omega \backslash \Phi|$ is odd.

Claim 1. $|\Omega|$ is odd. We need only to consider the case where G has no parallel edges since removing a pair of parallel edges from G does not change the parity of $|\Omega|$. The result is obvious if G is a tree. Now let u and v be two fixed vertices. Let $e = uw$ be any edge of G incident with u and w, where $w \neq v$. Consider the graph $G_e = G - e$. Then v and w are the only odd vertices in G_e. If e is a bridge of G, then w and v must be in the same component. By induction, the total number of $w - v$ trails in G_e is odd, denoted by $f(e)$.

Now assume that $e_1, e_2, \cdots, e_k$ are the edges in G incident with u but not v. So $k = d(u)$ or $k = d(u) - 1$. Note that $k = d(u) - 1$ if and only if u and v are adjacent in G. If $k = d(u)$, then the number of $u - v$ trails in G is

$$|\Omega| = f(e_1) + f(e_2) + \cdots + f(e_k);$$

otherwise, $k = d(u) - 1$ and the number of $u - v$ trails in G is

$$|\Omega| = f(e_1) + f(e_2) + \cdots + f(e_k) + 1.$$

As $d(u)$ is odd and $f(e_i)$ is odd for each i, $|\Omega|$ is odd.

Claim 2. $|\Phi|$ is even. Suppose that $T = w_0 w_1 \cdots w_k$, where $k \geq 3$, $u = w_0$ and $v = w_k$, is an $u - v$ trail that is not a path. Then some vertex in T, say w_i, is repeated in T, and so a circuit Q containing w_i in T is formed. The existence of such a Q gives rise to two $u - v$ trails in pair by reversing the sequence of vertices in Q. Thus, trails in Φ always occur in pairs, and so $|\Phi|$ is even.

It now follows from the above claims that $|\Omega - \Phi|$, which equals $|\Omega| - |\Phi|$, is odd.

$\square$

Suppose now that G is a connected Eulerian multigraph. Let $e = uv$ be an edge in G. Then u and v are the only two odd vertices in $G - e$. Note that $G - e$ is connected (why?). By Lemma 4.2.5, the number of $u - v$ paths in $G - e$ is odd. Toida then deduced from this the following necessary condition for a connected multigraph to be Eulerian.

(A) If G is a connected Eulerian multigraph, then any edge in G is contained in an odd number of cycles.

Example 4.2.6. The following graph G is Eulerian, and it can be checked that each edge in G is contained in an odd number of cycles. For instance, the edge ux is contained in the following 5 cycles: $uxvu$, $uxvwu$, $uxvwyu$, $uxvzwu$, $uxvzwyu$. You may want to find the number of cycles containing the edge uv.

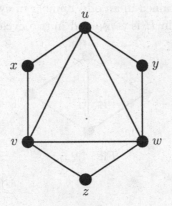

Fig. 4.2.4

How about the converse of (A)? That is, if every edge in a connected multigraph G is contained in an odd number of cycles, is G then Eulerian?

The answer is 'yes'. Indeed, by using a '**matroid**' approach, McKee (1984) showed that the converse is also true. Thus, we have another characterization of Eulerian multigraphs. The proof given below is by contradiction.

Theorem 4.2.7. *Let G be a connected multigraph. Then G is Eulerian if and only if every edge in G is contained in an odd number of cycles.*

Proof. It remains to prove the sufficiency. Assume that every edge in G is contained in an odd number of cycles. We shall prove that G is Eulerian. Suppose G is not Eulerian. Then, by Theorem 4.2.2, G contains an odd vertex w. Let $\sigma(e)$ denote the number of cycles containing an edge e.

Consider the sum

$$s = \sum \left(\sigma(e) \mid e \text{ incident with } w \right).$$

On one hand, by assumption, each term $\sigma(e)$ is odd and, as $d(w)$ is odd, the number of terms in the summation is also odd. Thus the sum s is odd.

On the other hand, as every cycle containing e also contains another edge e' incident with w, the sum s counts every cycle containing w twice. This implies that the sum s is even, a contradiction.

$\square$

Example 4.2.8. The following graph H is not Eulerian, and thus **not** every edge in H is contained in an odd number of cycles. Indeed, it is easy to see that every edge in H is contained in two cycles.

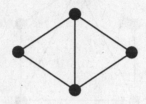

Fig. 4.2.5

4.3   Semi-Eulerian Multigraphs

A multigraph is Eulerian if it contains an Euler circuit, that is, a closed walk that passes through each edge once and only once. We now relax the 'closed' condition for such a walk. A multigraph is **semi-Eulerian** if it contains an **open** Euler trail, that is, an open walk that passes through each edge once and only once.

Example 4.3.1. Though none of the multigraphs in Question 4.1.3 is Eulerian, careful checking shows that each of them does possess an **open** walk which passes through each of its edges once and exactly once as shown in Fig. 4.3.1. They all are non-Eulerian but semi-Eulerian. Since they are not Eulerian, by Theorem 4.2.2, they contain odd vertices. Since the number of odd vertices in any multigraph, by Corollary 1.3.9, must be even, each of them contains at least two odd vertices. How many odd vertices are there in each of the multigraphs? What kinds of vertices are initial vertices and terminal vertices of such open walks?

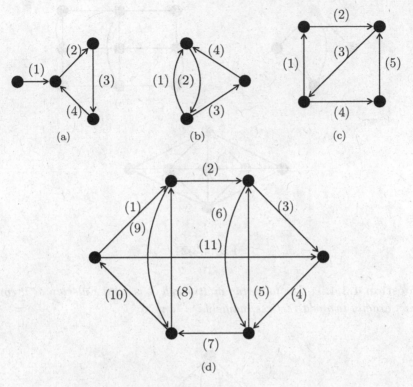

Fig. 4.3.1

We now state without proof the following result.

Theorem 4.3.2.   *Let G be a connected multigraph. Then G is semi-Eulerian if and only if G has exactly two odd vertices. Moreover, if G is semi-Eulerian, then the two odd vertices in G are the initial and terminal vertices of any Euler trail in G.*

□

Note: The multigraphs considered in Lemma 4.2.5 are thus semi-Eulerian multigraphs.

Question 4.3.3.   *In each of the following multigraphs, find the number of odd vertices. Which ones are semi-Eulerian?*

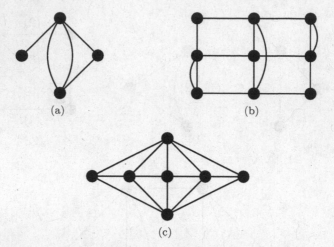

Fig. 4.3.2

Question 4.3.4.   *The following multigraph G is semi-Eulerian as it contains exactly two odd vertices, namely, 'v' and 'x'.*

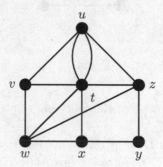

Fig. 4.3.3

    (i) *Form a new multigraph G^* by adding to G a new edge joining 'v' and 'x'. Is G^* Eulerian?*

    (ii) *Find an Euler circuit W in G^* (note that the new edge 'vx' is contained in W).*

    (iii) *Delete 'vx' (or 'xv') from W. Can you find an open Euler trail of G from the resulting sequence of edges?*

The above procedure of forming G^* from G could be useful in proving Theorem 4.3.2 (see Problem 4.3.7).

Example 4.3.5. The following multigraph G contains exactly 4 odd vertices, namely, b, c, x and z. Can you partition the edge set $E(G)$ into two trails Q_1 and Q_2 such that Q_1 is a $b - z$ trail while Q_2 is a $c - x$ trail in G?

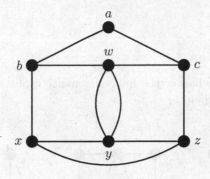

Fig. 4.3.4

The answer is 'yes'. A partition $\{Q_1, Q_2\}$ of $E(G)$ is shown in Fig. 4.3.5, where $Q_1 = bwyz$ and $Q_2 = cabxywczx$.

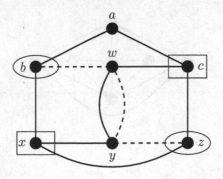

Fig. 4.3.5

A more general result which extends Theorem 4.3.2 is stated below (see Problem 4.3.16). Remember that the number of **odd** vertices in any multigraph is always **even**.

Theorem 4.3.6.   *Let G be a non-Eulerian multigraph. Then the $2k$ odd vertices in G can be arbitrarily divided into k pairs $\{u_i, v_i\}$'s and the edge set $E(G)$ can be partitioned into k trails Q_i's, such that Q_i is a $u_i - v_i$ trail for each $i = 1, 2, \cdots, k$.*

$\square$

Exercise for Section 4.3

1. Determine whether the following multigraphs are Eulerian, semi-Eulerian or neither:

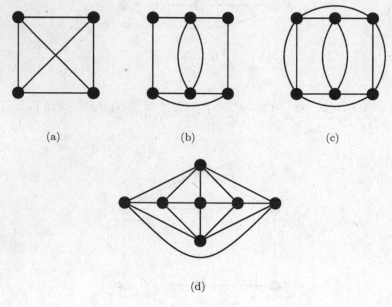

(a)   (b)   (c)

(d)

Fig. 4.3.6

2. Let G be the multigraph considered in Problem 4.1.1. Does G contain a circuit with 7 edges? Justify your answer.

3. Let G be an Eulerian multigraph of size m. Can G contain a circuit with $m - 1$ edges? Justify your answer.

4. Determine if each of the following statements is true:

   (i) If G is an Eulerian graph, then G contains no cut-vertices.
   (ii) If G is an Eulerian graph, then G contains no bridges.
   (iii) If G is an Eulerian graph of odd order and $\overline{G}$ is connected, then $\overline{G}$ is also an Eulerian graph.

5. Which $K(p, q)$, $p \geq q \geq 1$, are semi-Eulerian?
6. The following multigraph is semi-Eulerian as it contains exactly two odd vertices, namely, x and z.

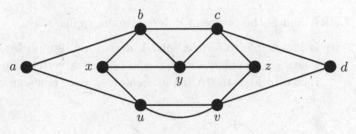

Fig. 4.3.7

   (i) Form a new multigraph G^* by adding to G a new edge joining x and z. Is G^* Eulerian?
   (ii) Find an Euler circuit W in G^*.
   (iii) Delete xz (or zx) from W. Can you find an open Euler trail of G from the resulting sequence of edges?

7. Prove Theorem 4.3.2.
8. Two halls are partitioned into small rooms for an exhibition event in two different ways as shown in Fig. 4.3.8(a) and (b), where A is the entrance and B is the exit.

   (i) Is it possible for a visitor to have a route which enters at A, passes through each door once and exactly once, and exits at B in partition (a)?
   (ii) Explain why such a route is not available in partition (b). Which door should be closed to ensure the existence of such a route?

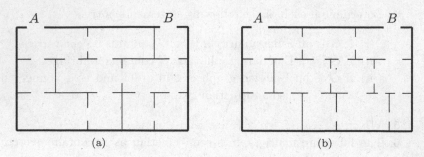

(a)   (b)

Fig. 4.3.8

9. Let G_1 and G_2 be two semi-Eulerian multigraphs.

   (i) Is it possible to form a semi-Eulerian multigraph by adding
       a new edge joining a vertex u in G_1 and a vertex v in G_2 as
       shown in Fig. 4.3.9? If the answer is 'yes', how can this be
       done?

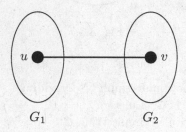

Fig. 4.3.9

   (ii) Is it possible to form an Eulerian multigraph by adding two
        new edges, each of which joins a vertex in G_1 and a vertex in
        G_2? If the answer is 'yes', how can this be done?

10. Let G_1 and G_2 be two connected multigraphs having $2p$ and $2q$
    odd vertices respectively, where $1 \leq p \leq q$. We wish to form an
    Eulerian multigraph from G_1 and G_2 by adding new edges, each of
    which joins a vertex in G_1 and a vertex in G_2. What is the least
    number of edges that should be added? How can this be done?

11. The following graph H is not Eulerian. What is the least number
    of new edges that should be added to H so that the resulting multi-
    graph becomes Eulerian? In how many ways can this be done?

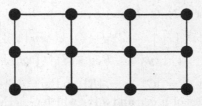

Fig. 4.3.10

12. Let G be a semi-Eulerian multigraph of order 8 and size 18, and with $\delta(G) = 3$ and $\Delta(G) = 6$. Assume that G contains exactly two vertices of degree 6. How many vertices of degree 3 does G have? Justify your answer and construct one such multigraph.

13. Let G be a graph which contains $K(5,6)$ as a spanning subgraph.

   (i) If G is semi-Eulerian, find the minimum size of G, and construct one such extremal semi-Eulerian graph G.

   (ii) If G is Eulerian, find the minimum size of G, and construct one such extremal Eulerian graph G.

14. Let G be a multigraph which contains $K(5,7)$ as a spanning subgraph.

   (i) Assume that G is semi-Eulerian. Can G be simple? If 'yes', find the minimum size of G, and construct one such extremal semi-Eulerian graph G.

   (ii) Assume that G is Eulerian. Prove that G cannot be simple. Find the minimum size of G, and construct one such extremal Eulerian multigraph G.

15. Let G be a $(8,10)$-Eulerian graph.

   (i) Let k be the maximum possible value of $\Delta(G)$. Determine k and construct all such G with $\Delta(G) = k$.

   (ii) Suppose $\Delta(G) = 4$.

      (a) Determine the number of vertices of degree 4 in G.

      (b) Assume further that no two vertices of degree 4 are adjacent. Construct all such G.

16. Prove Theorem 4.3.6.

17. Let G be a graph such that $V(G)$ is the disjoint union of four

non-empty sets P, Q, R and S, and

$$E(G) = \{xy \mid x \in P, y \in Q\} \cup \{xy \mid x \in Q, y \in R\}$$
$$\cup \{xy \mid x \in R, y \in S\} \cup \{xy \mid x \in S, y \in P\}.$$

Let $p = |P|$, $q = |Q|$, $r = |R|$ and $s = |S|$. Determine all the possible values of p, q, r and s for which G is Eulerian. Justify your answer.

18. Let G be a $(9, 20)$-Eulerian graph with $\Delta(G) = 6$. Assume that G has at most two vertices of degree 6. Find $\delta(G)$ and the number of vertices of degree $\delta(G)$ in G.

4.4   Finding Euler Circuits and Trails

Characterizations of Eulerian multigraphs and semi-Eulerian multigraphs were respectively presented in Section 4.2 and Section 4.3. In this section, we shall introduce two efficient algorithms, namely, Hierholzer's algorithm (1871) and Fleury's algorithm (1883), for constructing their Euler circuits or Euler trails.

The idea for the first algorithm is essentially the same as the argument given in the proof of $(3) \Rightarrow (1)$ in Theorem 4.2.2.

Algorithm 4.4.1. (Hierholzer's algorithm) Let G be an Eulerian multigraph.

Step 1: Choose an arbitrary vertex v in G. Starting from v, we obtain a circuit T_1 in G, terminating at v. If the edges in G are all in T_1, stop. Otherwise, set $i := 1$ and go to Step 2(a).

Step 2(a): Choose an arbitrary vertex u in T_i which is incident with an edge **not** in T_i, and extend from it to obtain a circuit T' in $G - E(T_i)$, terminating at u.

Step 2(b): Combine T_i and T' to obtain an enlarged circuit T_{i+1} in G. If the edges in G are all in T_{i+1}, stop. Otherwise, increase the value of i by 1 and return to Step 2(a).

Example 4.4.2.
Apply Algorithm 4.4.1 to find an Euler circuit in the following Eulerian multigraph.

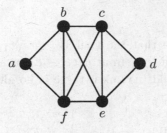

Fig. 4.4.1

Starting with a, we have T_1, T_2, T_3 as shown in Fig. 4.4.2.

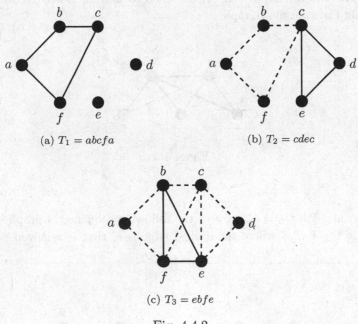

(a) $T_1 = abcfa$

(b) $T_2 = cdec$

(c) $T_3 = ebfe$

Fig. 4.4.2

An Euler circuit is $abcdebfecfa$.

Algorithm 4.4.3. (Fleury's algorithm) Let G be an Eulerian multigraph.

Step 1: Choose an arbitrary vertex v_0 in G and set $W_0 := v_0$. Set $i := 0$,

$G_i := G$, $E_i := \emptyset$.

Step 2(a): Suppose that the trail $W_i = v_0 e_1 v_1 ... e_i v_i$ has been constructed. Select an edge e_{i+1} from $E(G) - E_i$ such that (i) $e_{i+1} = v_i v_{i+1}$ for some vertex v_{i+1} and (ii) unless there is no alternative, e_{i+1} is not a bridge of G_i.

Step 2(b): Update $W_{i+1} := W_i e_{i+1} v_{i+1}$, $E_{i+1} := E_i \cup \{e_{i+1}\}$. Remove the edge e_{i+1} from G_i, together with any isolated vertices in G_i. If the resulting graph has no more edges, stop. Otherwise, let the resulting graph be G_{i+1}, increase the value of i by 1 and return to Step 2(a).

Example 4.4.4.   Apply Algorithm 4.4.3 to find an Euler circuit in the following Eulerian multigraph.

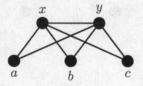

Fig. 4.4.3

Starting with $G_0 = G$, $v_0 = a$, the following sequence of graphs shows G_{i-1} for $i = 1, ..., 7$ where the dashed edge is e_i that is removed to form G_i.

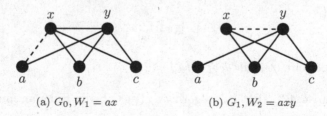

(a) $G_0, W_1 = ax$ (b) $G_1, W_2 = axy$

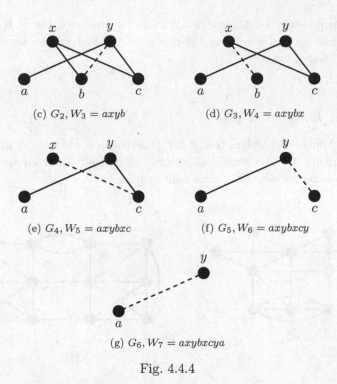

(c) $G_2, W_3 = axyb$

(d) $G_3, W_4 = axybx$

(e) $G_4, W_5 = axybxc$

(f) $G_5, W_6 = axybxcy$

(g) $G_6, W_7 = axybxcya$

Fig. 4.4.4

An Euler circuit of G is $W_7 = axybxcya$.

Remark 4.4.5.

(1) If we wish to find an Euler trail in a semi-Eulerian multigraph, the two algorithms discussed above can still be used. For Algorithm 4.4.3, we simply start at one of the odd vertices in the multigraph and the algorithm will end at the other odd vertex. For Algorithm 4.4.1, one method is to 'artificially' add one edge between the two odd vertices in the multigraph. This creates an Eulerian multigraph where Algorithm 4.4.1 can proceed. From the resulting Euler circuit, removal of the artificially added edge will result in an Euler trail as desired.

(2) We discuss briefly on the complexity of both algorithms. Although the graph traversal in Algorithm 4.4.3 is linear in the number of edges, we also need to consider the complexity of the bridge detection step. Using efficient algorithms for this step, the overall

complexity of Algorithm 4.4.3 can be improved but it is still significantly slower than Algorithm 4.4.1, which can be performed in
linear time.

Exercise for Section 4.4

1. Apply both Algorithm 4.4.1 (Hierholzer's algorithm) and Algorithm 4.4.3 (Fleury's algorithm) to find an Euler circuit in each
   of the following Eulerian multigraphs:

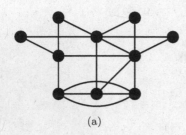 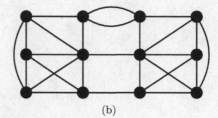

   (a)   (b)

Fig. 4.4.5

2. Identify two odd vertices in the following semi-Eulerian multigraph.
   Then apply Algorithm 4.4.3 (Fleury's algorithm) to find an Euler
   trail in the multigraph.

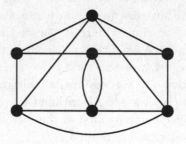

Fig. 4.4.6

4.5   The Chinese Postman Problem

Example 4.5.1.   Mr. Tan is a postman in a small town. There is a house located along each road in the town and there is a post-box in front of each house. Figure 4.5.1 shows the town with 13 roads and 8 road junctions, where the length of each road is labeled on the edge. Mr. Tan starts his workday everyday at the post office (located at P) where he collects the mail that he needs to distribute to all the 13 houses. There is no particular order that Mr. Tan needs to deliver the mail but as the length of each road is different, Mr. Tan would like to minimize the total distance he needs to travel before finishing his work for the day and returning to the post office.

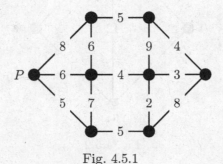

Fig. 4.5.1

Clearly, Mr. Tan's minimization problem would be easy to solve if the weighted graph G in Fig. 4.5.1 is Eulerian. If that was the case, Mr. Tan would just need to find an Euler circuit starting from vertex P which would allow him to pass through each road once and only once, deliver the letters to each of the 13 houses and return to P.

Unfortunately, G is not Eulerian since it has odd vertices, so in order for Mr. Tan to deliver all the letters and return to P, he would need to pass through some of the roads more than once. The problem now is: which ones?

Problems like the one in Example 4.5.1 were first proposed and studied by the Chinese mathematician Guan Meigu (see Kwan (1960)). Such problems are now affectionately known as **The Chinese Postman Problem** (abbreviation CPP).

The Chinese Postman Problem. Let G be a connected weighted multigraph. Find a closed walk $W = v_0 e_1 v_1 e_2 ... e_q v_q$ $(v_q = v_0)$ in G which contains all the edges of G such that $w(W) = \sum_{i=1}^{q} w(e_i)$ is the minimum.

Example 4.5.2.   Let G be the following weighted graph.

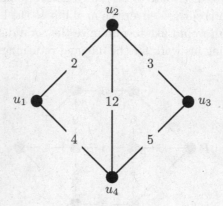

Fig. 4.5.2

Consider two closed walks $W_1 = u_2 u_1 u_4 u_2 u_4 u_3 u_2$ and $W_2 = u_2 u_1 u_4 u_2 u_3 u_4 u_1 u_2$ that both contain all edges of G. We have $w(W_1) = 38$ and $w(W_2) = 32$. Indeed, we may, by listing *all* possible closed walks containing all edges of G and comparing their weights, check that $w(W_2)$ is one with the minimum weight. Clearly, this way of solving the CPP by listing is inefficient if the order and size of G are large.

We now present two different algorithms that solve the CPP. The first algorithm is due to Guan Meigu himself.

Algorithm 4.5.3. (Guan's algorithm) Let G be a connected weighted multigraph.

Step 1: Pair off all odd vertices in G. For each pair $\{a, b\}$ of odd vertices, find an $a - b$ path P_{ab}. For each edge e in P_{ab}, duplicate e in G. Let the resulting multigraph be G_1.

Step 2: For all edges e in G, if e has at least 2 duplicate edges in G_1, delete an even number of them. Let the resulting multigraph be G_2.

Step 3: A cycle C in G is said to be *bad* if, in G_2, the sum of the weights of the duplicate edges in C exceeds $\frac{1}{2}w(C)$. If there is a bad cycle C in G, perform the following *complementary operation* with respect to C in G_2: delete all existing duplicate edges in C but duplicate each of the other edges in C. Repeat Step 3 until there are no more *bad* cycles. Let the resulting multigraph be G_3.

Step 4: Construct an Euler circuit in G_3.

Example 4.5.4. Solve the CPP for the multigraph G of Fig. 4.5.3 by applying Algorithm 4.5.3.

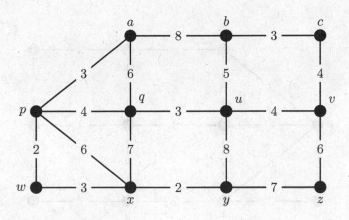

Fig. 4.5.3: G

Step 1. The four odd vertices a, b, v, y are paired off (randomly) as $\{a, v\}$ and $\{b, y\}$. For $\{a, v\}$, the $a - v$ path $abuv$ is chosen; for $\{b, y\}$, the $b - y$ path buy is chosen. Edges ab, bu, uv, bu and uy are duplicated to form G_1 of Fig. 4.5.4. Note that G_1 is an Eulerian multigraph.

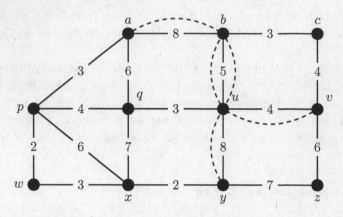

Fig. 4.5.4: G_1

Step 2. Since the edge bu in G has 2 duplicate edges in G_1, we delete both of them, resulting in G_2 of Fig. 4.5.5. Again, G_2 is an Eulerian multigraph.

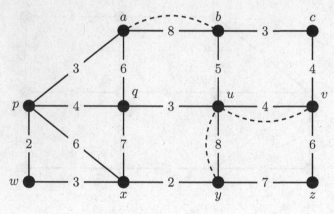

Fig. 4.5.5: G_2

Step 3. Consider the cycle $C : abcvuyxwpa$ in G. It is a bad cycle since $w(\{ab, vu, uy\}) = 20 > \frac{37}{2} = \frac{1}{2}w(C)$. We thus perform the complementary operation with respect to C on G_2, resulting in another Eulerian multigpaph G_3 of Fig. 4.5.6.

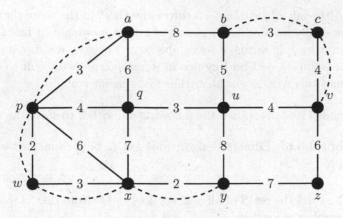

Fig. 4.5.6: G_3

Step 4. It is checked in Fig. 4.5.6 that there is no more bad cycle. We thus obtain an Euler circuit in G_3: *abcbuvcvzyxyuqapwxwpxqpa* of weight 98, which solves the CPP for G of Fig. 4.5.3.

Remark 4.5.5.

(1) It is interesting to note that the essential steps of Algorithm 4.5.3 had been written in the form of a chinese poem by Guan Meigu himself. We reproduce the poem below.

先分奇偶点
奇点对对联
联线不重叠
重叠要改变
边上联线长
不能过半圈

(2) Although Algorithm 4.5.3 solves the CPP, in the sense that it successfully finds an Euler circuit of minimum weight, it has a serious drawback in terms of its complexity. The step which requires the identification of **bad cycles** in a multigraph is difficult to execute and this renders the algorithm to be inefficient.

Edmonds (1965) proposed the following algorithm to solve the CPP.

Algorithm 4.5.6. (Edmonds' algorithm) Let G be a connected weighted multigraph.

Step 1: Find the set S of all odd vertices in G. Note that S must have an even number of vertices.

Step 2(a): Compute the distance (in the weighted graph G) $d_G(a, b)$, between each pair of vertices a, b in S.

Step 2(b): Form a weighted complete graph with vertex set S, where each edge ab, $a, b \in S$, is assigned the weight $w(ab) = d_G(a, b)$.

A **pairing** of S is a partition of S into 2-element subsets. The weight $w(M)$ of a pairing M of S is defined by

$$w(M) = \sum \left(w(xy) \mid \{x, y\} \in M \right).$$

A pairing M^* of S is **minimum** if $w(M^*) \leq w(M)$ for any pairing M of S.

Step 2(c): Find a minimum pairing M^* of S.

Step 3: For each pair $\{x, y\}$ in M^*, duplicate each edge contained in a shortest $x - y$ path in G. Let G_1 be the resulting Eulerian multigraph.

Step 4: Construct an Euler circuit in G_1.

Example 4.5.7. Solve the CPP for the same multigraph in Example 4.5.4 by applying Algorithm 4.5.6.

Step 1. The set of odd vertices is $S = \{a, b, y, v\}$. It is easy to find the following distances between the vertices in S.

Step 2(a). We have

$$d(a, b) = 8, \text{ a shortest } a - b \text{ path is } ab;$$
$$d(a, y) = 10, \text{ a shortest } a - y \text{ path is } apwxy;$$
$$d(a, v) = 13, \text{ a shortest } a - v \text{ path is } aquv;$$
$$d(b, y) = 13, \text{ a shortest } b - y \text{ path is } buy;$$
$$d(b, v) = 7, \text{ a shortest } b - v \text{ path is } bcv;$$
$$d(y, v) = 12, \text{ a shortest } y - v \text{ path is } yuv.$$

Step 2(b). We form the following weighted K_4 with vertex set S and the distances between the vertices as weights.

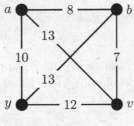

Fig. 4.5.7

Step 2(c). There are three possible pairings M_i, $i = 1, 2, 3$, that we can form, and they are shown in Fig. 4.5.8 with the respective weights $w(M_i)$.

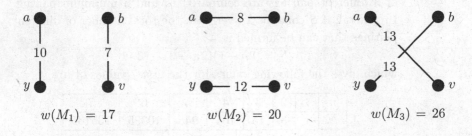

$$w(M_1) = 17 \qquad w(M_2) = 20 \qquad w(M_3) = 26$$

Fig. 4.5.8

Clearly M_1 is the minimum pairing, so we duplicate the edges ap, pw, wx, xy (forming the shortest $a - y$ path) and bc, cv (forming the shortest

$b - v$ path), resulting in the multigraph G_1 of Fig. 4.5.9.

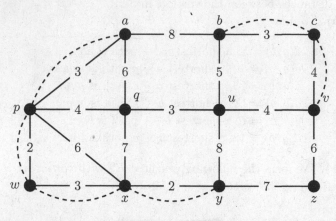

Fig. 4.5.9: G_1

G_1 is the same multigraph as G_3 in Example 4.5.4, so by applying Algorithm 4.4.3 to G_1, we would obtain an Euler circuit in G_1 with weight equal to 98 also.

Remark 4.5.8.

(1) Although Algorithm 4.5.6 does not involve identifying bad cycles like in Algorithm 4.5.3, its crucial step is the determination of a minimum pairing from the weighted complete graph with vertex set S. In Example 4.5.7, where S contains 4 vertices, the weights of 3 different pairings are compared to find a minimum pairing. In general, if S contains $2n$ vertices, the total number of different pairings that can be formed is
$$N = (2n - 1)(2n - 3) \cdots 3 \cdot 1$$
which gives the following values for the first 7 values of n:

n	1	2	3	4	5	6	7
N	1	3	15	105	945	10395	135135

As the value of N increases rapidly, this makes Algorithm 4.5.6 computationally difficult.

(2) However, Edmonds and Johnson (1973) came up with a technique that finds a minimum pairing efficiently. This modification to Algorithm 4.5.6 allows the CPP to be solved efficiently.

(3) The Chinese Postman Problem has found applications in various areas such as police patrol scheduling, routings of garbage and refuse collection, street sweepers, the spraying of roads with salt-grit to prevent ice formation, and many others. To reflect real-life situations more closely, many variations as well as extensions and generalizations of the CPP have been proposed and studied by numerous researchers from various disciplines such as mathematics, computer science, operations research, management science etc.

Exercise for Section 4.5

1. In Algorithm 4.5.6 (Edmond's algorithm), let ab, cd be two edges in a minimum pairing involving the odd vertices. Prove that the shortest paths P_1 from a to b and P_2 from c to d do not have any edge in common.

2. For each of the following weighted graphs, apply Algorithm 4.5.6 (Edmond's algorithm) to find a closed walk with minimum weight that contains all the edges. Repeat the problem using Algorithm 4.5.3 (Guan Meigu's algorithm).

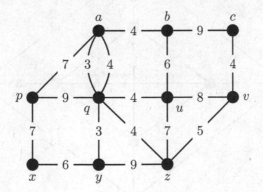

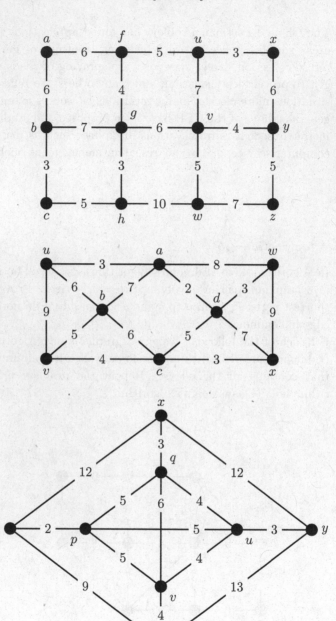

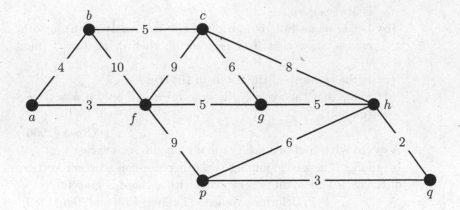

Fig. 4.5.10

4.6 Problem Set IV

1. Let G be an Eulerian multigraph. Show that $c(G - v) \leq \frac{1}{2}d(v)$ for each vertex v in G.

2. Find the necessary and sufficient conditions on p, q and r such that the complete tripartite graph $K(p, q, r)$ is Eulerian.

3. Let G be an Eulerian graph of size m that contains a spanning subgraph isomorphic to $C_k \,\square\, P_r$, where $k \geq 3$ and $r \geq 2$. Show that $m \geq 2kr$.

4. Let $G = O_n + H$, where H is a graph. Find the necessary and sufficient conditions for G to be Eulerian.

5. Find the necessary and sufficient conditions for the join $G = H_1 + H_2$, where H_1 and H_2 are graphs, to be Eulerian.

6. Let G be a non-trivial connected multigraph. For $A \subset V(G)$, let $e(A, V(G)\backslash A)$ denote the number of edges in G having an end in A and the other in $V(G)\backslash A$ (see Problem 2.3.29). Show that G is Eulerian if and only if $e(A, V(G)\backslash A)$ is even for every proper subset A of $V(G)$.

7. **[Graceful graphs]**

   (a) Given any $n \geq 3$ non-negative integers $a_i, i = 1, 2, \cdots, n$, prove that the sum

   $$|a_1 - a_2| + |a_2 - a_3| + \cdots + |a_{n-1} - a_n| + |a_n - a_1|$$

is always even.

(b) Let G be an Eulerian graph of size m. Applying (a), or otherwise, show that if G is graceful, then $m \equiv 0$ or 3 (mod 4).

(c) Is the converse of the result in (b) true?

(d) Show that the cycle C_n is graceful if and only if $n \equiv 0$ or 3 (mod 4).

$$\text{(Rosa (1966))}$$

8. **A cycle with a chord** is a graph obtained from a cycle C_n, $n \geq 4$, by adding a new edge joining two otherwise non-adjacent vertices of the cycle. Show that every cycle with a chord is graceful.

$$\text{(Delorme, Maheo, Thuillier, Koh and Teo (1980))}$$

9. Let $C_n(p,q)$ denote the graph obtained from a cycle C_n by adding three chords ab, bc and ca as shown in Fig. 4.6.1, such that, counting clockwise, there are p vertices between a and b, q vertices between b and c and 1 vertex between c and a, where $n = p + q + 4$ and $p \geq q \geq 1$. Show that the graph $C_n(p,q)$ is graceful if and only if $n \equiv 0$ or 1 (mod 4).

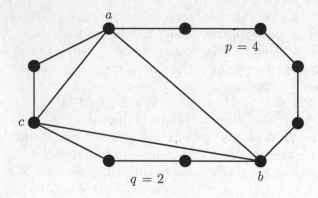

Fig. 4.6.1: $C_{10}(4,2)$

$$\text{(Koh, Rogers, Teo and Yap (1980))}$$

10. Given a connected multigraph G, let $\mu(G)$ denote the number of subsets of $E(G)$, including the empty set, each of which is contained in some spanning tree of G. For instance, in the graph G shown in Fig. 4.6.2, the empty set, every set of single edge and every set of

2 edges are such subsets. All sets of 3 edges, except those forming C_3 (namely, $wxyw$ and $wyzw$), are such subsets. As any spanning tree of G is of size 3, any set with at least 4 edges is not counted. Thus $\mu(G) = 1 + \binom{5}{1} + \binom{5}{2} + (\binom{5}{3} - 2) = 24.$

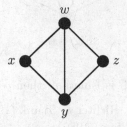

Fig. 4.6.2

(a) Evaluate $\mu(G)$ if G is

    (i) a tree of order n, $n \geq 2$;

    (ii) the cycle C_n, $n \geq 3$;

    (iii) an edge-gluing of C_p and C_q, $p \geq q \geq 3$, as shown in Fig. 4.6.3;

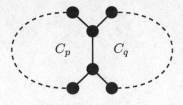

Fig. 4.6.3

    (iv) the complete bipartite graphs $K(3,2)$ and $K(4,2)$;

    (v) the complete graphs K_4 and K_5.

(b) Let G be a connected multigraph. Show that G is Eulerian if and only if $\mu(G)$ is odd.

                       (Shank (1979) and Fleischner (1989))

11. Given a connected multigraph G, let $pc(G)$ denote the number of partitions of $E(G)$ into cycles. Thus, by Theorem 4.2.2, $pc(G) \geq 1$ if and only if G is Eulerian.

(a) Evalute $pc(G)$, where G is the following multigraph.

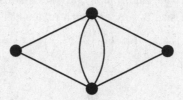

Fig. 4.6.4

(b) Show that if G is Eulerian, then $pc(G)$ is odd.

     (Halberstam, Richter & Shank (1984, unpublished); see also Bondy and Halberstam (1986))

Chapter 5

Hamiltonian Graphs and The Traveling Salesman Problem

5.1 Around the World and Spanning Cycles of a Graph

In 1857, Sir William Rowan Hamilton (1805–1865), who was then a Royal Astronomer of Ireland, invented a game called the **Icosian Game** (see Fig. 5.1.1(a)). The game involved a regular dodecahedron (see Fig. 5.1.1(b)) on which the 20 vertices were labeled using the names of 20 cities in the world. As discussed in Section 1.3 (see Fig. 1.3.10 and Fig. 1.3.11), the dodecahedron can be represented by a 3-regular graph as shown in Fig. 5.1.1(c). The object of the game was to travel 'Around the World' by finding a 'walk' in the dodecahedron which visited each city once and exactly once, starting and terminating at the same city.

(a)

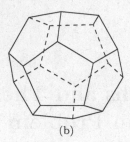 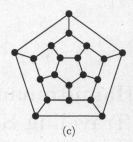

(b)   (c)

Fig. 5.1.1

One such walk is shown in Fig. 5.1.2. It is called a **spanning cycle** of the graph of Fig. 5.1.1(c). It is a cycle because the walk is closed and, other than the starting and terminating vertices, no vertex is repeated. It is spanning because all the 20 vertices of the graph are included in the walk.

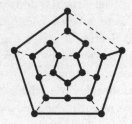

Fig. 5.1.2

A connected graph of order $n \geq 3$ is called a **Hamiltonian graph** if it contains a spanning cycle. If G is a Hamiltonian graph, then any spanning cycle of G is called a **Hamiltonian cycle** of G.

Remark 5.1.1. While the Eulerian notion applies to multigraphs, we confine ourselves to (simple) graphs for the Hamiltonian notion, because if a cycle contains an edge uv, then no other parallel edge of uv can be contained in the cycle.

Question 5.1.2.

(i) Is every cycle Hamiltonian?

(ii) If a graph G of order n is Hamiltonian, what is the least possible value of $e(G)$?

(iii) Is $K_n, n \geq 3$, Hamiltonian?

(iv) Which $K(p,q)$, where $p \geq q \geq 2$, are Hamiltonian?

(v) Which of the graphs in Fig. 5.1.3 are Hamiltonian?

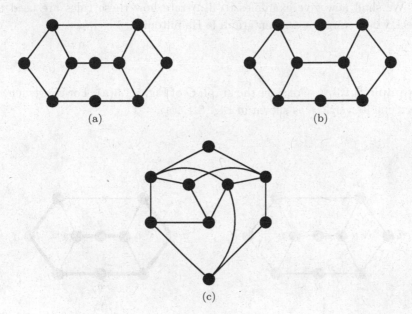

(a)   (b)

(c)

Fig. 5.1.3

Some useful rules for checking if a graph is Hamiltonian

Any vertex of degree two (if it exists) has an important role to play in checking whether a given graph G is Hamiltonian or not. This is because if C is a Hamiltonian cycle of G and v is a vertex of degree two in G, then in order to visit v along C, the two edges incident with v must both be contained in C, one for entering v and one for leaving v.

This observation together with two relevant ones are stated below:

(i) If v is a vertex of degree two in G, then the two edges incident with v are required to be contained in any Hamiltonian cycle of G.

(ii) During construction, if a non-spanning cycle is formed by those required edges, then G is not Hamiltonian.

(iii) If, during construction, two edges incident with a vertex are required, then all the other edges (if there exist) incident with the vertex should be deleted from further consideration.

We shall now give examples to illustrate how these rules are used to check whether or not a given graph is Hamiltonian.

Example 5.1.3.   Consider the graph G of Fig. 5.1.3(a). For convenience, we name its vertices as shown in Fig. 5.1.4(a).

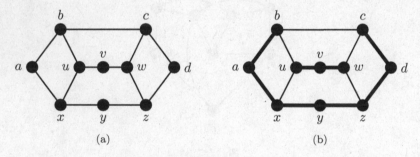

(a)   (b)

Fig. 5.1.4

There are four vertices of degree two in G, namely, a, d, v and y. If G has a Hamiltonian cycle C, then by rule (i), the eight edges $ab, ax, cd, dz, uv, vw, xy$ and yz are required edges in $E(C)$ as indicated in bold in Fig. 5.1.4(b).

It is now obvious that the set of eight required edges can be expanded to form such a C by adding bu and cw. Thus G is a Hamiltonian graph.

Example 5.1.4.   Consider the graph G of Fig. 5.1.3(b). Again, we name its vertices as shown in Fig. 5.1.5(a).

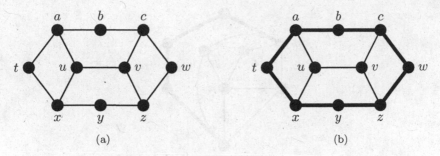

Fig. 5.1.5

Note that G has four vertices of degree two, namely, b, t, w and y. By rule (i), if G has a Hamiltonian cycle C, then the eight edges $ab, bc, at, tx, cw, wz, xy$ and yz are required edges in $E(C)$ as indicated in bold in Fig. 5.1.5(b).

Finally observe that these required edges by themselves form a cycle (a C_8), which is non-spanning. Thus, by rule (ii), G is not Hamiltonian.

Example 5.1.5. Consider the graph G of Fig. 5.1.3(c) and name its vertices as shown in Fig. 5.1.6(a).

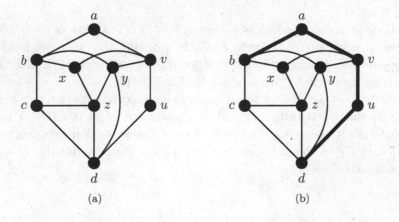

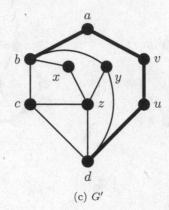

(c) G'

Fig. 5.1.6

There are two vertices of degree two in G, namely, a and u. Thus, if G has a Hamiltonian cycle C, by rule (i), the four edges ab, va, vu and ud are required edges in $E(C)$ as indicated in bold in Fig. 5.1.6(b).

Look at the vertex v, which is of degree four. As two of its incident edges (namely, va and vu) are required edges, by rule (iii), the other two of its incident edges (namely, vx and vy) should be deleted from further consideration. Let the resulting graph be G', as shown in Fig. 5.1.6(c).

Notice that the vertex x in G' is now of degree two. Thus, by rule (i), the edges xb amd xz are required edges in $E(C)$ as indicated in bold in Fig. 5.1.7(a).

Look at vertex b, which is of degree four in G'. As two of its incident edges, namely, ba and bx are in $E(C)$, by rule (iii), the other two incident edges, namely, bc and by should be deleted from further consideration. Let the resulting graph be G'', as shown in Fig. 5.1.7(b).

Notice that the vertex c (or y) is now of degree two in the resulting graph G''. By rule (i), the edges cz and cd are required edges in $E(C)$. Finally observe that these required edges form by themselves a non-spanning cycle (indeed, a C_8) as shown in bold in Fig. 5.1.7(c). Thus, by rule (ii), G is not Hamiltonian.

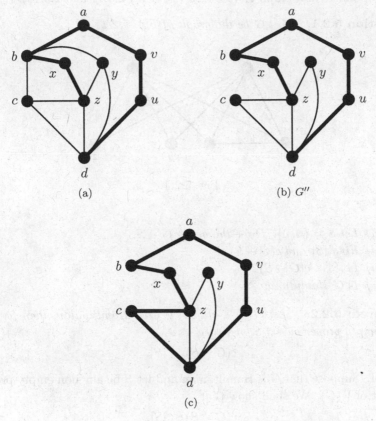

Fig. 5.1.7

5.2 A Necessary Condition for a Graph to be Hamiltonian

Theorem 4.2.2 provides us with two nice characterizations of Eulerian multigraphs. In contrast to this, up till now, no good characterization of Hamiltonian graphs has been found. Indeed, the problem of characterizing Hamiltonian graphs is very hard, and is considered as one of the major unsolved problems in Graph Theory.

In this section, we shall establish a result which gives a necessary condition for a graph to be Hamiltonian. As we shall witness, this result is very useful to show that certain graphs are not Hamiltonian.

Recall in Section 1.4 that the notation $c(H)$ is used to denote the number of components of a graph H (see also Problems 2.3.11 and 2.3.12).

Question 5.2.1.   *Let G be the graph of Fig. 5.2.1.*

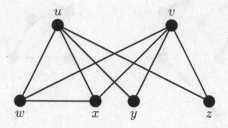

Fig. 5.2.1

  (i) Let $S = \{u, v\}$. Draw the graph $G - S$.
  (ii) Find $|S|$ and $c(G - S)$.
  (iii) Is $|S| < c(G - S)$?
  (iv) Is G Hamiltonian?

Theorem 5.2.2.   *Let G be a graph. If G is Hamiltonian, then for any non-empty proper subset S of $V(G)$,*

$$c(G - S) \leq |S|.$$

Proof. Suppose that G is Hamiltonian and let S be any non-empty proper subset of $V(G)$. We shall show that

$$c(G - S) \leq |S|.$$

Since G is Hamiltonian, G contains a Hamiltonian cycle, say, C. Two observations are in order:

  (1) As C is a spanning cycle of G, $C - S$ is a spanning subgraph of $G - S$, and we have (see Problem 2.3.11)

$$c(C - S) \geq c(G - S).$$

  (2) As C is a cycle and $S \subset V(C)$, we have (see Problem 2.3.12)

$$c(C - S) \leq |S|.$$

Now, combining the inequalities in (1) and (2), we have

$$c(G - S) \leq c(C - S) \leq |S|,$$

as required.

□

Remark 5.2.3. By Theorem 5.2.2,

$$c(G - S) \leq |S| \text{ for any non-empty proper subset } S \text{ of } V(G)$$

is a necessary condition for G to be Hamiltonian. Thus, given a graph G, if we can find a subset S of $V(G)$ such that

$$c(G - S) > |S|,$$

then we can conclude by Theorem 5.2.2 that G is not Hamiltonian.

Example 5.2.4. Consider the graph G of Fig. 5.1.5(a). We have proven in Example 5.1.4 that G is not Hamiltonian. Let us now apply Theorem 5.2.2 to draw the same conclusion.

Let $S = \{a, c, x, z\}$. Then $G - S$ is shown in Fig. 5.2.2.

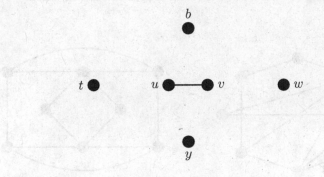

Fig. 5.2.2

Check that $|S| = 4$ and $c(G - S) = 5$, and we have $c(G - S) > |S|$. We thus conclude by Theorem 5.2.2 that G is not Hamiltonian.

Question 5.2.5. *Consider the graph of Fig. 5.1.6(a). Find a set S of vertices in G such that $c(G - S) > |S|$, and thus conclude that G is not Hamiltonian.*

Question 5.2.6. *A vertex w in a connected graph G is called a **cut-vertex** (see Problem 2.3.18) if $G - w$ is disconnected. Can G be Hamiltonian if G contains a cut-vertex? Why?*

Question 5.2.7. *The necessary condition given in Theorem 5.2.2 for a graph G to be Hamiltonian is, however, not sufficient. That is, the following implication is, in general, not true:*

If $c(G - S) \leq |S|$ for any non-empty proper subset S of $V(G)$, then G is Hamiltonian.

Find a graph G such that $c(G - S) \leq |S|$ for any non-empty proper subset S of $V(G)$, but G is not Hamiltonian. (See Problem 5.2.7.)

Exercise for Section 5.2

1. Determine whether the following graphs are Hamiltonian.

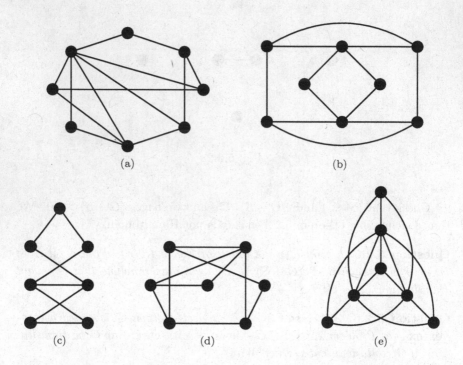

(a)   (b)

(c)   (d)   (e)

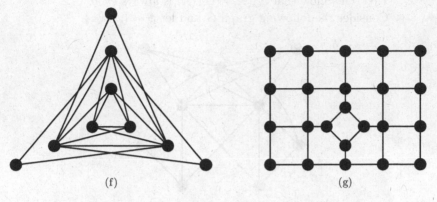

(f)   (g)

Fig. 5.2.3

2. Determine whether the following $m \times n$ rectangular grids, namely, $P_m \square P_n$ are Hamiltonian.

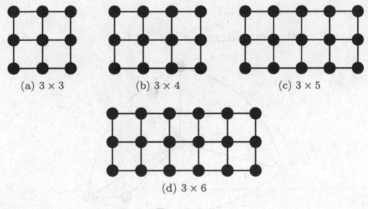

(a) 3×3 (b) 3×4 (c) 3×5

(d) 3×6

Fig. 5.2.4

3. Show that the graph $P_m \square P_n$ is Hamiltonian if and only if either m or n is even.

4. Construct all non-isomorphic 3-regular Hamiltonian graphs of order 4, 6 and 8 respectively.

5. Let G be a Hamiltonian graph with a Hamiltonian cycle C. For any non-empty proper subset A of $V(G)$, let $e_C(A, V(G) \backslash A)$ denote the number of edges in C having one end in A and the other in

$V(G)\backslash A$. Show that $e_C(A, V(G)\backslash A)$ is always even.

6. Consider the following graph G and let $S = \{x, y, z\}$.

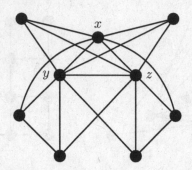

Fig. 5.2.5

    (i) Draw the graph $G - S$.

   (ii) Find $|S|$ and $c(G - S)$.

  (iii) Is $|S| < c(G - S)$?

  (iv) Is G Hamiltonian?

7. Let H be the graph shown in Fig. 5.2.6.

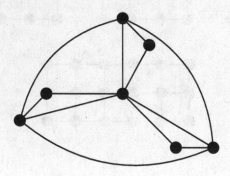

Fig. 5.2.6

    (i) Verify that $c(H - S) \leq |S|$ for all non-empty proper subsets S of $V(H)$.

  (ii) Is H Hamiltonian?

  (iii) Is the converse of Theorem 5.2.2 true?

8. Let G be a Hamiltonian (n, m)-graph such that $m = n + 2$ and $n \geq 5$.

   (i) Prove that $2 \leq d(x) \leq 4$ for each vertex x in G.
   (ii) If G is also semi-Eulerian, what can be said about the structure of G?

9. Let G be a semi-Eulerian and Hamiltonian $(12, 17)$-graph with $\delta(G) = 2$ and $\Delta(G) = 4$.

   (i) How many vertices of degree 3 can G have?
   (ii) How many vertices of degree 2 does G have?
   (iii) Construct one such graph G.
   (iv) Assume that the odd vertices are adjacent in G, but no two vertices of degree 2 are adjacent in G. What can be said about the structure of the subgraph induced by the set of vertices of degree 4 in G?

10. Let H be a semi-Eulerian and Hamiltonian $(7, 12)$-graph with a Hamiltonian cycle C. Assume that $\delta(H) = 2$ and $\Delta(H) = 5$, and that H has exactly two vertices of degree 2.

    (i) Find the number of vertices of degree 4 and the number of vertices of degree 5 in H.
    (ii) Assume that the two vertices of degree 2 are adjacent in C. Construct all such graphs H.

11. Let G be a Hamiltonian bipartite graph of order 8.

    (i) Explain why $\delta(G) \geq 2$ and $\Delta(G) \leq 4$.
    (ii) Assume further that G is Eulerian and $\Delta(G) = 4$. What can be said about the structure of G?

12. Denote each of the following graphs by H and answer the following questions for each H.

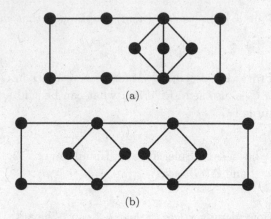

(a)

(b)

Fig. 5.2.7

(i) Is H Hamiltonian? Why?

(ii) Let $m(H)$ denote the minimum number of new edges that need to be added to H so that the resulting graph H^* is Hamiltonian (note that $V(H^*) = V(H)$). Find the value of $m(H)$ and justify your answer.

(iii) Construct two such Hamiltonian graphs H^* obtained by adding $m(H)$ new edges to H.

5.3 Sufficient Conditions for a Graph to be Hamiltonian

Theorem 5.2.2 gives a necessary condition for a graph G to be Hamiltonian, but this condition is not sufficient. We now look for a condition which is sufficient for G to be Hamiltonian: that is, we wish to find a condition $(\#)$, say, so that if G satisfies $(\#)$, then G is Hamiltonian.

Complete graphs K_n, $n \geq 3$, are Hamiltonian. This fact, however, does not surprise us because these graphs are the most 'dense' (i.e., containing the most number of edges for a given n), and so the task of forming a Hamiltonian cycle is easy. Must a graph G be that 'dense' in order to possess a Hamiltonian cycle? Certainly not! But then, what is the 'least density' expected of G to ensure the existence of a Hamiltonian cycle in G?

One of the most relevant quantities to measure the 'density' of a graph is the degree of a vertex. From this perspective, Dirac (1952a) proved the following very first significant result on Hamiltonian graphs.

Theorem 5.3.1.   *Let G be a graph of order $n \geq 3$. If $d(v) \geq \frac{n}{2}$ for each vertex v in G, then G is Hamiltonian.*

□

Question 5.3.2.   *Let H be a 51-regular graph of order 102. Is H Hamiltonian? Why?*

Question 5.3.3.   *A graph H has its degree sequence $(7,7,6,6,5,5,4,4)$. Is H Hamiltonian? Why?*

Question 5.3.4.   *Consider the following graph H, and let $n = v(H)$.*

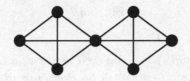

Fig. 5.3.1

(i) *Is it true that $d(v) \geq \frac{n}{2}$ for each vertex v in H?*
(ii) *Is it true that $d(v) \geq \frac{n-1}{2}$ for each vertex v in H?*
(iii) *Is H Hamiltonian?*

Question 5.3.5.   *Is the converse of Theorem 5.3.1 true? That is, if G is a Hamiltonian graph of order n, is it true that $d(v) \geq \frac{n}{2}$ for each vertex v in G?*

The condition imposed on a graph G in Theorem 5.3.1 to ensure that G contains a Hamiltonian cycle is based on the degree of each individual vertex in G. Ore (1960) found another sufficient condition which applies to pairs of non-adjacent vertices rather than single ones as shown in Theorem 5.3.6.

Theorem 5.3.6.   *Let G be a graph of order $n \geq 3$. If*

$$d(u) + d(v) \geq n$$

for every pair of non-adjacent vertices u and v in G, then G is Hamiltonian.

□

Question 5.3.7.   *Is the converse of Theorem 5.3.6 true? That is, if G is a Hamiltonian graph of order $n \geq 3$, is it true that $d(u) + d(v) \geq n$ for every pair of non-adjacent vertices u, v in G?*

Question 5.3.8.   *Consider the following graph G, and let $n = v(G)$.*

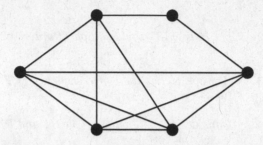

Fig. 5.3.2

(i)   *Is it true that $d(v) \geq \frac{n}{2}$ for each vertex v in G?*
(ii) *Is it true that $d(u)+d(v) \geq n$ for every pair of non-adjacent vertices u, v in G?*
(iii) *Can you conclude that G is Hamiltonian by Theorem 5.3.6?*
(iv) *Can you conclude that G is Hamiltonian by Theorem 5.3.1?*

Question 5.3.9.   *Consider the following graph G, and let $n = v(G)$.*

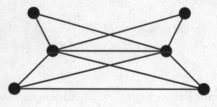

Fig. 5.3.3

(i)   *Is it true that $d(u)+d(v) \geq n$ for every pair of non-adjacent vertices u, v in G?*
(ii) *Is G Hamiltonian?*

The closure of a graph

Given a graph G, we shall now define a new graph from G call the *closure* of G, which was introduced by Bondy and Chvátal (1976). We then proceed to show the equivalence of the hamiltonicity of G and its closure, and derive Theorem 5.3.1 and Theorem 5.3.6 from this result.

Let G be a graph of order $n \geq 3$. The **closure** $cl(G)$ of G is the graph obtained from G by recursively joining pairs of non-adjacent vertices u and v with $d(u) + d(v) \geq n$ until no such pair exists.

Example 5.3.10. Consider the graph G shown in Fig. 5.3.4(a) with $n = 5$. The two vertices u and v are non-adjacent, and $d(u) + d(v) = 5 = n$, so we join u and v with an edge, resulting in the graph G' shown in Fig. 5.3.4(b). Note that G' has no further pair of non-adjacent vertices u, v such that $d(u) + d(v) \geq 5$. Thus the closure of G is G'.

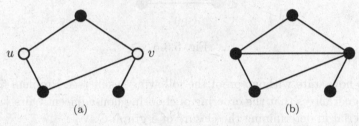

(a)   (b)

Fig. 5.3.4

Example 5.3.11. Consider the graph G shown in Fig. 5.3.5(a) with $n = 5$. Once again, the two vertices u and v are non-adjacent, and $d(u) + d(v) = 5 = n$, so we join u and v with an edge, resulting in the graph G' shown in Fig. 5.3.5(b). Subsequently, Fig. 5.3.5(c) to Fig. 5.3.5(e) show the resulting graphs with edges added between non-adjacent vertices u, v satisfying $d(u) + d(v) \geq n$. The closure of G is thus the complete graph of order 5, K_5, as shown in Fig. 5.3.5(e).

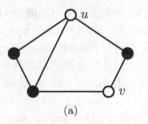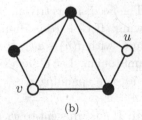

(a)   (b)

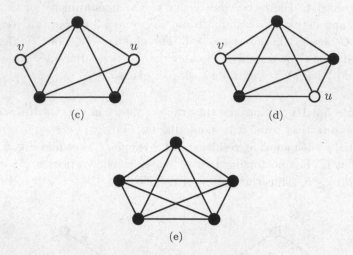

Fig. 5.3.5

We now state without proof the following result (see Problem 5.3.10), which guarantees that the ordering of choosing non-adjacent pairs $\{u, v\}$ is immaterial in determining the closure of a graph G.

Theorem 5.3.12.   *If G_1 and G_2 are two graphs obtained from G by recursively joining pairs of non-adjacent vertices u, v such that $d(u) + d(v) \geq n$, until no such pair exists, then $G_1 = G_2$.*

$\square$

We shall now establish the following result due to Bondy and Chvátal (1976).

Theorem 5.3.13.   *Let G be a graph of order $n \geq 3$. Then G is Hamiltonian if and only if $cl(G)$ is Hamiltonian.*

Proof. The necessity is trivial. To prove the sufficiency, it suffices to prove the following: If $G + uv$ is Hamiltonian, where u and v are non-adjacent vertices in G with $d(u) + d(v) \geq n$, then G is also Hamiltonian. Suppose G is not Hamiltonian. Let C be a spanning cycle of $G + uv$. Then $uv \in E(C)$ and $C - uv$ is a spanning $u - v$ path in G. Rename the vertices so that $C - uv : u = w_1 w_2 \cdots w_n = v$.

   Claim: If u is adjacent to w_k ($k = 2, 3, \cdots, n-1$), then v is not adjacent to w_{k-1}.

Otherwise, $w_1 \cdots w_{k-1} w_n w_{n-1} \cdots w_k w_1$ is a spanning cycle in G, a contradiction.

Assume $d(u) = p$. By the above claim, $d(v) \leq (n-1) - p$. But then $d(u) + d(v) \leq p + n - 1 - p = n - 1$, a contradiction. Thus G is Hamiltonian and the proof is complete.

$\square$

Corollary 5.3.14. *Let G be a graph of order $n \geq 3$. If $cl(G)$ is complete, then G is Hamiltonian.*

$\square$

Question 5.3.15. *Let us return to Theorem 5.3.6. Assume the condition in the statement holds, that is, $d(u) + d(v) \geq n$ for each pair of non-adjacent vertices u and v in G. What can you say about $cl(G)$? Can you conclude that G is Hamiltonian? (See Problem 5.3.9.)*

Question 5.3.16. *Does Theorem 5.3.1 follow from Theorem 5.3.6? (See Problem 5.3.3.)*

Exercise for Section 5.3

1. The following graph H is Hamiltonian.

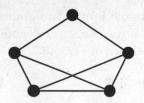

Fig. 5.3.6

    (i) Does the Hamiltonicity of H follow from Theorem 5.3.1?

    (ii) Does the Hamilotinicity of H follow from Theorem 5.3.6?

2. A graph G has $(8, 8, 8, 7, 7, 7, 6, 5, 5, 5)$ as its degree sequence. Is G Hamiltonian? Why?

3. Let G be a graph of order $n \geq 3$. The sufficient condition given by Dirac in Theorem 5.3.1 states that

$$\text{(D)} \quad d(v) \geq \frac{n}{2} \quad \text{for each } v \text{ in } V(G).$$

The sufficient condition given by Ore in Theorem 5.3.6 states that

(O) $d(u)+d(v) \geq n$ for every pair of u, v in $V(G)$ with $uv \notin E(G)$.

(1) Which of the following implications is true?

    (i) (D) $\Rightarrow$ (O);

    (ii) (O) $\Rightarrow$ (D).

(2) Which of the following implications is true?

    (i) Theorem 5.3.1 $\Rightarrow$ Theorem 5.3.6;

    (ii) Theorem 5.3.6 $\Rightarrow$ Theorem 5.3.1.

4. Let G be an (n, m)-graph with $n \geq 3$.

    (i) Assume that there exist two non-adjacent vertices u and v in G such that $d(u)+d(v) \leq n-1$. Show that $m \leq \binom{n-1}{2}+1.$

    (ii) Deduce that if $m \geq \binom{n-1}{2}+2$, then G is Hamiltonian.

5. Construct a non-Hamiltonian graph of order $n \geq 3$ with size $\binom{n-1}{2}+1$.

6. For each odd integer $n \geq 3$, construct a non-Hamiltonian graph G of order n such that $\delta(G) = \frac{n-1}{2}$.

7. Let $G + H$ denote the **join** of two graphs G and H (see Problem 2.5.6). For a positive integer r, denote by rK_2 the union of r disjoint edges as shown in Fig. 5.3.7.

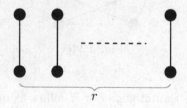

Fig. 5.3.7

    (i) Determine whether the join $(3K_2)+O_7$ is Hamiltonian. Justify your answer.

    (ii) Determine whether the join $(4K_2)+O_7$ is Hamiltonian. Justify your answer.

8. Find the closure of the following graphs.

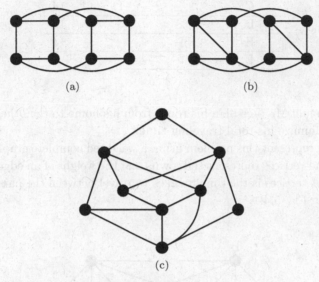

(a)   (b)

(c)

Fig. 5.3.8

9. Show that if G is a graph satisfying the conditions in Theorem 5.3.6, then $cl(G)$ is complete, and hence G is Hamiltonian.
10. Prove Theorem 5.3.12.

5.4   The Traveling Salesman Problem

Example 5.4.1.   Mr. Lee is a salesman living in town A. Every day, he needs to travel from his house to all the other 5 towns (B to F) to promote his products before returning home in the evening. The 6 towns are well connected by public transport and there are direct buses between each pair of towns. The time (in minutes) it takes for Mr. Lee to travel between each pair of towns is shown in the table below.

Between	A	B	C	D	E	F
A	–	13	13	12	17	13
B	–	–	13	14	15	16
C	–	–	–	11	14	14
D	–	–	–	–	15	15
E	–	–	–	–	–	14
F	–	–	–	–	–	–

How should Mr. Lee plan his route from his home to the other 5 towns so as to minimize his total traveling time?

We can represent this problem using a weighted complete graph of order 6, where the vertices represent the towns and the weight of an edge between each pair of vertices is the time it takes to travel between the pair of towns as shown in Fig. 5.4.1.

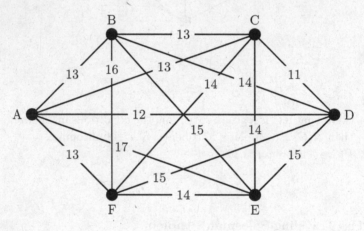

Fig. 5.4.1

Clearly, G has many spanning cycles, each representing a route that Mr. Lee can take. The total traveling time would be the sum of all the weights of edges in a particular spanning cycle. The question now reduces to finding a spanning cycle with the **minimum weight**.

> **The Traveling Salesman Problem**. (TSP) Let G be a weighted complete graph of order n. Find a spanning cycle C in G such that
>
> $$w(C) = \sum_{e \in E(C)} w(e)$$
>
> is the minimum among all spanning cycles of G.

Remark 5.4.2.

(1) Unlike the shortest path problem, minimum spanning tree problem, chinese postman problem, there are currently no efficient algorithms for solving the traveling salesman problem.

(2) Research questions relating to the traveling salesman problem are still being worked on by many mathematicians and computer scientists. Interested readers may visit http://www.tsp.gatech.edu/.

(3) Some instances of the traveling salesman problem have the weighted complete graphs G satisfying the triangle inequality

$$w(xy) \le w(xz) + w(zy)$$

for any vertices x, y, z in G.

We will next introduce three *approximation* algorithms that produce spanning cycles of *low* weight (not necessarily the lowest).

A **greedy algorithm** is one that optimizes the choices at each step, that is, making local optimal choices, without regard to previous choices or the subsequent effect of that choice. An example of a greedy algorithm is the Kruskal's algorithm (Algorithm 3.6.3) introduced for the Minimum Spanning Tree Problem (see Section 3.6). The following algorithm, called the **nearest neighbor** algorithm, is another example of a greedy algorithm.

Algorithm 5.4.3. (Nearest neighbor algorithm for TSP) Let G be a weighted complete graph.

Step 1: Choose an arbitrary vertex v_1 in G and set $i := 1$ and $S_1 := \{v_1\}$.

Step 2(a): Suppose that the path $P_i = v_1 v_2 ... v_i$ has been constructed. Select a vertex y from $V(G) - S_i$ such that

$$w(v_i y) \le w(v_i x) \quad \text{for all } x \in V(G) - S_i.$$

Set $v_{i+1} := y$, $P_{i+1} := v_1 v_2 ... v_i v_{i+1}$, $S_{i+1} := S_i \cup \{v_{i+1}\}$.

Step 2(b): If $i + 1 < n$, increase the value of i by 1 and return to Step 2(a). If $i + 1 = n$, then the spanning cycle obtained is

$$C = v_1 v_2 ... v_n v_1.$$

Example 5.4.4. Using vertex A as the starting vertex, apply Algorithm 5.4.3 on the weighted complete graph from Example 5.4.1 to find a spanning cycle of low weight.

The successive choices of vertices following A are as follows.

(a) A $\rightarrow$ D since $w(AD)$ is the minimum among all $w(AX)$, X $\in$ {B, C, D, E, F}.

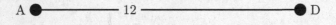

Fig. 5.4.2

(b) A $\rightarrow$ D $\rightarrow$ C since $w(DC)$ is the minimum among all $w(DX)$, X $\in$ {B, C, E, F}.

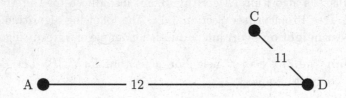

Fig. 5.4.3

(c) A $\rightarrow$ D $\rightarrow$ C $\rightarrow$ B since $w(CB)$ is the minimum among all $w(CX)$, X $\in$ {B, E, F}.

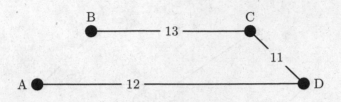

Fig. 5.4.4

(d) $A \to D \to C \to B \to E$ since $w(BE)$ is the minimum among all $w(BX)$, $X \in \{E, F\}$.

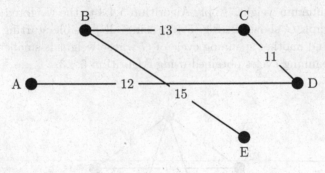

Fig. 5.4.5

(e) We have only one choice from vertex E, which is $E \to F$. This completes the spanning cycle $A \to D \to C \to B \to E \to F \to A$. The weight of this cycle is 78.

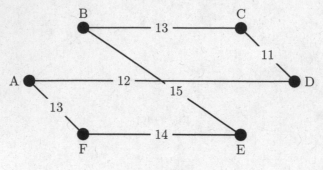

Fig. 5.4.6

Remark 5.4.5.

(1) The spanning cycle obtained by using Algorithm 5.4.3 depends on the starting vertex. Try Example 5.4.1 again using Algorithm 5.4.3, starting from vertex B.

(2) Generally speaking, even if we consider all possible starting vertices, Algorithm 5.4.3 may not necessarily yield a spanning cycle of minimum weight. Apply Algorithm 5.4.3 to the weighted complete graph G shown in Fig. 5.4.7, using all possible starting vertices. Find another spanning cycle of G whose weight is smaller than all spanning cycles obtained using Algorithm 5.4.3.

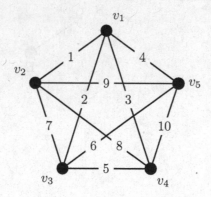

Fig. 5.4.7

Recall that the subgraph induced by a set F of edges is denoted by $[F]$.

Algorithm 5.4.6. (Minimum edge algorithm for TSP) Let G be a weighted complete graph.

Step 1: Choose an edge e_1 of minimum weight.

Step 2(a): Suppose $e_1, e_2, ..., e_i$, $1 \le i \le n-2$, have already been chosen. Select e_{i+1} from $E(G) - \{e_1, e_2, ..., e_i\}$ such that
      (i) $[\{e_1, e_2, ..., e_{i+1}\}]$ contains no cycles;
      (ii) no vertex in $[\{e_1, e_2, ..., e_{i+1}\}]$ is of degree 3;
      (iii) $w(e_{i+1})$ is minimum subject to (i) and (ii).

Step 2(b): Increase the value of i by 1. If $i < n - 1$, return to Step 2(a). If $i = n - 1$, join the two ends of the path $[\{e_1, e_2, ..., e_{n-1}\}]$ to form a spanning cycle.

Example 5.4.7. Apply Algorithm 5.4.6 to the weighted complete graph in Example 5.4.1.

The sequence of edges chosen using Algorithm 5.4.6 is CD, AD, AB, CE, EF and BF. The spanning cycle with weight 80 is shown in Fig. 5.4.8.

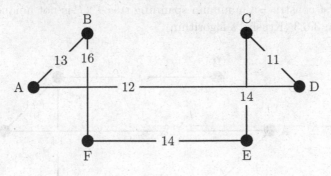

Fig. 5.4.8

Remark 5.4.8. Note that Algorithm 5.4.6 is also a greedy algorithm.

The last approximation algorithm, due to Christofides (1976), uses the notions of a minimum spanning tree, minimum pairing (see Algorithm 4.5.6) and Eulerian multigraph.

Algorithm 5.4.9. (Christofides' algorithm for TSP) Let G be a weighted complete graph.

Step 1: Construct a minimum spanning tree T of G.

Step 2(a): Let S be the set of odd vertices in T. Form the weighted induced subgraph $[S]$ of G. Note that $[S]$ is also a weighted complete graph with even number of vertices.

Step 2(b): Find a minimum pairing M^* of S. Add the edges of M^* to T to produce an Eulerian multigraph $\hat{T}$.

Step 3: Find an Euler circuit $W : u_1 u_2 ... u_k (= u_1)$ in $\hat{T}$.

Step 4: Transform W into a spanning cycle as follows. Suppose that u_i is the first vertex to be repeated in W. Let u_j be the first vertex *after* u_i that is not repeated in W. Replace the trail '$u_{i-1} u_i .... u_j$' by the edge $u_{i-1} u_j$. Repeat this process until a spanning cycle C is formed.

Example 5.4.10. Apply Algorithm 5.4.9 to the weighted complete graph in Example 5.4.1.

We first construct a minimum spanning tree T (T is not unique), using Algorithm 3.6.3 (Kruskal's algorithm).

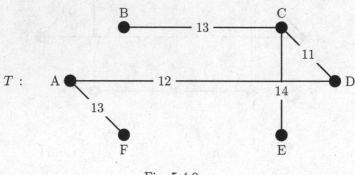

Fig. 5.4.9

$S = \{$B, C, E, F$\}$ is the set of odd vertices in T. The weighted complete graph induced by S is shown in Fig. 5.4.10.

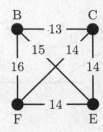

Fig. 5.4.10

There are 3 possible pairings M_i, $i = 1, 2, 3$ of S that we can form, namely $M_1 = \{BC, EF\}$, $M_2 = \{BF, CE\}$ and $M_3 = \{BE, CF\}$. It can be easily checked that the minimum pairing is M_1 (with a weight of 27).

We now add the edges BC and EF to T and obtain an Eulerian multi-graph $\hat{T}$.

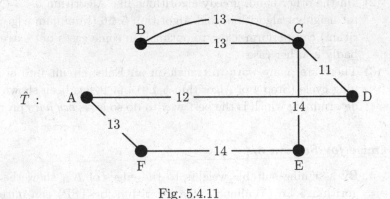

Fig. 5.4.11

An Euler circuit in $\hat{T}$ is AFECBCDA. The first vertex to be repeated is C and the first vertex after C that is not repeated is B, so we replace the trail ECB by EB. There are no more repeated vertices and the resulting spanning cycle is AFEBCDA, whose weight is 78.

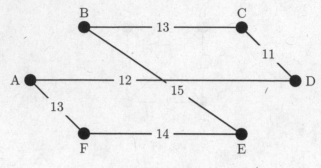

Fig. 5.4.12

Remark 5.4.11.

(1) Let G be a weighted complete graph and $\alpha = \min\{w(C) \mid C$ is a spanning cycle of $G\}$. If C^* is a spanning cycle obtained by Algorithm 5.4.9 (Christofides' algorithm), it has been established that $\alpha \le w(C^*) \le \frac{3}{2}\alpha$. This means that Algorithm 5.4.9 cannot be 'that far off' from the actual minimum.

(2) On the other hand, greedy algorithms like Algorithm 5.4.3 (Nearest neighbor algorithm) and Algorithm 5.4.6 (Minimum edge algorithm) can perform 'close to optimal' in some cases but extremely badly in other cases.

(3) There are many ways to transform an Euler circuit into a spanning cycle (Step 4 of Algorithm 5.4.9) but it has been shown that determining which is the best way to do so is a *very hard* problem.

Exercise for Section 5.4

1. By assigning suitable weights to the edges of K_4, show that Algorithm 5.4.6 (Minimum edge algorithm for TSP) can output a solution that is really bad.

2. For each of the problems below, the weighted complete graph is given. Find an approximate solution for the TSP by applying respectively (i) Algorithm 5.4.3 (Nearest neighbor algorithm for TSP) (ii) Algorithm 5.4.6 (Minimum edge algorithm for TSP) and (iii) Algorithm 5.4.9 (Christofides' algorithm for TSP).

   (a) The weighted complete graphs K_6 with weight matrices given below:

	(1)	(2)	(3)	(4)	(5)	(6)
(1)	0	3	5	4	6	6
(2)	3	0	5	6	5	7
(3)	5	5	0	5	6	5
(4)	4	6	5	0	4	3
(5)	6	5	6	4	0	7
(6)	6	7	5	3	7	0

	(1)	(2)	(3)	(4)	(5)	(6)
(1)	0	3	3	2	7	3
(2)	3	0	3	4	5	5
(3)	3	3	0	1	4	4
(4)	2	4	1	0	5	5
(5)	7	5	4	5	0	4
(6)	3	5	4	5	4	0

	(1)	(2)	(3)	(4)	(5)	(6)
(1)	0	5	5	4	6	4
(2)	5	0	4	3	7	7
(3)	5	4	0	2	3	6
(4)	4	3	2	0	5	8
(5)	6	7	3	5	0	3
(6)	4	7	6	8	3	0

(b) The weighted complete graph K_9 with weight matrix given below:

	(1)	(2)	(3)	(4)	(5)	(6)	(7)	(8)	(9)
(1)	0	6	5	4	4	2	6	8	5
(2)	6	0	5	3	8	7	3	6	3
(3)	5	5	0	8	10	7	8	11	8
(4)	4	3	8	0	9	6	6	9	6
(5)	4	8	10	9	0	7	11	13	10
(6)	2	7	7	6	7	0	8	10	7
(7)	6	3	8	6	11	8	0	9	6
(8)	8	6	11	9	13	10	9	0	9
(9)	5	3	8	6	10	7	6	9	0

5.5   Problem Set V

1. A path in a graph G is called a **Hamiltonian path** of G if it
   includes all the vertices in G.

   (i) Is the following graph Hamiltonian? Does it contain a Hamiltonian path?

   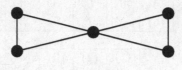

   Fig. 5.5.1

   (ii) Prove that if G is a graph of order $n \geq 2$ such that $\delta(G) \geq$
   $\frac{n-1}{2}$, then G contains a Hamiltonian path.

2. Show that the Petersen graph is non-Hamiltonian, but contains a
   Hamiltonian path.

3. Let G be a graph which contains a Hamiltonian path. Show that
   $c(G - S) \leq |S| + 1$ for every proper subset S of $V(G)$.

4. Construct a cubic non-Hamiltonian graph G which has the following property: for any two non-adjacent vertices x, y in G, there is
   a Hamiltonian $x - y$ path.

5. Let H be a Hamiltonian graph of order 11 with $\delta(H) = 3$. Assume
   further that H is semi-Eulerian having an open Euler $u - v$ trail.

   (a) Find the number of odd vertices in H.
   (b) Show that $e(H) \geq 21$.
   (c) Is $H - u$ always connected?
   (d) Is $H - \{u, v\}$ always connected?
   (e) Assume that $d(u) = d(v) = 3$, and u and v are adjacent. Show
   that $H - \{u, v\}$ is connected.

6. Let r be a positive integer.

   (a) Prove that the complete 3-partite graph $K(r, 2r, 3r)$ is Hamiltonian.
   (b) Is the graph $K(r, 2r, 3r + 1)$ Hamiltonian?

7. Let H be a spanning subgraph of a graph G. Which of the following
   statements is/are true?

   (a) If H is Eulerian, then G is Eulerian.

   (b) If H is semi-Eulerian, then G is semi-Eulerian.

   (c) If H is Hamiltonian, then G is Hamiltonian.

8. Let $L(G)$ be the line graph of G (see Problem II.11).

   (a) Prove that if G is Eulerian, then $L(G)$ is Hamiltonian.

   (b) Prove that if G is Eulerian, then so is $L(G)$.

   (c) Is the converse of (b) true?

   (d) Prove that if G is Hamiltonian, then so is $L(G)$.

   (e) Is the converse of (d) true?

9. Let $G \,\square\, H$ be the Cartesian product of the graphs G and H (see Problem II.12).

   (a) Show that if G and H are Eulerian, then so is $G \,\square\, H$.

   (b) Is the converse of the result in (a) true?

   (c) Show that the graph $C_k \,\square\, P_r$ is Hamiltonian, where $k \geq 3$ and $r \geq 1$.

   (d) Show that if G and H are Hamiltonian, then so is $G \,\square\, H$.

   (e) Is the converse of the result in (d) true?

   (f) For $n > 1$, which n-cube graphs are Eulerian? Hamiltonian?

10. Let G be a $2k$-regular graph of order $4k + 1$, where $k \geq 1$.

   (a) Show that

     (i) $\mathrm{diam}(G) = 2$;

     (ii) G contains no cut-vertex.

   (b) Assume that G contains a cycle C_{4k}. Prove that G is Hamiltonian.

11. Let p be an odd integer with $p \geq 3$.

   (a) Applying Theorem 5.3.1 (Dirac's theorem), or otherwise, show that the complete graph K_{2p-1} contains a $(p-1)$-regular spanning subgraph.

   (b) Let G be a graph of order $2p$. Show that either G or $\overline{G}$ contains $K(1, p)$ as a subgraph.

   (c) Is the result of (b) still true if G is of order $2p - 1$?

12. A graph G of order $n \geq 3$ is said to be **almost Hamiltonian** if G is not Hamiltonian but $G + e$ is Hamiltonian for each edge in $\overline{G}$.

   (a) Construct an almost Hamiltonian graph containing a cut-vertex.

   (b) Construct an almost Hamiltonian graph containing no cut-vertex.

(c) Let G be an almost Hamiltonian graph containing no cut-vertex, and assume that there exists a vertex w in G such that $N(w) = \{u, v\}$. Show that $\min\{d(u), d(v)\} \geq 4$.

13. Let G be a graph of order $n \geq 3$ with degree sequence $d_1 \leq d_2 \leq \cdots \leq d_n$. Assume that the following implication holds:

$$d_j \leq j < \frac{n}{2} \Rightarrow d_{n-j} \geq n - j.$$

Show that $cl(G)$ is complete, and so G is Hamiltonian.

(Chvátal (1972))

14. The following sufficient condition for a graph G of order n to be Hamiltonian was established by Fan Geng-Hua: G contains no cut-vertices and $\max\{d(u), d(v)\} \geq \frac{n}{2}$ for all vertices u, v in G with $d(u, v) = 2$.

(a) Verify that the following graphs satisfy Fan's condition. Find also their closures.

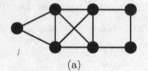 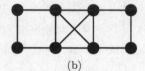

(a)   (b)

Fig. 5.5.2

(b) For each $n = 4k$, where $k \geq 3$, construct a graph H of order n that satisfies Fan's condition but $cl(H)$ is not complete.

(see Fan (1984))

15. [k-**Hamiltonian graphs**] Let n and k be integers such that $0 \leq k \leq n - 3$. A graph G of order n is said to be k-**Hamiltonian** if the removal of any r vertices from G, where $0 \leq r \leq k$, results in a Hamiltonian graph.

(a) Is it true that every k-Hamiltonian graph is also $(k - 1)$-Hamiltonian?

(b) For each $n \geq 5$, construct a graph of order n which is 1-Hamiltonian but not 2-Hamiltonian.

(c) Let G be a graph of order n such that $d(u) + d(v) \geq n + k$ for any pair of non-adjacent vertices u, v in G, where $0 \leq k \leq n - 3$. Show that G is k-Hamiltonian.

16. **[Hamiltonian-connected graphs]** A graph G is said to be **Hamiltonian-connected** if any two distinct vertices u, v of G are joined by a spanning $u - v$ path.

    (a) Show that every Hamiltonian-connected graph of order at least 3 is Hamiltonian.

    (b) Give an example to show that the converse of (a) is false.

    (c) Show that, for any tree T, the cube T^3 (for definition, see Problem II.9) is Hamiltonian-connected.

    (d) Deduce from (c) that the cube of any connected graph is Hamiltonian-connected.

    (see Sekanina (1960) and Karaganis (1968))

    Note: Fleischner (1974) proved a deep result that the square of any connected graph without cut-vertices is Hamiltonian.

17. Let G be a connected graph of order at least 4. Show that G^3 is 1-Hamiltonian. Give an example of G to show that G^3 may not be 2-Hamiltonian.

    (Chartrand and Kapoor (1969))

18. Let G be a connected graph of order at least 5. Show that

    (a) if w is an end-vertex of G, then $G^3 - w = (G - w)^3$;

    (b) if x is not a cut-vertex of G, then $G^3 - \{x, y\}$ is Hamiltonian, for any other vertex y in G.

    (Koh and Teo (1989))

Chapter 6

Connectivity

6.1 Motivation

Consider the following four graphs of order five:

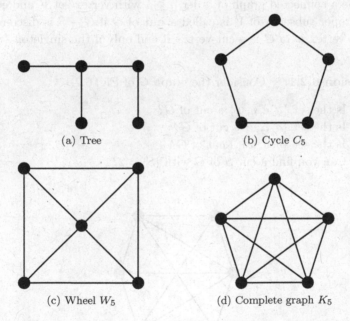

(a) Tree

(b) Cycle C_5

(c) Wheel W_5

(d) Complete graph K_5

Fig. 6.1.1

One of their common features is that they are 'connected'. On the other hand, it is also easy to see that the 'strengths' of their connectedness are different. For instance, for the tree (a), we need only to remove one of

237

its vertices to destroy its connectedness. For the cycle (b) and the wheel (c), we need respectively two and three of their vertices to do so. The connectedness of the complete graph (d) is the 'strongest' in the sense that no matter how many vertices (from one to four) we remove from it, the resulting subgraph is still connected.

How about the 'edge-deletion' version of the above discussion?

In this chapter, we shall introduce an important parameter, called connectivity, which measures how 'strong' the connectedness of a given graph is. Two versions of the notion, namely, the vertex-connectivity and the edge-connectivity, will be presented. Our emphasis is on the vertex-connectivity, and the main objective is to study the structure and properties of graphs that have attained a certain level of vertex-connectivity.

6.2   Vertex-connectivity and Edge-connectivity

Let G be a connected graph of order $n \geq 1$ with vertex set V and edge set E. A proper subset S of V is called a **cut** of G if $G - S$ is disconnected. Thus, a vertex w in G is a cut-vertex if and only if the singleton $\{w\}$ is a cut of G.

Discussion 6.2.1.   Consider the graph G of Fig. 6.2.1

   (i) Is the set $\{a, c, v, w\}$ a cut of G?
  (ii) Is the set $\{a, c, v\}$ a cut of G?
 (iii) Is the set $\{b, u, w\}$ a cut of G?
 (iv) Can you find a cut S of G with $|S| = 2$?

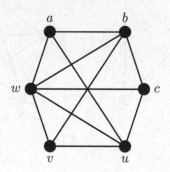

Fig. 6.2.1

Vertex-connectivity

The **vertex-connectivity** (or simply the **connectivity**) of G, denoted by $\kappa(G)$, is defined by

$$\kappa(G) = \begin{cases} n-1, & \text{if } G \cong K_n; \\ \min\{|S| \mid S \text{ is a cut of } G\}, & \text{if } G \not\cong K_n. \end{cases}$$

In other words, $\kappa(G)$ is the minimum number of vertices in G whose removal results in either a disconnected graph or a trivial graph (a singleton). Any cut S of G with $|S| = \kappa(G)$ is called a **minimum** cut of G.

Remark 6.2.2.

(1) When we remove any r vertices $(1 \le r \le n-1)$ from the complete graph K_n, the resulting subgraph is never disconnected; it is a trivial graph if $r = n-1$. By definition, $\kappa(K_n) = n - 1$.

(2) We define, for convention, $\kappa(G) = 0$ if G is disconnected.

Example 6.2.3. For the four graphs in Fig. 6.1.1, we have:

(a) $\kappa(G) = 1$, (b) $\kappa(C_5) = 2$, (c) $\kappa(W_5) = 3$, and (d) $\kappa(K_5) = 4$.

Discussion 6.2.4. Consider the graph G of Fig. 6.2.2.

(i) Find a cut S of G with $|S| = 3$.
(ii) Does G contain any cut S with $|S| = 2$?
(iii) Evaluate $\kappa(G)$.

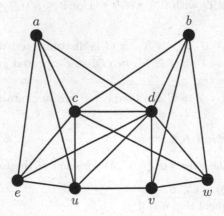

Fig. 6.2.2

Edge-connectivity

A subset F of E is called an **edge-cut** of G if $G - F$ is disconnected. Thus, an edge f in G is a bridge of G if and only if the singleton $\{f\}$ is an edge-cut of G. The **edge-connectivity** of G, denoted by $\kappa'(G)$, is defined by

$$\kappa'(G) = \begin{cases} 0, & \text{if } v(G) = 1; \\ \min\{|F| \mid F \text{ is an edge-cut of } G\}, & \text{if } v(G) \geq 2. \end{cases}$$

That is, $\kappa'(K_1) = 0$, and if $n \geq 2$, then $\kappa'(G)$ is the minimum number of edges in G whose removal results in a disconnected graph. Any edge-cut F of G with $|F| = \kappa'(G)$ is called a **minimum** edge-cut of G. Likewise, we define $\kappa'(G) = 0$ if G is disconnected.

Example 6.2.5.

(1) For the four graphs in Fig. 6.1.1, we have:

    (a) $\kappa'(G) = 1$;   (c) $\kappa'(W_5) = 3$;
    (b) $\kappa'(C_5) = 2$;   (d) $\kappa'(K_5) = 4$.

Note that the values of κ' and κ are the same for all these four cases.

(2) For the graph G of Fig. 6.2.2, however, while $\kappa(G) = 3$, we have $\kappa'(G) = 4$.

(3) The graphs G with $0 \leq \kappa(G) \leq 1$ or $0 \leq \kappa'(G) \leq 1$ are characterized below.

    (a) $\kappa(G) = 0 \Leftrightarrow G \cong K_1$ or G is disconnected $\Leftrightarrow \kappa'(G) = 0$.

    (b) $\kappa(G) = 1 \Leftrightarrow G \cong K_2$ or G is a connected graph containing a cut-vertex.

    (c) $\kappa'(G) = 1 \Leftrightarrow G$ is a connected graph containing a bridge.

Exercise for Section 6.2

1. Evaluate the values of $(\kappa, \kappa', \delta)$ for each of the following graphs:

    (i) the cycle $C_n, n \geq 3$,
    (ii) the wheel $W_n, n \geq 4$,
    (iii) the complete graph $K_n, n \geq 1$,
    (iv) the complete bipartite graph $K(p, q)$, where $1 \leq p \leq q$,
    (v) the graph $K_n - e$, where $n \geq 3$ and e is an edge in K_n.

(vi)

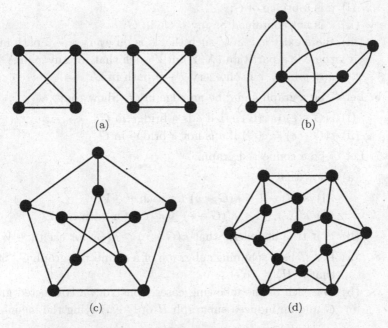

(a)   (b)

(c)   (d)

Fig. 6.2.3

2. Let $z(G)$ denote the number of vertices which are not cut-vertices of a non-trivial connected graph G. Show that $z(G) \geq 2$. Characterize G such that $z(G) = 2$.

3. Which of the following statements are true?

   (i) If G is an Eulerian graph, then $\kappa(G) \geq 2$.
   (ii) If G is an Eulerian graph, then $\kappa'(G) \geq 2$.
   (iii) If G is a Hamiltonian graph, then $\kappa(G) \geq 2$.

4. Let G be a connected graph and w a vertex in G. Prove that the following statements are equivalent:

   (i) w is a cut-vertex of G;
   (ii) there exist u, v in $V - \{w\}$ such that w is on every $u - v$ path in G;
   (iii) there is a partition (X, Y) of $V - \{w\}$ such that for any x in X and any y in Y, w is on every $x - y$ path in G.

5. Let G be a connected graph and e an edge in G. Prove that the following statements are equivalent:

   (i) e is a bridge of G;

   (ii) e is not contained in any cycle in G;

   (iii) there exist u, v in V such that e is on every $u - v$ path in G;

   (iv) there is a partition (X, Y) of V such that for any x in X and any y in Y, e is on every $x - y$ path in G.

6. Let G be a graph and e be an edge in G. Show that

   (i) $c(G - e) = c(G) + 1$ if e is a bridge in G;

   (ii) $c(G - e) = c(G)$ if e is not a bridge in G.

7. Let G be a connected graph.

   (a) Show that

   (i) $\kappa(G) - 1 \leq \kappa(G - v)$ for each $v \in V$;

   (ii) $\kappa'(G) - 1 \leq \kappa'(G - e) \leq \kappa'(G)$ for each $e \in E$.

   (b) Is it true in general that $\kappa(G - v) \leq \kappa(G)$ for each $v \in V$?

8. (a) Let H be a spanning subgraph of a connected graph G. Show that $\kappa(H) \leq \kappa(G)$.

   (b) For each of the following cases, construct a connected graph G and an induced subgraph H of G satisfying the inequality

   (i) $\kappa(H) < \kappa(G)$;

   (ii) $\kappa(H) > \kappa(G)$.

9. Let G be a connected graph. An edge-cut F of G is said to be **minimal** if no proper subset of F is an edge-cut of G. Clearly, every minimum edge-cut of G is also a minimal edge-cut of G. Show that if F is a minimal edge-cut of G, then $G - F$ consists of two components.

6.3   The Triple $(\kappa, \kappa', \delta)$

By checking all the graphs G shown in the previous sections, inclusive of Exercise 6.2, it is observed that the inequalities $\kappa(G) \leq \kappa'(G) \leq \delta(G)$ always hold. Indeed, we are going to establish in this section the validity of the inequalities.

Theorem 6.3.1.   (Whitney (1932)) *For any graph G of order $n \geq 1$,*
$$0 \leq \kappa(G) \leq \kappa'(G) \leq \delta(G) \leq n - 1.$$

Proof. The first and the last inequalities are trivial. We may assume that G is connected and $n \geq 2$.

We first show that $\kappa'(G) \leq \delta(G)$. Let $u \in V$ be such that $d(u) = \delta(G)$. Consider the set $F = \{e \in E \mid e$ is incident with $u\}$. Clearly, F is an edge-cut of G, and so by definition, we have

$$\kappa'(G) \leq |F| = d(u) = \delta(G).$$

We next show that $\kappa(G) \leq \kappa'(G)$. If $G \cong K_n$, $\kappa(G) = \kappa'(G) = n - 1$, and the result follows. Assume now that $G \not\cong K_n$ (and so $n \geq 3$). Let F be a minimum edge-cut of G. Then, by Problem 6.2.9, $G - F$ consists of two components G_1 and G_2, say, of order r and $n - r$, respectively, where $1 \leq r \leq n - 1$. Note that every edge in G which joins a vertex in G_1 to a vertex in G_2 must be in F.

Claim: There exist $u \in V(G_1)$ and $v \in V(G_2)$ such that $uv \notin E$.

Otherwise, every vertex in G_1 is adjacent to every vertex in G_2, and we have $|F| = r(n - r)$. Observe that $n - 1 \geq \delta(G) \geq \kappa'(G) \geq |F| = r(n - r) \geq n - 1$. Thus, we have $\delta(G) = n - 1$, and so $G \cong K_n$, which is a contradiction.

Now, for each edge in F, let $w(e)$ be a vertex incident with e but $w(e) \notin \{u, v\}$. Let $S = \{w(e) \mid e \in F\}$. Then $|S| \leq |F|$ and, by the above claim which ensures that there will be at least one vertex in each of G_1 and G_2 after S has been removed, S is a cut of G. We thus have $\kappa(G) \leq |S| \leq |F| = \kappa'(G)$, as required.

$\square$

Theorem 6.3.1 says that for any graph G, the three parameters $\kappa(G)$, $\kappa'(G)$ and $\delta(G)$ satisfy the inequalities: $\kappa(G) \leq \kappa'(G) \leq \delta(G)$. Can these restrictions be improved? The answer is 'no' as shown in Theorem 6.3.2.

Theorem 6.3.2. (Chartrand and Harary (1968)) *For any triple (p, q, r) of positive integers with $p \leq q \leq r$, there exists a graph G such that $\kappa(G) = p$, $\kappa'(G) = q$ and $\delta(G) = r$.*

Proof. Given the triple (p, q, r), let H_1 and H_2 each be the complete graph K_{r+1}, $A \subseteq V(H_1)$ with $|A| = p$, and $B \subseteq V(H_2)$ with $|B| = q$. Construct a graph G by taking the disjoint union of H_1 and H_2, and adding q new edges joining B and A in such a way that each vertex in B is incident with exactly one new edge and each vertex in A is incident with at least one new edge. It can be verified that G is a desired graph.

$\square$

Example 6.3.3. For $(p, q, r) = (3, 4, 5)$, a graph G as described in the above proof is shown in Fig. 6.3.1.

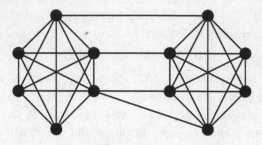

Fig. 6.3.1

Exercise for Section 6.3

1. Construct a graph G with $(\kappa(G), \kappa'(G), \delta(G)) = (2, 4, 6)$.
2. Let G be a graph with $\Delta(G) \leq 3$. Show that $\kappa(G) = \kappa'(G)$. Does the equality still hold if $\Delta(G) = 4$?
3. Let G be a graph of order $n \geq 3$. Show that if $\delta(G) \geq n - 2$, then $\kappa(G) = \delta(G)$. Does the equality still hold if $\delta(G) = n - 3$?
4. Let G be a graph of order $n \geq 3$. Show that if $\delta(G) \geq \frac{1}{2}n$, then $\kappa'(G) = \delta(G)$. Does the equality still hold if $\delta(G) = \lceil \frac{n}{2} - 1 \rceil$?
5. Suppose G is a k-regular graph with $\kappa(G) = 1$, where $k \geq 3$. What is the maximum possible value for $\kappa'(G)$?
6. Let G be a graph of order $n \geq 2$ with $\kappa(G) = p \geq 1$. Determine each of the following values, expressed in terms of n, p or q, where $G + H$ denotes the join of two graphs G and H.

   (a) $\kappa(G + O_q)$, $q \geq 1$;
   (b) $\kappa(G + C_q)$, $q \geq 3$;
   (c) $\kappa(G + K_q)$, $q \geq 2$.

7. Is there any relation among $\kappa(G + H)$, $\kappa(G)$ and $\kappa(H)$?
8. Let G be a connected graph with $\delta(G) \geq 2$. Show that $\kappa(L(G)) = 1$ if and only if $\kappa'(G) = 1$.
9. Let G be a graph of order $n \geq 2$. Show that

$$1 \leq \kappa(G) + \kappa(\overline{G}) \leq \kappa'(G) + \kappa'(\overline{G}) \leq n - 1.$$

6.4   *k*-connected Graphs

In this section, we shall introduce classes of graphs in terms of their connectivity. Let k be a positive integer. A non-trivial graph G is said to be **k-connected** if $\kappa(G) \geq k$. By convention, the graph K_1 is 0-connected.

Discussion 6.4.1.   Consider the following wheel W_6.

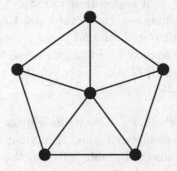

Fig. 6.4.1

Note that $\kappa(W_6) = 3$. Is it right to say that

- (i) W_6 is 1-connected?
- (ii) W_6 is 2-connected?
- (iii) W_6 is 3-connected?
- (iv) W_6 is 4-connected?

Observation 6.4.2.   Let G be a graph of order $n \geq 2$.

- (i) If G is k-connected, then $0 \leq k \leq n-1$.
- (ii) By convention, G is 0-connected if and only if it is disconnected.
- (iii) G is 1-connected if and only if it is connected.
- (iv) G is $(n-1)$-connected if and only if $G \cong K_n$.
- (v) If G is k-connected, where $k \geq 2$, then it is also $(k-1)$-connected.

Example 6.4.3.

- (i) The graph K_2 or any other connected graph containing a cut-vertex is 1-connected but not 2-connected.
- (ii) Every cycle C_n, $n \geq 3$, is k-connected, for $k = 1, 2$, but not 3-connected.

(iii) Every wheel W_n, $n \geq 4$, is k-connected, for $k = 1, 2, 3$, but not 4-connected.

Observation 6.4.4.   The following two observations are very useful in proving results relating to k-connectedness.

(i) Let G be a connected and non-complete graph. Then G is k-connected, $k \geq 1$, if and only if for all $S \subseteq V$ with $|S| = k - 1$, $G - S$ is still connected. (See Problem 6.4.10.)

(ii) Let G be a graph of order n and $\{v_1, v_2, ..., v_k\} \subseteq V$. Let G' be the graph obtained from G by adding a new vertex w and k new edges $wv_1, wv_2, ..., wv_k$. If G is k-connected, then G' is also k-connected. (See Problem 6.4.11.)

In what follows, we shall first state without proof a classic result known as Menger's theorem on connectivity, and then use it to establish Whitney's characterization of k-connected graphs. To end this section, we present a property of k-connected graphs due to Dirac.

Let G be a connected graph, and u, v be two non-adjacent vertices in G. A subset S of $V - \{u, v\}$ is said to **separate** u and v if $G - S$ is disconnected and u and v are in different components of $G - S$. A set $I = \{Q_1, Q_2, ..., Q_k\}$ of $u-v$ paths (each Q_i is a $u-v$ path) in G is said to be **internally-disjoint** if $V(Q_i) \cap V(Q_j) = \{u, v\}$ for all i, j with $1 \leq i < j \leq k$.

Discussion 6.4.5.   Let S be any set of vertices in G separating u and v, and I be any set of internally-disjoint $u - v$ paths in G. Is it true that $|I| \leq |S|$? Why?

Theorem 6.4.6.   (Menger (1927)) *Let G be a connected graph, and u, v be two non-adjacent vertices in G. Then the minimum number of vertices separating u and v is equal to the maximum number of internally-disjoint $u - v$ paths in G.*

$\square$

Example 6.4.7.   Consider the graph G of Fig. 6.4.2(a) with two non-adjacent vertices u and v. It can be checked that the minimum number of vertices separating u and v is 4, which is equal to the maximum number of internally-disjoint $u - v$ paths in G, as shown in Fig. 6.4.2(b).

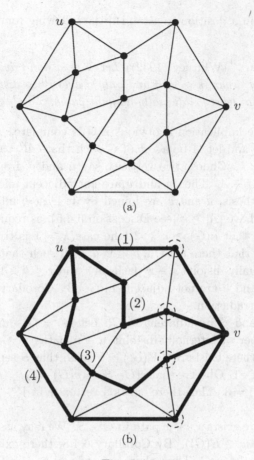

Fig. 6.4.2

Corollary 6.4.8. *Let u and v be two non-adjacent vertices in a connected graph G. Then*

$$\kappa(G) \leq \max\{|I| \mid I \text{ is a set of internally-disjoint } u-v \text{ paths}\}.$$

Proof. By definition, for any set S that separates u and v, $\kappa(G) \leq |S|$. Thus, by Theorem 6.4.6,

$$\kappa(G) \leq \min\{|S| \mid S \text{ separates } u \text{ and } v\}$$
$$= \max\{|I| \mid I \text{ is a set of internally-disjoint } u-v \text{ paths}\}.$$

$\square$

We are now in a position to establish the following fundamental result on connectivity.

Theorem 6.4.9. (Whitney (1932)) *Let G be a non-trivial graph. Then G is k-connected, where $k \geq 1$, if and only if every two distinct vertices in G are joined by at least k internally-disjoint paths.*

Proof. ($\Leftarrow$) The implication is obvious if G is complete. Assume that G is not complete, and let A be a cut of G such that $|A| = \kappa(G)$. We shall show that $|A| \geq k$. Choose two vertices, say u and v, from two different components of $G - A$. Then u and v are not adjacent and are separated by A. By hypothesis, u and v are joined by at least k internally-disjoint paths. We thus have $|A| \geq k$ (see Discussion 6.4.5), as required.

($\Rightarrow$) Assume that $\kappa(G) \geq k \geq 2$ (the case $k = 1$ is trivial). Suppose on the contrary that there exist $u, v \in V, u \neq v$, such that the maximum number of internally-disjoint $u - v$ paths is r and $r \leq k - 1$.

Case 1: u and v are not adjacent in G. By Corollary 6.4.8, $\kappa(G) \leq r \leq k - 1$, a contradiction.

Case 2: u and v are adjacent in G. Let $G' = G - uv$. Clearly, the maximum number of internally-disjoint $u - v$ paths in G' is $r - 1$. By Theorem 6.4.6, there exists $S \subset V(G') = V$ such that S separates u and v in G' and $|S| = r - 1$. Observe that $v(G - S) = v(G) - |S| \geq (k+1) - (r-1) \geq (k+1) - (k-2) = 3$. Thus there exists a vertex w in $V - S$ other than u or v.

Claim. There exists a $u - w$ path in $G' - S$. We may assume that $uw \notin E(G')$ (and so $uw \notin E(G)$). By Corollary 6.4.8, there exist k internally-disjoint $u - w$ paths in G. Thus, there are at least $k - 1$ internally-disjoint $u - w$ paths in G'. Since $|S| = r - 1 \leq k - 2$, at least one of these $u - w$ paths is disjoint from S.

Likewise, there exists a $v - w$ path in $G' - S$. By combining a $u - w$ path and a $v - w$ path in $G' - S$, we obtain a $u - v$ walk in $G' - S$. This, however, contradicts the fact that S separates u and v in G'.

$\square$

Example 6.4.10. The graph G of Fig. 6.4.2 is 4-connected and it can be verified that every two distinct vertices in G are joined by at least 4 internally-disjoint paths.

Corollary 6.4.11. *Let G be a graph with $v(G) \geq 3$. The following statements are equivalent:*

(i) *G is 2-connected.*

(ii) *Every two vertices in G are joined by at least two internally-disjoint paths.*

(iii) *Every two vertices in G are contained in a common cycle.*

(iv) *Every two edges in G are contained in a common cycle.*

(v) *Any vertex together with any edge are both contained in a common cycle.*

□

A connected graph of order $n \geq 2$ is called a **block** if it is K_2 when $n = 2$ or it is 2-connected when $n \geq 3$. Thus, Corollary 6.4.11 (see also Problem 6.4.14) provides us with a number of characterizations of blocks.

Let G be a connected graph of order $n \geq 3$. A subgraph H of G is called a **block** of G if H is itself a block and H is not properly contained in any subgraph which is a block. Clearly, every block of G must be an induced subgraph of G. Figure 6.4.3 shows a connected graph having 3 cut-vertices and 5 blocks. It is noted that while any disconnected graph is a disjoint union of *components*, any connected graph with cut-vertices is formed by *blocks* which meet at cut-vertices.

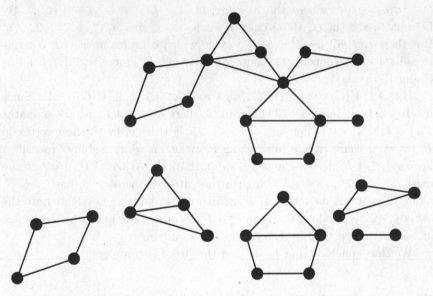

Fig. 6.4.3

By Corollary 6.4.11, every two vertices of a 2-connected graph are contained in a common cycle. We shall now establish the following extension.

Theorem 6.4.12.   (Dirac (1960)) *Let G be a k-connected graph, where $k \geq 2$. Then every set of k vertices is contained in a common cycle in G.*

To prove the above theorem, we shall make use of the following result (see Problem 6.4.8).

Lemma 6.4.13.   *Let G be a k-connected graph and $u, v_1, v_2, ..., v_k$ be any $k+1$ vertices in G. Then there exists a $u-v_i$ path Q_i, for each $i = 1, 2, ..., k$, in G such that $V(Q_i) \cap V(Q_j) = \{u\}$ whenever $i \neq j$.*

$\square$

Proof of Theorem. We assume that $k \geq 3$. Let A be a set of k distinct vertices. Choose a cycle C such that $|A \cap V(C)| = h$ is maximum (note that $h \geq 2$ by Corollary 6.4.11). Our aim is to show that $h = k$. Suppose on the contrary that $h < k$.

   Claim. $|V(C)| \geq h + 1$. If not, then $|V(C)| = h$. We may write $C : a_1 a_2 ... a_h a_1$, where the a_i's are in A. Let $a \in A - V(C)$. By Lemma 6.4.13, in G, there exist h such $a - a_i$ paths $Q_i, i = 1, 2, ..., h$. But then $a_1 Q_1 a Q_2 a_2 a_3 ... a_h a_1$ is a cycle which contains at least $h + 1$ vertices in A. This contradicts the maximality of 'h'. Thus, $|V(C)| \geq h + 1$ as claimed.

   Let $A \cap V(C) = \{w_1, w_2, ..., w_h\}$ and let $w_{h+1} \in V(C) - A$. Since $h + 1 \leq k$, by Lemma 6.4.13 again, in G, there exist $h + 1$ such $a - w_i$ paths $P_i, i = 1, 2, ..., h + 1$. For each $i = 1, 2, ..., h + 1$, let v_i be the first vertex in P_i (as we traverse from a to w_i along P_i) which is contained in C (possibly, $v_i = w_i$). Let P_i' denote the $a - v_i$ subpath of P_i. As $|A \cap V(C)| = h$, there exist $i, j = 1, 2, ..., h + 1, i \neq j$, such that at least one $v_i - v_j$ path, say R, along C contains no vertex of A as an interior vertex. Let R' denote the other $v_i - v_j$ path along C. Then $P_i' R' P_j'$ forms a cycle in G which contains at least $h + 1$ vertices in A, a contradiction again.

   We thus conclude that $h = k$ and the proof is complete.

$\square$

Remark 6.4.14. By Corollary 6.4.11, the converse of Theorem 6.4.12 is true if $k = 2$. Is the converse true if $k \geq 3$? Obviously, no. The cycle

C_n, where $n \geq k \geq 3$, is the simplest counter-example.

Exercise for Section 6.4

1. Consider each of the following graphs H.

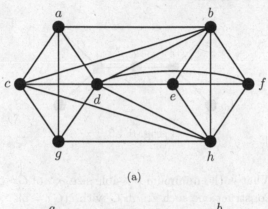

(a)

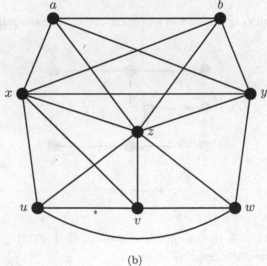

(b)

Fig. 6.4.4

Answer the following.

(a) Evaluate $\kappa(H)$, $\kappa'(H)$ and $\delta(H)$.
(b) Find a minimum cut of H.

(c) Find a minimum edge-cut of H.

(d) What is the maximum number of edges that can be removed from H such that the resulting spanning subgraph H' of H is still $\kappa(H)$-connected?

2. Let G be a graph with $\kappa'(G) \geq 3$ which contains the following graph as a spanning tree:

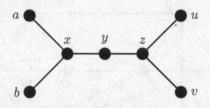

Fig. 6.4.5

(a) What is the minimum possible size m^* of G?

(b) Construct one such graph G with $e(G) = m^*$.

3. A graph G contains the following graph as a spanning subgraph.

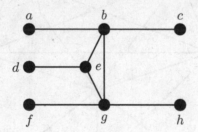

Fig. 6.4.6

(a) What is the minimum size m' of G if $\kappa'(G) \geq 2$? Construct one such graph G of size m'.

(b) What is the minimum size m^* of G if $\kappa'(G) \geq 3$? Construct one such graph G of size m^*.

4. **k-connected graphs of fixed order with minimum size**

(a) Let G be a k-connected graph of order n, where $n > k \geq 2$. Show that $e(G) \geq \lceil \frac{kn}{2} \rceil$.

(b) Construct

   (i) a 4-connected $(10, 20)$-graph;
   (ii) a 5-connected $(10, 25)$-graph;
   (iii) a 5-connected $(11, 28)$-graph.

Note. General constructions of k-connected graphs of order n with minimum size $\left\lceil \frac{kn}{2} \right\rceil$ were given in Harary (1962).

5. Let G be a 2-connected graph of order $n \geq 5$ with $\text{diam}(G) = 2$. Murty (1969) proved that $e(G) \geq 2n - 5$. For each $n \geq 5$, construct one such graph G with $e(G) = 2n - 5$.

6. Is it true that if G is a graph with $\kappa(G) = k \geq 1$, then every two vertices u and v in G are joined by exactly k internally-disjoint $u - v$ paths?

7. Let G be a connected graph of order $n \geq 2$. Show that

$$n \geq 2 + \kappa(G)(\text{diam}(G) - 1).$$

8. Let G be a graph.

   (a) Show that if G is k-connected, then for any $k + 1$ vertices $u, v_1, v_2, ..., v_k$ in G, there exists a $u - v_i$ path Q_i, for each $i = 1, 2, ..., k$, in G such that $V(Q_i) \cap V(Q_j) = \{u\}$ whenever $i \neq j$.
   (b) Is the converse of (a) true?

9. Let G be a connected graph of order n. Two paths in G are said to be **disjoint** if they have no vertex in common.

   (a) Show that if G is 3-connected and $n \geq 6$, then for any two disjoint 3-element subsets A and B of V, there exist three pairwise disjoint paths in G, each connecting a vertex in A to a vertex in B.
   (b) Is the converse of (a) true?
   (c) Generalize the above results.

10. Let G be a non-complete graph and k be a positive integer. Prove that G is k-connected if and only if for all $S \subseteq V(G)$ with $|S| \leq k - 1$, $G - S$ is connected.

11. Let G be a graph and A be a set of k vertices in G, where $k \geq 1$. Let G' be the graph obtained from G by adding a new vertex w and k new edges wv, where v is in A.

   (a) Show that if G is k-connected, then G' is also k-connected.
   (b) Is the converse of (a) true?

12. Find the connectivity of the following graph, and use it to verify Theorem 6.4.9.

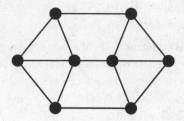

Fig. 6.4.7

13. Prove Corollary 6.4.11.

14. Let G be a graph with $v(G) \geq 3$. Prove that the following statements are equivalent:

  (i) G is 2-connected.
  (ii) For every two vertices u and v, and every edge e in G, there exists a $u - v$ path containing e.
  (iii) For every two vertices u and v, and every edge e in G, there exists a $u - v$ path not containing e.
  (iv) For every three vertices u, v, w in G, there exists a $u - v$ path in G containing w.
  (v) For every three vertices u, v, w in G, there exists a $u - v$ path in G not containing w.

15. Let G be a connected graph with cut-vertices. A block of G is called an **end-block** if it contains only one cut-vertex of G. Thus, the graph of Fig. 6.4.3 contains three end-blocks (namely, the 1st, 4th and 5th from left to right). Show that G contains at least two end-blocks.

16. Let G be a graph in which every two blocks have a vertex in common. Show that all blocks in G have a vertex in common.

17. Let G be a k-connected graph, and let A and B be two disjoint subsets of V such that $|A| = 2$ and $|B| = k - 2$, where $k \geq 3$. Show that there exists a cycle C in G such that $A \subseteq V(C)$ and $B \cap V(C) = \emptyset$. Generalize the above result.

18. Let G be a k-connected graph where $k \geq 2$. By Theorem 6.4.12, any set of k vertices is contained in a common cycle. Can we have a stronger conclusion that any set of $k + 1$ vertices is contained in

a common cycle?

19. Let G be a connected graph. Show that G is Eulerian if and only if every block of G is Eulerian.

6.5   Problem Set VI

1. A graph G contains the following $(9, 10)$-graph H as a spanning subgraph.

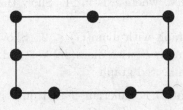

Fig. 6.5.1

(i) Is $\kappa(G) \geq 2$?

(ii) What is the minimum size m^* of G if $\kappa(G) \geq 3$? Construct one such graph G of size m^*.

(iii) What is the minimum size m' of G if G is an Eulerian 3-connected graph? Construct one such graph G of size m'.

2. Let G be a 2-connected graph with $\delta(G) \geq 3$. Show that there exists an edge e in G such that $G - e$ is 2-connected.

3. Let G be a 2-connected graph. Suppose that there exists a vertex x in G such that $N(x) = \{u, v\}$ and $uv \in E(G)$. Show that either $G \cong K_3$ or $G - uv$ is 2-connected.

4. Let G be a 2-connected graph. Show that one of the following statements holds:

(i) $G \cong K_3$;

(ii) there exists a vertex x in G such that $d(x) = 2$ and the two neighbors of x are not adjacent;

(iii) $G - e$ is 2-connected for some edge e in G.

5. Let G be a graph of order at least three. Show that G is 2-connected if and only if $G \cong K_3$ or G can be obtained from K_3 by applying a finite sequence of the following operations:

(i) replace an edge uv by a $u - v$ path;

(ii) add a new edge joining two non-adjacent vertices.

6. Show that the center $C(G)$ of a connected graph G lies within a single block of G.

7. Let G be a 2-connected graph of order $n \geq 3$. Assume that G contains no even cycle as a subgraph. Show that n is odd and $G \cong C_n$.

8. Let G be a graph and A, a set of k vertices in G, where $k \geq 1$. Let G^* be the graph obtained from G by adding a new vertex w and k new edges wv, where v is in A. Show that if $\kappa'(G) \geq k$, then $\kappa'(G^*) \geq k$.

9. Let G be a graph with $\mathrm{diam}(G) \leq 2$. Show that $\kappa'(G) = \delta(G)$. Construct a graph H with $\mathrm{diam}(H) = 3$ and $\kappa'(H) < \delta(H)$.

10. Let G be a connected graph.

(i) Show that G is Eulerian if and only if $|F|$ is even for any minimal edge-cut F in G.

(ii) Deduce that $\kappa'(G)$ is even if G is Eulerian. (See Problem 6.2.9.)

(iii) Is the converse of (ii) true?

11. Let G be a 2-connected graph which contains no $K(1,3)$ or $K(1,3) + e$ as shown in Fig. 6.5.2 as an induced subgraph. Show that G is Hamiltonian.

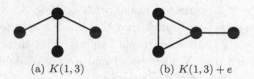

(a) $K(1,3)$ (b) $K(1,3) + e$

Fig. 6.5.2

(Goodman and Hedetniemi (1974))

12. Let G be a k-Hamiltonian (n, m)-graph (see Problem V.15). Show that

(i) $\kappa(G) \geq k + 2$;
(ii) $m \geq \frac{1}{2}n(k + 2)$.

13. Let G be an incomplete graph of order $n \geq 3$. Show that $\kappa(G) \geq 2(\delta(G) + 1) - n$. For each $n \geq 3$, construct a graph G of order n for which the equality holds.

14. Let G be a graph of order $n \geq 3$, and k an integer with $1 \leq k \leq n-1$. Show that if $\delta(G) \geq \frac{1}{2}(n+k-2)$, then G is k-connected.

15. Let G be a connected graph. Denote by $b(G)$ the number of blocks of G, and for any vertex v in G, denote by $b(v)$ the number of blocks of G which contain v. Show that

$$b(G) = 1 + \sum_{v \in V} \left(b(v) - 1 \right).$$

16. Let G be a connected graph. Denote by $cv(G)$ the number of cut-vertices of G, and for each block B of G, denote by $cv(B)$ the number of cut-vertices of G which are contained in B. Show that

$$cv(G) = 1 + \sum \left(cv(B) - 1 \mid B \text{ is a block of } G \right).$$

17. Let G be a 3-connected graph. Show that for any two vertices u and v in G, there exist two internally disjoint $u-v$ paths of different lengths in G. Is the result still true if G is only 2-connected?

18. Show that for each $n \geq 1$, $\kappa(Q_n) = \kappa'(Q_n) = \delta(Q_n) = n$.

Chapter 7

Independence, Matching and Covering

7.1 Introduction

Let G be a graph with vertex set V and edge set E. We first name a natural relation among 'vertices' in G. Two vertices are said to be **independent** if they are not adjacent in G; and a set A of vertices in G is called an **independent set** if every two vertices in A are independent in G. Next, we define also a natural relation among 'edges'. Two edges are said to be **independent** if they are not adjacent, that is, they are not incident with a common vertex; and a set M of edges in G is called a **matching** if every two edges in M are independent in G.

Now, we introduce two relations between 'vertices' and 'edges' in G. A set Q of vertices in G is called a **vertex cover** of G if every edge in G is incident with a vertex in Q. A set W of edges in G is called an **edge cover** of G if every vertex in G is incident with an edge in W.

In this chapter, we shall study these four notions and discuss some basic interrelationships among them. We begin in Section 7.2 with the motivation for the study of the notion of matchings, but will confine it to bipartite graphs first. We then proceed to present in Section 7.3 the celebrated Hall's theorem on the existence of a 'largest matching' in *bipartite* graphs. The application of Hall's theorem to the study of the existence of a system of distinct representatives for a family of sets is given in Section 7.4. In Section 7.5, we study the notion of matchings in *general* graphs. Berge's characterization theorem on 'maximum matchings' and Tutte's fundamental theorem on the existence of a 'perfect matching' in any general graph are presented in this section. In Section 7.6, the four quantities, namely, the independence number, matching number, vertex covering number and edge covering number of G, arising respectively from the above four notions, are introduced in order; and their interrelationships linked by the well-known Gallai's identities form the core of this final section.

7.2   Matchings in Bipartite Graphs

In Example 1.2.3 (see also Section 3.1), the following bipartite graph which models a job-application situation is shown:

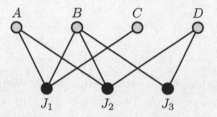

Fig. 7.2.1

Now we ask: Is it possible to assign each applicant to a job for which he or she applies?

The answer is clearly 'no' as there are more applicants than jobs available. Indeed, the best we can do is to assign three applicants to three jobs in a way such as the following:

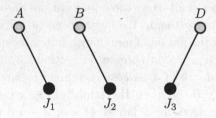

Fig. 7.2.2

The bipartite graph in Fig. 7.2.3 shows the acquaintance relationship between four men (M's) and five women (W's), where a vertex representing a man is adjacent to a vertex representing a woman if and only if the man is acquainted with the woman.

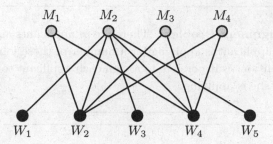

Fig. 7.2.3

Again, we ask: Is it possible to marry off the four men in such a way that each man marries a woman he is acquainted with?

Though now we have more women than men, after some tries, we start to believe that this is also not possible. Indeed, the only women whom M_1, M_3 and M_4 are acquainted with are W_2 and W_4; and it is thus not possible to marry off the four men according to the condition.

For this case, the best we can do is to marry off three of the men in a way such as the following:

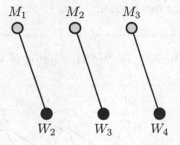

Fig. 7.2.4

The examples above are instances respectively of the following two well-known problems.

The Assignment Problem. There are m applicants and n jobs, and each applicant is applying for a number of these jobs. Under what conditions is it possible to assign each applicant to a job for which he/she is applying?

The Marriage Problem. There are m men and n women, and each man is acquainted with a certain number of the women. Under what conditions is it possible to marry off these m men in such a way that each man marries a woman he is acquainted with?

In this and the next section, we shall study these problems and provide solutions to them using bipartite graphs as models.

Figure 7.2.2 depicts an assignment for three applicants to three jobs, and this assignment is represented by a set of three edges in which no two edges have an end in common. The same applies to Fig. 7.2.4. These motivate us to introduce the following:

Let G be a graph. A non-empty set M of edges in G is called a **matching** of G if no two edges in M are adjacent; that is, every two edges in M are independent.

Question 7.2.1. *Consider the bipartite graph G with bipartition (X, Y) as shown in Fig. 7.2.5.*

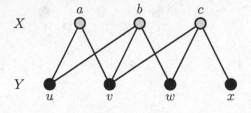

Fig. 7.2.5

(i) *Is {au} a matching?*

(ii) *Is {av, cv} a matching?*

(iii) *Is {au, bv, cw} a matching?*

(iv) *Two matchings are different if they are different as sets. Find three different matchings with three edges.*

(v) *Can you find in G a matching with four edges?*

Question 7.2.2.

(i) *Let G be a graph of order n ≥ 2, and let M be a matching in G. At most how many edges can M contain?*

(ii) *Find a tree of order 10 which contains a matching with 5 edges.*

(iii) *Find a connected graph of order 6 in which every matching can contain only one edge.*

Question 7.2.3. *Consider the bipartite graph G with bipartition (X, Y) as shown in Fig. 7.2.6.*

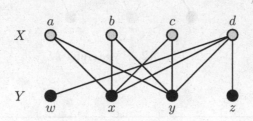

Fig. 7.2.6

(i) *Find in G a matching M with three edges, i.e., $|M| = 3$.*

(ii) *Does G contain a matching M with $|M| = 4 = |X|$? Why?*

In the bipartite graph G of Fig. 7.2.5, there exists a matching M such that $|M| = |X|$, i.e., every vertex in X is incident with an edge in M. This is, however, not the case in the bipartite graph of Fig. 7.2.6.

Let M be a matching of G. A vertex v in G is said to be M-**saturated** if v is incident with an edge in M. Suppose now that G is a bipartite graph with bipartition (X, Y). We say that M is a **complete matching from X to Y** if every vertex in X is M-saturated (that is, $|M| = |X|$).

Thus, in the bipartite graph of Fig. 7.2.5, the matching M shown in Fig. 7.2.7 is a complete matching from X to Y (but not a complete matching from Y to X); whereas there is no complete matching from X to Y in the bipartite graph of Fig. 7.2.6.

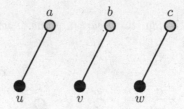

Fig. 7.2.7: Matching M

Let G be a graph. A matching M in G is said to be **perfect** if every vertex in G is M-saturated.

In particular, if G is bipartite with bipartition (X, Y), then M is **perfect** if $|X| = |M| = |Y|$.

Question 7.2.4.   *Two bipartite graphs are shown in Fig. 7.2.8(a) and (b).*

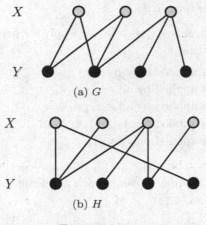

(a) G

(b) H

Fig. 7.2.8

(i) *Does G have a complete matching from X to Y?*
(ii) *Does G have a perfect matching?*
(iii) *Does H have a complete matching from X to Y?*
(iv) *Does H have a perfect matching?*

Remark 7.2.5. Let G be a bipartite graph with bipartition (X, Y).

(1) If G contains a complete matching from X to Y, then $|X| \le |Y|$; but the converse is not true.
(2) If G contains a perfect matching, then $|X| = |Y|$; but again the converse is not true.
(3) If G contains a complete matching M from X to Y, then M is perfect when and only when $|X| = |Y|$.

It is now clear that the assignment problem and the marriage problem as stated at the beginning of this section can be re-formulated using graph terminology as follows:

Problem. Let G be a bipartite graph with bipartition (X, Y). Under what conditions is there a complete matching from X to Y?

We shall provide a solution to this problem in the next section.

Exercise for Section 7.2

1. Five applicants $A_1, A_2, \cdots, A_5$ apply for five jobs $J_1, J_2, \cdots, J_5$. It is known that

   (i) J_1 is applied only by A_2,
   (ii) J_2 is applied by all except A_4,
   (iii) J_3 is applied by all except A_2,
   (iv) J_4 is applied by A_2 and A_4, and
   (v) J_5 is applied only by A_4.

   (a) Draw a bipartite graph that models the situation.
   (b) Is it possible to assign each applicant to a job for which he/she applies?

2. In the preceding problem, suppose that the applicant A_5 changes his/her mind and applies for J_5 instead of J_2.

   (a) Draw a bipartite graph that models the situation.
   (b) Is it possible to assign each applicant to a job for which he/she applies?

3. Five applicants apply to work in a company. There are six jobs available: $J_1, J_2, \cdots, J_6$. Applicant A is qualified for jobs J_2 and J_6; B is qualified for jobs J_1, J_3 and J_4; C is qualified for jobs J_2, J_3 and J_6; D is qualified for jobs J_1, J_2 and J_3; E is qualified for all jobs except J_4 and J_6.

   (a) Draw a bipartite graph that models the situation.
   (b) Is it possible to assign each applicant to a job for which he/she is qualified?

4. Five men $M_1, M_2, \cdots, M_5$ and five women $W_1, W_2, \cdots, W_5$ have and only have the following acquaintance relationships between them:

   (i) each of W_1, W_2 and W_3 is acquainted with all the men,
   (ii) each of M_1 and M_5 is acquainted with all the women.

   (a) Draw a bipartite graph that models the situation.
   (b) Is it possible to marry off these five men in such a way that each man marries a woman he is acquainted with?
   (c) If M_1 insists on marrying W_1, is it possible to marry off the remaining ones in such a way that each man marries a woman he is acquainted with?

5. Consider the following set of codewords:

$$X = \{ab, abc, cd, bcd, de\}.$$

We wish to transmit these codewords as messages. Instead of transmitting the whole codeword, we transmit a single letter which is contained in it, as its representative. Can this be done in such a way that the five codewords can be recovered uniquely from their five respective representatives?

6. A school has vacancies for seven teachers, one for each of the subjects Chemistry, English, French, Geography, History, Mathematics and Physics. There are seven applicants for the vacancies and all are qualified to teach more than one subject. The applicants and their subjects are listed in the table below.

   (a) Draw a bipartite graph to represent this situation.
   (b) Determine the maximum number of (suitably qualified) teachers the school can employ.

Applicants	Subjects qualified
Miss Adilah	Mathematics, Physics
Miss Boey	Chemistry, English, Mathematics
Miss Doraisamy	Chemistry, French, History, Physics
Mr. Richmond	English, French, History, Physics
Mr. Singh	Chemistry, Mathematics
Mr. Tan	Mathematics, Physics
Mr. Wong	English, Geography, History

7. Consider the following bipartite graph G with bipartition (X, Y):

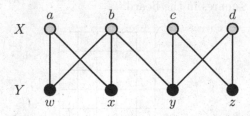

Fig. 7.2.9

(i) Is $\{bx, cy\}$ a matching?
(ii) Is $\{ax, by, cy\}$ a matching?

(iii) Is $\{ax, by, cz\}$ a matching?

(iv) Is $\{ax, bw, cz, dy\}$ a matching? a perfect matching?

(v) Is there any perfect matching that contains the edge 'by'?

(vi) Find the number of perfect matchings in G.

8. For $n \geq 3$, find the number of perfect matchings of the cycle C_n.

9. For $n \geq 1$, find the number of perfect matchings of the graph $K(n, n)$.

10. For $n \geq 2$, find the number of perfect matchings of the graph K_n.

11. Consider the following bipartite graph G with bipartition (X, Y):

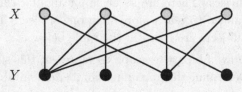

Fig. 7.2.10

(i) Does G contain a complete matching from X to Y?

(ii) Does G contain a perfect matching?

(iii) Find two matchings M and M' in G with $|M| = |M'| = 3$.

(iv) How many matchings M are there in G with $|M| = 3$?

12. Consider the 6×6 grid-board shown in Fig. 7.2.11 whose upper left and lower right corner squares are removed. You are given 17 dominoes, each covering exactly two adjacent squares (squares that have an edge in common) of the board. Can you use them to cover the 34 squares in the board?

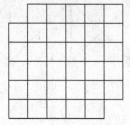

Fig. 7.2.11

13. Prove that the following 3-regular graph (triple flyswat) does not have a perfect matching, but does have a matching with seven edges.

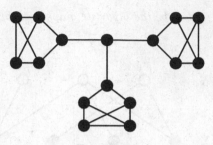

Fig. 7.2.12

14. Let T be a tree of order $n \geq 2$ and M, a matching in T.

   (i) What is the largest possible value for $|M|$? Construct one T which contains one such M having its $|M|$ attaining this largest value.

   (ii) What is the least possible value for $|M|$? Construct one T which contains one such M having its $|M|$ attaining this least value.

15. The 20 members of a local tennis club have scheduled exactly 14 two-person games among themselves, with each member playing in at least one game. Prove that within this schedule there must be a set of 6 games with twelve distinct players. (USAMO, 1989)

16. Generalize the result in the preceding problem by replacing '20 members' by 'n members', and '14 games' by 'm games'.

7.3 Hall's Theorem

In what follows, we will write, for simplicity, V for $V(G)$ and E for $E(G)$.

Before providing a solution to the problem stated at the end of Section 7.2, let us first introduce the following notation.

Let G be a graph. For a set S of vertices in G, let $N(S)$ denote the set of vertices in G which are adjacent to some vertex in S; i.e.,

$$N(S) = \{v \in V \mid v \text{ is adjacent to some vertex in } S\}.$$

It is clear that

$$N(S) = \bigcup_{a \in S} N(a).$$

Question 7.3.1. *Let G be the bipartite graph with bipartition (X, Y) as shown in Fig. 7.3.1.*

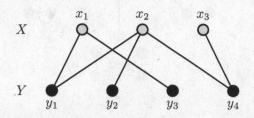

Fig. 7.3.1

(i) Complete the following table:

S	$N(S)$
$\{x_1\}$	$\{y_1, y_3\}$
$\{x_2\}$	
$\{x_3\}$	
$\{x_1, x_2\}$	
$\{x_1, x_3\}$	$\{y_1, y_3, y_4\}$
$\{x_2, x_3\}$	
X	

(ii) Does there exist a complete matching from X to Y?

Question 7.3.2. *Consider the bipartite graph G with bipartition (X, Y) as shown in Fig. 7.3.2.*

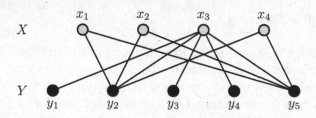

Fig. 7.3.2

(i) Let $S = \{x_1, x_2, x_4\}$. Find $N(S)$.

(ii) Is there a complete matching from X to Y in G? Why?

We are now in a position to establish the following classic result on matchings in bipartite graphs due to the English algebraist Hall (1935).

Theorem 7.3.3. *Let G be a bipartite graph with bipartition (X, Y). Then G contains a complete matching from X to Y if and only if $|S| \leq |N(S)|$ for every subset S of X.*

Proof. $(\Rightarrow)$ Suppose on the contrary that there exists $S \subseteq X$ such that $|S| > |N(S)|$. Then it is clear that there is no complete matching from X to Y in G.

$(\Leftarrow)$ Assume that in G,

$$|S| \leq |N(S)| \text{ for all } S \subseteq X. \tag{$*$}$$

We shall show that G contains a complete matching from X to Y by induction on $|X|$. The above statement is obviously true if $|X| = 1$. Assume that it is true when $|X| \leq k - 1$. Consider now that $|X| = k$, where $k \geq 2$.

Case (1). $|S| + 1 \leq |N(S)|$ for all $S \subset X$ and $S \neq \emptyset$.

Let $x \in X$. Then there exists $y \in Y$ such that $xy \in E(G)$. Let $G' = G - \{x, y\}$. Clearly, G' satisfies $(*)$ for all $S \subseteq X - \{x\}$. Thus, by the induction hypothesis, G' contains a complete matching M' from $X - \{x\}$ to $Y - \{y\}$. It follows that $M' \cup \{xy\}$ is a complete matching from X to Y.

Case (2). There exists $S_0 \subset X$, $S_0 \neq \emptyset$ such that $|S_0| = |N(S_0)|$.

(i) Consider the subgraph $G_0 = [S_0 \cup N(S_0)]$ of G. Observe that G_0 is also bipartite, and satisfies $(*)$ (with X replaced by S_0). By the induction hypothesis, G_0 contains a complete matching M_0 from S_0 to $N(S_0)$.

. (ii) Consider $G' = G - (S_0 \cup N(S_0))$.

Claim: G' satisfies $(*)$ (with X replaced by $X \backslash S_0$).

If not, then there exists $S_1 \subseteq X \backslash S_0$ such that $|S_1| > |N(S_1)|$ in G'. But then $|S_0 \cup S_1| > |N(S_0 \cup S_1)|$ in G, a contradiction.

Thus, the above claim holds, and by the induction hypothesis, G' contains a complete matching M' from $X \backslash S_0$ to $Y \backslash N(S_0)$.

Now, combining (i) and (ii), we obtain a matching M, namely $M_0 \cup M'$, which is a complete matching from X to Y in G.

$\square$

As an immediate consequence of Theorem 7.3.3, we have:

Corollary 7.3.4.   *Let G be a bipartite graph with bipartition (X, Y) such that $|X| = |Y|$. Then G has a perfect matching if and only if $|S| \le |N(S)|$ for every subset S of X.*

$\square$

Example 7.3.5.   In Question 7.3.1, the bipartite graph G with bipartition (X, Y) as depicted in Fig. 7.3.1 has a complete matching from X to Y by Theorem 7.3.3 since, as shown in the table therein, $|S| \le |N(S)|$ for every subset S of X.

On the other hand, in Question 7.3.2, the bipartite graph G with bipartition (X, Y) as depicted in Fig. 7.3.2 has no complete matching from X to Y by Theorem 7.3.3 as there exists a subset S of X, namely, $S = \{x_1, x_2, x_4\}$, such that $|S| > |N(S)|$.

A special family of bipartite graphs which always contain perfect matchings is given below (see Problem 7.3.6).

Corollary 7.3.6.   *Let G be a bipartite graph with bipartition (X, Y). If there exists a positive integer k such that $d(y) \le k \le d(x)$ for all y in Y and x in X, then G has a complete matching from X to Y.*

$\square$

Corollary 7.3.7.   *Every k-regular bipartite graph, where $k \ge 1$, always contains a perfect matching.*

$\square$

Exercise for Section 7.3

1. Let G be the bipartite graph you constructed in Problem 7.2.1. Let $S = \{A_1, A_3, A_5\}$. Find $N(S)$. Is $|N(S)| < |S|$? What conclusion can you draw from Theorem 7.3.3?

2. Let G be the bipartite graph with bipartition (X, Y) as shown in Problem 7.2.7.

   (a) Complete the following table:

S	$N(S)$
$\{a\}$	
$\{b\}$	
$\{c\}$	
$\{d\}$	
$\{a,b\}$	
$\{a,c\}$	
$\{a,d\}$	
$\{b,c\}$	
$\{b,d\}$	
$\{c,d\}$	
$\{a,b,c\}$	
$\{a,b,d\}$	
$\{a,c,d\}$	
$\{b,c,d\}$	
X	

    (b) Is it true that $|S| \le |N(S)|$ for all $S \subseteq X$?

    (c) What conclusion can you draw from Corollary 7.3.4?

3. Let G be the bipartite graph with bipartition (X, Y) as shown in Problem 7.2.11.

    (a) Find a subset S of X such that $|S| > |N(S)|$.

    (b) What conclusion can you draw from Theorem 7.3.3?

4. Consider the following bipartite graph G with bipartition (X, Y):

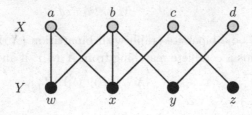

Fig. 7.3.3

    (a) Let $S = \{a, b\}$. Find $N(S)$.

    (b) Let E_1 be the set of edges in G incident with some vertex in S. Find E_1.

   (c) Let E_2 be the set of edges in G incident with some vertex in $N(S)$. Find E_2.

   (d) Is $E_1 \subseteq E_2$?

5. Let G be a bipartite graph with bipartition (X, Y). For $S \subseteq X$, let E_1 be the set of edges in G incident with some vertex in S, and let E_2 be the set of edges in G incident with some vertex in $N(S)$. Is it true in general that $E_1 \subseteq E_2$? Why?

6. Let G be a bipartite graph with bipartition (X, Y). Assume that there exists a positive integer k such that $d(y) \le k \le d(x)$ for each vertex y in Y and each vertex x in X.

   Let $S \subseteq X$, and denote by E_1 the set of edges in G incident with some vertex in S, and by E_2 the set of edges in G incident with some vertex in $N(S)$ (see the preceding problem).

   (a) Show that $k|S| \le |E_1| \le |E_2| \le k|N(S)|$.

   (b) Deduce from Theorem 7.3.3 that G has a complete matching from X to Y.

   (c) Deduce from (b) that every k-regular bipartite graph with $k \ge 1$ has a perfect matching.

7. Let G be a bipartite graph with bipartition (X, Y). Let $\Delta(G) = k \ge 1$ and

$$X^* = \{x \in X \mid d(x) = k\}.$$

Assume that X^* is not empty. Determine whether the following statement ($\#$) is true and justify your answer.

   ($\#$) *G contains a matching M such that every vertex in X^* is M-saturated.*

8. Let G be a bipartite graph with bipartition (X, Y). Prove that G contains a complete matching from X to Y if and only if

$$|X \backslash N(T)| \le |Y \backslash T|$$

for all $T \subseteq Y$.

9. Let G be a bipartite graph. Prove that G contains a perfect matching if and only if $|S| \le |N(S)|$ for all $S \subseteq V(G)$.

10. Let G be a connected bipartite graph with bipartition (X, Y), where $|X| \ge 2$ and $|Y| \ge 2$. Prove that the following statements are equivalent:

   (a) Each edge of G is contained in a perfect matching of G.

(b) $|X| = |Y|$ and $|S| < |N(S)|$ for all $S \subset X$ with $S \neq \emptyset$.

(c) $G - \{x, y\}$ has a perfect matching for any $x \in X$ and $y \in Y$.

11. Let G be a bipartite graph with bipartition (X, Y) such that $|X| - |Y| = p \geq 1$. Form a larger bipartite graph G^* with bipartition $(X, Y \cup Y^*)$, where $|Y^*| = p$, such that
    (1) G^* contains G as an induced subgraph, and
    (2) every vertex in Y^* is adjacent to every vertex in X.
    Prove that G^* has a perfect matching if and only if G has a matching with $|Y|$ edges.

12. Let G be a bipartite graph with bipartition (X, Y), and let k be an integer such that $1 \leq k \leq |X|$. Show that G contains a matching M with $|M| = k$ if and only if

$$|S| \leq |N(S)| + |X| - k$$

    for all $S \subseteq X$.

13. Let G be a bipartite graph with bipartition (X, Y). Assume that $d(x) \geq 6$ and $d(y) \leq 8$ for all $x \in X$ and $y \in Y$. Prove that G contains a matching M with $|M| \geq \frac{3}{4}|X|$.

14. Let G be a bipartite graph with bipartition (X, Y). For $S \subseteq X$, the **deficiency** $\rho(S)$ of S is defined as

$$\rho(S) = |S| - |N(S)|.$$

    Assume that $d(x) \geq 3$ and $d(y) \leq 4$ for all $x \in X$ and $y \in Y$. Show that

$$|X| \geq 4\rho(S),$$

    for all $S \subseteq X$.

15. Let G be a bipartite graph with bipartition (X, Y). Let A, B be subsets of X. Show that
    (a) $N(A \cup B) = N(A) \cup N(B)$;
    (b) $N(A \cap B) \subseteq N(A) \cap N(B)$;
    (c) $|N(A \cup B)| + |N(A \cap B)| \leq |N(A)| + |N(B)|$;
    (d) $\rho(A \cup B) + \rho(A \cap B) \geq \rho(A) + \rho(B)$. (See Problem 7.3.14.)

    Define the **deficiency** $\rho(G)$ of G as

$$\rho(G) = \max\{\rho(A) \mid A \subseteq X\}.$$

    Show that

    (e) if $\rho(A) = \rho(B) = \rho(G)$, then $\rho(A \cap B) = \rho(G)$.

16. Let G be a bipartite graph with bipartition (X, Y). Show that the maximum number of vertices in X that can be saturated by a matching in G is equal to $|X| - \rho(G)$. (For the definition of $\rho(G)$, see Problem 7.3.15.)

17. Let k and n be positive integers with $k \leq n$. A $k \times n$ matrix with entries from $\{1, 2, \cdots, n\}$ is called a **Latin rectangle** if each 'i' in $\{1, 2, \cdots, n\}$ appears exactly once in each row and at most once in each column. A $k \times n$ Latin rectangle is called a **Latin square** of order n if $k = n$.

    Consider the 3×5 Latin rectangle $L = \begin{pmatrix} 1\ 2\ 3\ 4\ 5 \\ 5\ 1\ 2\ 3\ 4 \\ 4\ 5\ 1\ 2\ 3 \end{pmatrix}$.

    Define a bipartite graph G with bipartition (X, Y) associated with L as follows:

    (i) $X = \{1, 2, 3, 4, 5\}$,
    (ii) $Y = \{C_1, C_2, C_3, C_4, C_5\}$, where C_i is the ith column of L,
    (iii) 'i' in X is adjacent to 'C_j' in Y if and only if 'i' does not appear in 'C_j'.

    (a) Draw the diagram of G.
    (b) What is the degree of each vertex in X?
    (c) What is the degree of each vertex in Y?
    (d) Is G 2-regular?
    (e) Does G contain a perfect matching?
    (f) Display a perfect matching in G if your answer to (e) is 'yes'.
    (g) Use the perfect matching obtained in (f) to append a new row to L to form a 4×5 Latin rectangle L'.
    (h) Expand L' to form a Latin square of order 5.

18. Consider the 2×6 Latin rectangle $L = \begin{pmatrix} 1\ 2\ 3\ 4\ 5\ 6 \\ 3\ 6\ 4\ 5\ 2\ 1 \end{pmatrix}$.

    Define, likewise, the bipartite graph G with bipartition (X, Y) associated with L as shown in the preceding problem by replacing X by $\{1, 2, 3, 4, 5, 6\}$, and Y by $\{C_1, C_2, C_3, C_4, C_5, C_6\}$, where C_i is the ith column of L.

    (a) Draw the diagram of G.
    (b) What is the degree of each vertex in X?
    (c) What is the degree of each vertex in Y?
    (d) Is G 4-regular?
    (e) Does G contain a perfect matching?

(f) Display a perfect matching in G if your answer to (e) is 'yes'.

(g) Use the perfect matching obtained in (f) to append a new row to L to form a 3×6 Latin rectangle L'.

7.4 System of Distinct Representatives

In this section, we shall introduce the notion of a system of distinct representatives for a family of finite sets, and prove a classic result about the existence of such a system by applying Hall's theorem, i.e., Theorem 7.3.3.

We begin with the following example.

Example 7.4.1. In the math department of a university, there are five staff committees with their executives (excluding their Head) elected as shown below:

$$\begin{aligned}
&\text{Colloquium(C):} &&\{a, b\}, \\
&\text{Library(L):} &&\{b, c, d\}, \\
&\text{Research(R):} &&\{a, b\} \\
&\text{Sport(S):} &&\{d, e\} \\
&\text{Teaching(T):} &&\{b, e\}
\end{aligned}$$

The Head would like to call a meeting where each committee is represented by one executive and different committees must be represented by distinct representatives. Can this be done?

By trial and error, it is not difficult to see that it can be done. Indeed, one possible solution is shown below (where '$\triangleright$' indicates 'representing'):

$$a \triangleright C, \quad c \triangleright L, \quad b \triangleright R, \quad d \triangleright S \text{ and } e \triangleright T.$$

Note that C and R have the same executives.

The above example provides an instance for the following important notion in Combinatorics.

Let W be a non-empty set, and $S_1, S_2, \cdots, S_m$ be non-empty finite subsets (*but not necessarily distinct*) of W. A **system of distinct representatives (SDR)** for the family $(S_1, S_2, \cdots, S_m)$ is a sequence of m elements $(a_1, a_2, \cdots, a_m)$ such that $a_i \in S_i$ for each $i = 1, 2, \cdots, m$, and $a_i \neq a_j$ whenever $i \neq j$.

Thus, in Example 7.4.1, we see that the sequence (a, c, b, d, e) is an SDR for the family (C, L, R, S, T). Note that it is allowable that $S_i = S_j$ for some distinct i and j; for instance, we have $C = \{a, b\} = R$ in Example 7.4.1.

Example 7.4.2.   Let W be the set of natural numbers. Consider the family of subsets of W in each of the following cases:

    (i) $S_1 = \{1, 2\}, S_2 = \{2, 3\}, S_3 = \{3, 4\}, S_4 = \{4, 5\}$ and $S_5 = \{1, 5\}$.
   (ii) $S_1 = \{1, 2\}, S_2 = \{2, 3\}, S_3 = \{3, 4, 5\}, S_4 = \{1, 3\}$ and $S_5 = \{1, 2, 3\}$.

Does the family $(S_1, S_2, \cdots, S_5)$ have an SDR?

    (i) The family $(S_1, S_2, \cdots, S_5)$ has an SDR, for instance, $(1, 2, 3, 4, 5)$.
   (ii) The family $(S_1, S_2, \cdots, S_5)$ does not have any SDR. Why? One way to argue is as follows: Observe that

$$S_1 \cup S_2 \cup S_4 \cup S_5 = \{1, 2, 3\}.$$

As there are more sets (4 sets) than members (3 only), it is clear that no SDR for the family could exist.

In general, given a family of m sets $(S_1, S_2, \cdots, S_m)$, if there exist k of these sets, where $1 \leq k \leq m$, whose union has less than k members, then it is obvious that the family does not have any SDR. That is, if there exists some $I \subseteq \{1, 2, \cdots, m\}$, I non-empty, such that

$$\left| \bigcup_{i \in I} S_i \right| < |I|,$$

then the family does not have any SDR. In other words, if $(S_1, S_2, \cdots, S_m)$ has an SDR, then

$$\left| \bigcup_{i \in I} S_i \right| \geq |I|,$$

for any subset I of $\{1, 2, \cdots, m\}$.

Is the converse true? That is, if $\left| \bigcup_{i \in I} S_i \right| \geq |I|$, for any subset I of $\{1, 2, \cdots, m\}$, is it true that the family $(S_1, S_2, \cdots, S_m)$ would have an SDR?

The answer to the above question is in the affirmative, and this positive answer, as shown in Theorem 7.4.3, was given by Hall (1935) also.

Theorem 7.4.3.   *Let W be a non-empty set, and let $(S_1, S_2, \cdots, S_m)$ be a family of non-empty finite subsets of W, where $m \geq 1$. Then the family*

$(S_1, S_2, \cdots, S_m)$ *has an SDR if and only if*

$$\left| \bigcup_{i \in I} S_i \right| \geq |I|$$

for any subset I of $\{1, 2, \cdots, m\}$.

The necessity of Theorem 7.4.3, as pointed out earlier, is trivial. We shall now apply Hall's matching theorem (Theorem 7.3.3) to prove the sufficiency of Theorem 7.4.3.

Proof. (Proof of the sufficiency of Theorem 7.4.3.)

Assume that $(S_1, S_2, \cdots, S_m)$ is a given family of non-empty finite subsets of W, where $m \geq 1$, satisfying the condition that

$$\left| \bigcup_{i \in I} S_i \right| \geq |I|,$$

for any subset I of $\{1, 2, \cdots, m\}$. We may assume that W is finite. Our aim is to show that the family $(S_1, S_2, \cdots, S_m)$ has an SDR.

To begin with, we form a bipartite graph G with bipartition (X, Y), where

$$X = \{S_1, S_2, \cdots, S_m\} \text{ and } Y = W,$$

such that 'S_i' in X and 'y' in Y are *adjacent* in G when and only when y is an *element* in S_i. (For instance, the bipartite graph G with bipartition (X, Y) corresponding to Example 7.4.1 is shown in Fig. 7.4.1, noting that Y is the set of staff members in the department.)

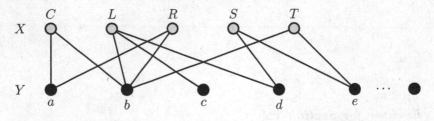

Fig. 7.4.1

Next, we shall show that the inequality $|A| \leq |N(A)|$ holds in G for every subset A of X. Thus, let A be a subset of $X = \{S_1, S_2, \cdots, S_m\}$. We may write $A = \{S_i \mid i \in I\}$ for some subset I of $\{1, 2, \cdots, m\}$. Note that $|A| = |I|$.

We may ask: what is $N(A)$ in this case? Well, by our definition of G, $N(A)$ consists of all elements of S_i, where 'i' is in I; that is,

$$N(A) = \bigcup_{i \in I} S_i.$$

Since the family $(S_1, S_2, \cdots, S_m)$ satisfies the condition that $\left| \bigcup_{i \in I} S_i \right| \geq |I|$, for any subset I of $\{1, 2, \cdots, m\}$, we then have

$$|N(A)| = \left| \bigcup_{i \in I} S_i \right| \geq |I| = |A|;$$

that is, $|A| \leq |N(A)|$ holds in G for every subset A of X, as required.

Accordingly, by Theorem 7.3.3, G possesses a complete matching from X to Y. Write this matching as

$$S_1 \triangleright y_1, S_2 \triangleright y_2, \cdots S_m \triangleright y_m.$$

It is now clear that the sequence $(y_1, y_2, \cdots, y_m)$ is an SDR for the family $(S_1, S_2, \cdots, S_m)$. (For the bipartite graph G shown in Fig. 7.4.1, a complete matching from X to Y exists, and is shown in Fig. 7.4.2, which in turn produces an SDR, namely, (b, c, a, d, e) for the family (C, L, R, S, T).)

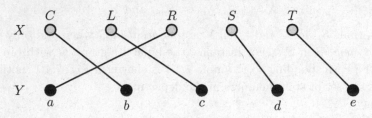

Fig. 7.4.2

$\square$

Exercise for Section 7.4

    1. Find an SDR for each of the following families of sets.

        (a) $(\{1, 2\}, \{2, 5\}, \{4\}, \{3, 5\})$;

        (b) $(\{1, 3\}, \{1, 3\}, \{2\}, \{1, 4\}, \{4, 5\})$;

        (c) $(\{5\}, \{1, 6\}, \{2, 3\}, \{1, 4\}, \{3\}, \{1, 4\})$.

2. For each of the following families of sets, determine whether it has an SDR. Justify your answers.

   (a) $(\{1\}, \{2,3\}, \{1,2\}, \{1,3\}, \{1,4,5\})$;
   (b) $(\{1,2\}, \{2,3\}, \{4,5\}, \{4,5\})$;
   (c) $(\{1,2\}, \{2,3\}, \{3,4\}, \{4,5\}, \{5,1\})$.

3. There are four clubs in a school with their committee members as shown below:

$$\text{Club (A):} \quad \{a,b\}$$
$$\text{Club (B):} \quad \{a,c,e\}$$
$$\text{Club (C):} \quad \{b,c\}$$
$$\text{Club (D):} \quad \{b,d\}$$

   Let $X = \{A, B, C, D\}$ and $Y = \{a,b,c,d,e\}$. We construct a bipartite graph G with bipartition (X, Y) as follows: A vertex (club) in X is adjacent to a vertex (person) in Y if and only if that person is a committee member of the club.

   (a) Draw the graph G.
   (b) Let $S = \{A\}$. Find $N(S)$ in G.
   (c) Let $S = \{A, B\}$. Find $N(S)$ in G.
   (d) Let $S = \{A, B, C\}$. Find $N(S)$ in G.
   (e) Does there exist a subset S of X such that $|S| > |N(S)|$?
   (f) Does there exist a complete matching from X to Y?
   (g) Display a complete matching M from X to Y if your answer to (f) is 'yes'.
   (h) Provide an SDR for the family (A, B, C, D) from M.

4. Let $S_1 = \{b,c\}$, $S_2 = \{a\}$, $S_3 = \{a,b\}$ and $S_4 = \{c,d\}$. Verify that the family (S_1, S_2, S_3, S_4) satisfies the condition stated in Theorem 7.4.3, and thus conclude that the family has an SDR. Provide also one such SDR.

5. Six teachers A, B, C, D, E and F are members of five committees. The memberships of the committees are $\{A, B, C\}$, $\{D, E, F\}$, $\{A, D, E, F\}$, $\{A, C, E, F\}$, and $\{A, B, F\}$. The activities of each committee are to be reviewed by a teacher who is not on the committee, and different committees are to be reviewed by different teachers. Can five distinct teachers be selected? If 'yes', show one such assignment.

6. Show that each of the following families of sets has **no** SDR by Theorem 7.4.3.

(a) $(\{1,2\},\{1\},\{3,4\},\{2\})$;

(b) $(\{1\},\{2,3\},\{1,4,5\},\{1,2\},\{1,3\})$;

(c) $(\{2,3\},\{2,3,4,5,6\},\{3,4\},\{4,5\},\{2,5\},\{2,4\})$.

7. For $n \geq 2$, let $S_1 = \{1\}$, $S_2 = \{1,2\}$ and for each $i = 3, \cdots, n$, let $S_i = \{1,2,\cdots,i\}$.

   (a) Show that the family $(S_1, S_2, \cdots, S_n)$ has an SDR.
   (b) How many different SDRs does $(S_1, S_2, \cdots, S_n)$ have?

8. For $n \geq 2$ and for each $i = 1, 2, \cdots, n-1$, let $S_i = \{i, i+1\}$, and $S_n = \{n, 1\}$.

   (a) Show that the family $(S_1, S_2, \cdots, S_n)$ has an SDR.
   (b) How many different SDRs does $(S_1, S_2, \cdots, S_n)$ have?

9. Find the number of SDRs for each of the following families, where n is a positive integer:

   (a) $(\{1,2\},\{1,3\},\{1,4\},\cdots,\{1,n\})$;
   (b) $(\{1,n+1\},\{2,n+2\},\cdots,\{n,2n\})$.

10. Let $S_1 = \{1,a\}$, $S_2 = \{1, 2a-1\}$, $S_3 = \{2, 4-a\}$ and $S_4 = \{2, a+1\}$, where $a \in \{1,2,3,4\}$. Find all possible values of 'a' for which the family (S_1, S_2, S_3, S_4) has an SDR.

11. There are 12 clubs at a junior college. It is known that each club has at least 3 members and no student is a member of four or more clubs. Prove that this family of 12 clubs has an SDR.

7.5   Maximum Matchings and Perfect Matchings

Let G be a graph. A matching M in G is said to be **maximum** if for any matching M' in G, $|M| \geq |M'|$. Clearly, every perfect matching in G is maximum, but not vice versa.

Example 7.5.1.   Figure 7.5.1 shows two graphs, each of which contains several maximum matchings, but none of them is perfect.

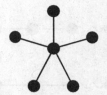

(a) Any single edge forms a maximum matching

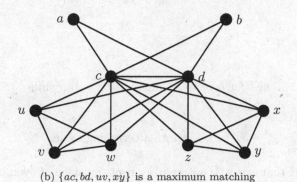

(b) $\{ac, bd, uv, xy\}$ is a maximum matching

Fig. 7.5.1

In this section, we shall present two results on matchings in general graphs. The first one, due to Berge (1957), characterizes matchings that are maximum; and the second one, established by Tutte (1947), characterizes graphs that possess a perfect matching.

To present Berge's result, we need to introduce the following concepts.

Let M be a matching in G. A path $v_1 v_2 \cdots v_p$ in G is said to be M-**alternating** if $v_i v_{i+1} \in M$ whenever $v_{i-1} v_i \notin M$ for each $i = 2, 3, \cdots, p - 1$. Likewise, a cycle $u_1 u_2 \cdots u_p u_1$ in G is said to be M-**alternating** if $u_i u_{i+1} \in M$ whenever $u_{i-1} u_i \notin M$ for each $i = 2, 3, \cdots, p$, where $u_{p+1} = u_1$.

Example 7.5.2. Consider the matching $M = \{bd, ce, fi, gj\}$ in the graph G shown in Fig. 7.5.2.

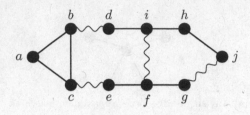

Fig. 7.5.2

Then

$abdifgjh$ is an M-alternating path; while

$bdifecb$ is an M-alternating cycle.

It is clear that every M-alternating cycle must be of even order.

Recall that the symmetric difference of two sets A and B, denoted by $A \triangle B$, is defined as $A \triangle B = (A \setminus B) \cup (B \setminus A)$. Note also that the notion of the subgraph $[F]$ of G induced by a set F of edges was defined in Section 3.6 (see Example 3.6.2).

Observation 7.5.3. Let M_1 and M_2 be two distinct matchings in a graph G. Then each component of the graph $[M_1 \triangle M_2]$ is either an M_i-alternating path or an M_i-alternating cycle for $i = 1, 2$.

Proof. Let $H = [M_1 \triangle M_2]$ and v be a vertex in H. Then v is M_1-saturated or M_2-saturated (or both). Thus, $1 \le d_H(v) \le 2$, and so each component of H is either a path or a cycle. It is clear that the resulting paths or cycles must be M_i-alternating, for $i = 1, 2$.

$\square$

Example 7.5.4. Figure 7.5.3(a) shows a graph G with two matchings M_1 and M_2 shown in (b) and (c) respectively. The graph $[M_1 \triangle M_2]$, which consists of two components, is shown in (d); note that the first one is an M_i-alternating cycle while the second one is an M_i-alternating path, for each $i = 1, 2$.

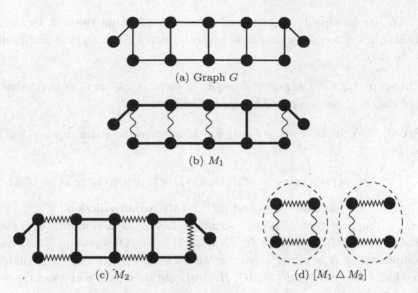

(a) Graph G

(b) M_1

(c) M_2 (d) $[M_1 \triangle M_2]$

Fig. 7.5.3

In Example 7.5.2, we showed an M-alternating path, that is, $abdifgjh$ in G with respect to the given matching M. Note that the two ends of this path, namely a and h, are not M-saturated. In this case, the *length* of this path has to be *odd* (indeed, 7) and, along the path, the number of edges in M (there are 3) is one less than the number of edges *not* in M (there are 4). This indicates that the current matching M is not maximum.

A $u - v$ path in M is said to be M-**augmenting** if it is M-alternating and the vertices u and v are *not* M-saturated.

Note that every M-augmenting path P is of odd length and

$$|E(P) \cap (E(G) \setminus M)| = |E(P) \cap M| + 1.$$

Question 7.5.5.

  (i) In Fig. 7.5.3(b), can you find an M_1-augmenting path in G? Is M_1 maximum?

 (ii) In Fig. 7.5.3(c), can you find an M_2-augmenting path in G? Is M_2 maximum?

We are now ready to state and prove the following theorem by Berge which gives a necessary and sufficient condition for a matching in a graph to be maximum.

Theorem 7.5.6.   *Let G be a graph. A matching M in G is maximum if and only if G contains no M-augmenting path.*

Proof. ($\Rightarrow$) Suppose there exists an M-augmenting path: $v_1 v_2 \cdots v_{2k}$ in G. Let

$$M' = (M \setminus \{v_2 v_3, v_4 v_5, \cdots, v_{2k-2} v_{2k-1}\}) \cup \{v_1 v_2, v_3 v_4, \cdots, v_{2k-1} v_{2k}\}.$$

Then M' is a matching in G and $|M'| > |M|$, a contradiction.

($\Leftarrow$) Suppose that M is not maximum. Let M' be a matching in G such that $|M'| > |M|$. Consider $H = [M \triangle M']$. By Observation 7.5.3, each component of H is either a path or an even cycle, with edges alternately in M and M'. Since $|M'| > |M|$, H must contain a path component, say a $u - v$ path, where u and v are M'-saturated. Clearly, this $u - v$ path is an M-augmenting path in G.

$\square$

We shall now turn our attention to the notion of perfect matchings. Any graph of odd order certainly contains no perfect matchings. As shown in Example 7.5.1, a graph may not have a perfect matching even if it is of even order. When does a graph contain a perfect matching? This is the question that we are going to discuss in what follows.

Let H be a graph. A component F of H is said to be *odd* if $v(F)$ is odd. Denote by $o(H)$ the number of odd components of H. Thus, for the graph H in Fig. 7.5.4, we have $o(H) = 3$.

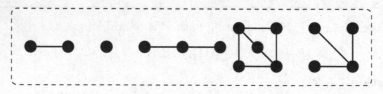

Fig. 7.5.4: H

Let us consider again the graph G shown in Fig. 7.5.1(b). How do we argue that G contains no perfect matchings? One way is given below. Take

$S = \{c, d\}$. Removing S from G, we obtain the subgraph $G - S$ as shown in Fig. 7.5.5:

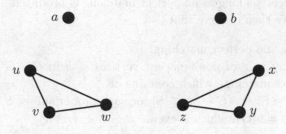

Fig. 7.5.5

Note that it consists of four components, and each of them is odd. Suppose G has a perfect matching M. Then the vertices a, b, one in $\{u, v, w\}$, and one in $\{x, y, z\}$ must be matched under M with four respective distinct vertices in S, which is not possible as $|S| = 2$. We thus conclude that G contains no perfect matching.

In general, we have:

Observation 7.5.7. Let G be a graph with vertex set V. If G contains a perfect matching, then

$$o(G - S) \le |S| \text{ for each proper subset } S \text{ of } V. \qquad (*)$$

Proof. Assume that G has a perfect matching M. Let S be a proper subset of V and consider $G - S$. Suppose $G - S$ has exactly k odd components, say $H_1, H_2, \cdots, H_k$. As H_i is odd, some vertex of H_i, say x_i, must be matched under M with a vertex, say y_i, in S. As M is a matching, the y_i's are distinct. Thus,

$$o(G - S) = k = |\{y_1, y_2, \cdots, y_k\}| \le |S|.$$
□

We have just seen in the above observation that the condition $(*)$ is a necessary condition for a graph G to have a perfect matching. Tutte (1947) proved that it is also a sufficient condition. We now establish this classic result, but the proof we present below was given by Lovász (1975).

Theorem 7.5.8. *A graph G with vertex set V has a perfect matching if and only if*

$$o(G - S) \le |S| \text{ for each proper subset } S \text{ of } V. \qquad (*)$$

Proof. ($\Leftarrow$) Suppose that G satisfies $(*)$ but has no perfect matching. Let G^* be a graph obtained from G by successively joining pairs of non-adjacent vertices, as long as no perfect matching is produced, until no such pair exists. We then observe that

    (1) G^* has no perfect matching,

    (2) for any pair of non-adjacent vertices a, b in G^*, $G^* + ab$ has a perfect matching which contains ab,

    (3) $o(G^* - S) \le o(G - S) \le |S|$ for each proper subset S of V, and

    (4) $v(G^*) = v(G)$, which is even.

Let W be the set of vertices of degree $v(G^*) - 1$ in G^*. By (1), $W \ne V(G^*)(= V)$.

Claim. Every component of $G^* - W$ is complete.

Suppose on the contrary that there is a component B of $G^* - W$ which is not complete. There there exist three vertices u, v and w in B such that u is adjacent to v, v is adjacent to w, but w is not adjacent to u (see Problem 1.4.12). Note that as v is not in W, there exists a vertex z in $G^* - W$ such that v and z are not adjacent. (See Fig. 7.5.6.)

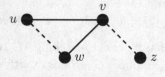

Fig. 7.5.6

By (2), $G^* + uw$ has a perfect matching M_1 and $G^* + vz$ has a perfect matching M_2. Clearly, uw is in $M_1 \backslash M_2$ and vz is in $M_2 \backslash M_1$.

Let $H = [M_1 \triangle M_2]$ in $G^* + uw + vz$. As $d_H(x) = 2$ for each vertex x in H, H is a disjoint union of even cycles, consisting of edges alternately in M_1 and M_2.

We have two cases to consider.

Case 1. uw and vz are in different components of H. (See Fig. 7.5.7(a).)

Let C be the cycle in H containing vz. Then the edges of M_1 in C together with the edges of M_2 not in C form a perfect matching in G^*, contradicting (1).

Case 2. uw and vz are in the same component, say C, of H. (See

Fig. 7.5.7(b).)

By symmetry of u and w, we may assume that the four vertices u, v, w and z occur in the following order: v, z, w, u on C. Then the edges of M_1 in the path $vz \cdots w$ along C, together with the edge vw and the edges of M_2 not in the path $vz \cdots w$ along C, would form a perfect matching in G^*, contradicting (1) again.

Thus each component of $G^* - W$ is complete, as claimed.

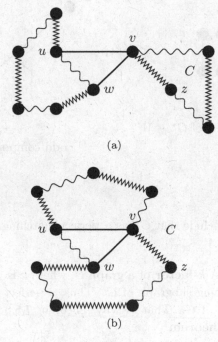

(a)

(b)

Fig. 7.5.7

By (3), the number of odd components of $G^* - W$ is at most $|W|$. By the above claim and the fact that each vertex in W is of degree $v(G^*) - 1$ in G^*, it is now easy to form a perfect matching in G^*, as shown in Fig. 7.5.8, which contradicts (1) again.

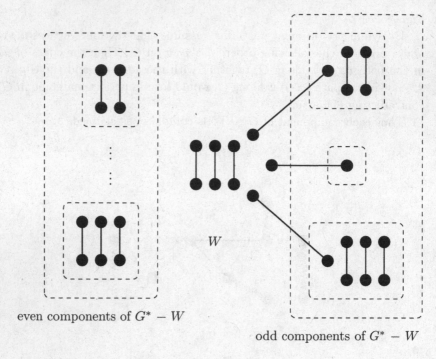

even components of $G^* - W$

odd components of $G^* - W$

Fig. 7.5.8

We therefore conclude that G has a perfect matching if G satisfies $(*)$.

□

Remark 7.5.9.   A **k-factor** of a graph G, where k is a positive integer, is a k-regular spanning subgraph of G. Thus, a perfect matching of G is precisely a 1-factor of G. That is why Theorem 7.5.8 is also known as **Tutte's 1-factor theorem**.

When G is a tree, Chungphaisan (1976) observed that the condition $(*)$ in Theorem 7.5.8 can be substantially simplified (see Problem 7.5.5).

Corollary 7.5.10.   *Let T be a tree. Then T contains a perfect matching if and only if $o(T - v) = 1$ for each vertex v in T.*

□

Exercise for Section 7.5

1. Let G be a graph and M a matching in G. Show that G contains a maximum matching that saturates all M-saturated vertices.
2. Consider the following graph G:

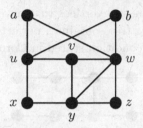

Fig. 7.5.9

   (a) Does G have a perfect matching?
   (b) Find six maximum matchings in G.
   (c) Is there any maximum matching containing the edge wy?

3. Consider the following graph G:

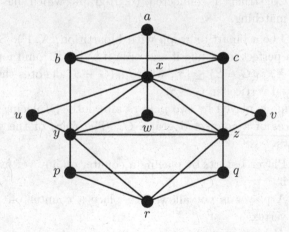

Fig. 7.5.10

   (i) Does G have a perfect matching?
   (ii) What is the size of a maximum matching in G?

(iii) Is there any maximum matching containing the set of edges $\{yz, pq\}$?

(iv) Find the number of maximum matchings containing the set of edges $\{ax, bc\}$.

4. Show that every tree has at most one perfect matching. Construct all trees with at most 8 vertices that have a perfect matching.

5. (a) The following tree contains a perfect matching. Verify that it satisfies the condition stated in Corollary 7.5.10.

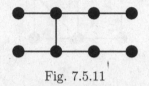

Fig. 7.5.11

(b) Prove Corollary 7.5.10.

6. Let G be a graph of order n and W a subset of V. Show that $n + |W| - o(G - W)$ is always even for all subsets W of V.

7. Let G be a connected $(6, m)$-graph.

(i) Show that if $m \geq 10$, then G has a perfect matching.

(ii) Construct a connected $(6, 9)$-graph which has no perfect matching.

8. Let G be a bipartite graph with bipartition (X, Y). Show that G has a perfect matching if and only if $|X| = |Y|$ and for any subset S of X, $i_0(G - S) \leq |S|$, where $i_0(G - S)$ denotes the number of isolated vertices in $G - S$.

9. A game is played by two players alternatively placing counters on vertices of a connected graph G. The rules of the game are as follows:

(1) Player 1 starts by placing a counter on any vertex to 'occupy' it.

(2) A player is not allowed to place a counter on an occupied vertex.

(3) At his turn, a player must place a counter on a vertex adjacent to the vertex just occupied by the player before.

(4) The last player to occupy a vertex wins.

Show that the player who starts first has a winning strategy if and only if G contains no perfect matching.

7.6   Independence, Covering and Gallai Identities

Let G be a graph of order n. Recall that a set A of vertices in G is said to be independent if no two vertices in A are adjacent in G. Trivially, every singleton $\{v\}$, where v is a vertex in G, is an independent set in G. Note also that A is independent in G if and only if $[A]$ is a complete subgraph in $\overline{G}$. The **independence number** of G, denoted by $\alpha(G)$, is defined by

$$\alpha(G) = \max\{|A| \mid A \text{ is an independent set in } G\}.$$

An independent set A in G is said to be **maximum** if $|A| = \alpha(G)$.

For the edge version, a set M of edges is a matching if no two edges in M are adjacent in G. Trivially, every singleton $\{e\}$, where e is an edge in G, is a matching in G. The **matching number** of G, denoted by $\alpha'(G)$, is defined by

$$\alpha'(G) = \max\{|M| \mid M \text{ is a matching in } G\}.$$

Thus, a matching M in G is **maximum** if $|M| = \alpha'(G)$.

Example 7.6.1.   The graph H of Fig. 7.6.1 is the Cartesian product $P_3 \,\square\, P_3$. It can be checked that $\alpha(H) = 5$ and $\alpha'(H) = 4$. It is noted that $A = \{a, c, e, g, i\}$ is the unique maximum independent set in H.

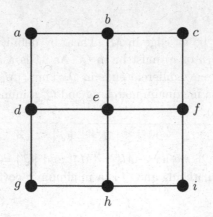

Fig. 7.6.1

In this final section, we shall introduce a notion, called covering (with vertex and edge versions), and shall see that it is very closely related to the notion of independence as shown in two famous identities, called the Gallai identities.

A set Q of vertices in G is called a **vertex cover**, or simply **v-cover** of G, if each edge in G is incident with a vertex in Q. Trivially, the whole vertex set V is a v-cover of G. The **vertex covering number**, or simply **v-covering number** of G, denoted by $\beta(G)$, is defined by

$$\beta(G) = \min\{|Q| \mid Q \text{ is a } v\text{-cover of } G\}.$$

A v-cover Q of G is said to be **minimum** if $|Q| = \beta(G)$.

Example 7.6.2. In the graph H of Fig. 7.6.1, $\beta(H) = 4$ and $Q = \{b, d, f, h\}$ is the unique minimum v-cover of H. Note also that the independent set $A = \{a, c, e, g, i\}$ and $Q = \{b, d, f, h\}$ are set-complementary in $V(H)$.

Let M be a matching and Q a v-cover of G. Two simple results linking M and Q are given below.

Observation 7.6.3.

   (i) $|M| \le |Q|$;
  (ii) If $|M| = |Q|$, then M is a maximum matching and Q is a minimum v-cover.

Proof.

   (i) Let $e = uv$ be an edge in M. Then, by definition of v-cover, one of its ends (u or v) must be in Q. As M is a matching, different edges in M have different ends in Q. Thus, $|M| \le |Q|$.
  (ii) Let M^* be a maximum matching and Q' a minimum v-cover in G. Then, by (i),

$$|M| \le |M^*| \le |Q'| \le |Q|.$$

As $|M| = |Q|$, we have $|M| = |M^*|$ and $|Q'| = |Q|$. Thus M is a maximum matching and Q is a minimum v-cover of G.

$\square$

It follows from the above observation (i) that

$$\alpha'(G) \le \beta(G). \tag{#}$$

The equality may hold. For instance, for the graph H of Fig. 7.6.1, we have $\alpha'(H) = 4 = \beta(H)$. This is, however, not true in general. Take any odd cycle C_{2k+1}. It can be checked that $\alpha'(C_{2k+1}) = k < k+1 = \beta(C_{2k+1})$ (see Problem 7.6.9). Until now, characterizations of graphs G for which $\alpha'(G) = \beta(G)$ have not yet been found. Nevertheless, we do have the following sufficient condition given by König (1931) and Egerváry (1931). Note that its proof given below makes use of Hall's theorem (Theorem 7.3.3).

Theorem 7.6.4. *For any bipartite graph G, $\alpha'(G) = \beta(G)$.*

Proof. By (#), we need only to show that $\beta(G) \leq \alpha'(G)$. Thus, let Q be a minimum v-cover of G (i.e., $|Q| = \beta(G)$). Our aim is to construct a matching M in G such that $|M| = |Q|$.

Let (X, Y) be a bipartition of G. Denote $R = Q \cap X$, $T = Q \cap Y$ and let $G_1 = [R \cup (Y \setminus T)]$ and $G_2 = [(X \setminus R) \cup T]$.

Observe that as Q $(= R \cup T)$ is a v-cover of G, G has no edges joining vertices between $X \setminus R$ and $Y \setminus T$.

We shall now apply Theorem 7.3.3 to prove the following:

Claim. G_1 has a complete matching from R to $Y \setminus T$.

Let $S \subseteq R$, and consider $N_{G_1}(S) \subseteq Y \setminus T$. If $|N_{G_1}(S)| < |S|$, then as $N_{G_1}(S)$ covers all edges incident to S that are not covered by T, the set $Q' = (R \setminus S) \cup N_{G_1}(S) \cup T$ is a v-cover of G and $|Q'| < |Q|$, a contradiction. Thus, $|S| \leq |N_{G_1}(S)|$ for all $S \subseteq R$, and so by Theorem 7.3.3, G_1 has a complete matching M_1 from R to $Y \setminus T$.

Likewise, G_2 has a complete matching M_2 from T to $X \setminus R$. Clearly, $M = M_1 \cup M_2$ is a matching in G with $|M| = |M_1| + |M_2| = |R| + |T| = |Q|$, as required. It follows that $\beta(G) = |Q| = |M| \leq \alpha'(G)$.

$\square$

Question 7.6.5. *Is it true that if $\alpha'(G) = \beta(G)$, then G must be bipartite? (See Problem 7.6.10.)*

We have just discussed relations between v-covers and matchings. Let us now present a relation between v-covers and independent sets. Example 7.6.2 shows an independent set $A = \{a, c, e, g, i\}$ and a v-cover $Q = \{b, d, f, h\}$ of the graph H of Fig. 7.6.1. Note that they are set-complementary in $V(H)$. Indeed, we have the following general result (see Problem 7.6.17).

Theorem 7.6.6. *Let G be a graph of order $n \geq 2$ and A, a subset of V. Then*

(i) *A is an independent set in G if and only if $V \setminus A$ is a v-cover of G;*

(ii) *A is a maximum independent set in G if and only if $V \setminus A$ is a minimum v-cover of G.*

$\square$

Corollary 7.6.7.   *For any graph G of order $n \geq 2$, $\alpha(G) + \beta(G) = n$.*

$\square$

We finally introduce the edge version of covering. Let G be a graph with $\delta(G) \geq 1$ (i.e., containing no isolated vertices). A set W of edges in G is called an **edge cover**, or simply e-**cover** of G, if every vertex in G is incident with an edge in W. Trivially, the whole edge set E is an e-cover of G. The **edge covering number**, or simply e-**covering number** of G, denoted by $\beta'(G)$, is defined by

$$\beta'(G) = \min\{|W| \mid W \text{ is an } e\text{-cover of } G \}.$$

An e-cover W of G is said to be **minimum** if $|W| = \beta'(G)$.

Discussion 7.6.8.

(i) Explain why we need to impose the condition that $\delta(G) \geq 1$ for the introduction of e-covers.

(ii) Find $(\alpha, \alpha', \beta, \beta')$ for each of the graphs in Fig. 7.6.2.

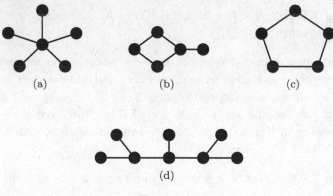

(a)   (b)   (c)

(d)

Fig. 7.6.2

(iii) Let W be a minimum e-cover in G. Consider the subgraph $[W]$ induced by W in G. Can $[W]$ contain a C_3? a P_4? What can you say about the structure of $[W]$?

(iv) Theorem 7.6.6(i) states that for a subset A of V, A is an independent set in G if and only if $V \setminus A$ is a v-cover of G. Assuming $\delta(G) \geq 1$, for the edge version, if M is a matching in G, is the set $E \setminus M$ an e-cover of G? If W is an e-cover of G, is $E \setminus W$ a matching in G?

Comparing with Theorem 7.6.6(i) for the vertex version, the relationship between matchings and e-covers is more complicated. Nevertheless, parallel to Corollary 7.6.7, we do have the following result due to Gallai (1959):

Theorem 7.6.9. *For any graph G of order n with $\delta(G) \geq 1$, $\alpha'(G) + \beta'(G) = n$.*

Proof. Let M be a matching such that $|M| = \alpha'(G)$. We shall construct an e-cover W with $|W| = n - \alpha'(G)$ by extending M. Let A be the set of vertices in G which are not M-saturated, and to each a in A, select an edge e_a incident to a (note that $\delta(G) \geq 1$). Since M is maximum, $e_a \neq e_b$ if a and b are distinct vertices in A. Let $W = M \cup \{e_a \mid a \in A\}$. Clearly, W is an e-cover of G and $|W| = |M| + |A| = \alpha'(G) + n - 2\alpha'(G)$. Thus, $\beta'(G) \leq |W| = n - \alpha'(G)$, and so

$$\alpha'(G) + \beta'(G) \leq n. \tag{1}$$

Let W be an e-cover of G such that $|W| = \beta'(G)$. We shall construct a matching M with $|M| = n - \beta'(G)$. Consider $[W]$. As W is minimum, $[W]$ contains no C_3 and no paths of length 3. Thus, W consists of k stars, where $k \geq 1$. Observe that $n = k + |W|$. Form M by picking an edge from each star in $[W]$. Then M is a matching with $|M| = k = n - |W| = n - \beta'(G)$. Thus, $\alpha'(G) \geq |M| = n - \beta'(G)$, and so

$$\alpha'(G) + \beta'(G) \geq n. \tag{2}$$

The result thus follows by (1) and (2).

$\square$

To end this chapter, we now summarize the main results in this section.

Let G be a graph of order $n \geq 2$ (assume $\delta(G) \geq 1$ when $\beta'(G)$ is involved). Then we have the **Gallai identities**:

$$\alpha(G) + \beta(G) = n = \alpha'(G) + \beta'(G)$$

and the inequalities:

$$\alpha'(G) \leq \beta(G) \text{ and } \alpha(G) \leq \beta'(G).$$

If G is bipartite, then

$$\alpha'(G) = \beta(G) \text{ and } \alpha(G) = \beta'(G).$$

Exercise for Section 7.6

    1. Consider each of the following graphs G:

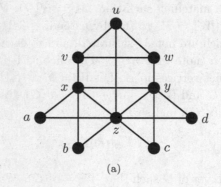

(a)

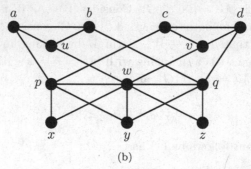

(b)

Fig. 7.6.3

(a) Does G have a perfect matching?

(b) What is the size of a maximum matching in G?

(c) Find in G

   (i) a maximum matching,

   (ii) a maximum independent set,

   (iii) a minimum v-cover, and

   (iv) a minimum e-cover.

   Hence find the values of $\alpha(G)$, $\alpha'(G)$, $\beta(G)$ and $\beta'(G)$.

2. Let $\mathcal{T}(n)$ be the family of all trees of order $n \geq 3$. Which T in $\mathcal{T}(n)$ has

   (i) the largest value of $\alpha(T)$?

   (ii) the largest value of $\alpha'(T)$?

   (iii) the least value of $\beta(T)$?

   (iv) the least value of $\beta'(T)$?

3. Let H be a subgraph of a graph G.

   (a) Show that

      (i) $\beta(H) \leq \beta(G)$;

      (ii) $\alpha'(H) \leq \alpha'(G)$.

   (b) Show that $\alpha(H) \leq \alpha(G)$ if H is an induced subgraph of G; and that this condition is sometimes necessary.

   (c) Assume that H and G contain no isolated vertices. Is it true that $\beta'(H) \leq \beta'(G)$?

4. Let G be a bipartite graph with bipartition (X, Y), where $|X| \geq |Y|$. Show that $\alpha(G) \geq |X|$. Is it true in general that $\alpha(G) = |X|$?

5. Prove that a graph G is bipartite if and only if $\alpha(H) \geq \frac{1}{2}v(H)$ for every subgraph H of G.

6. Let G be a graph with $\delta(G) \geq 1$. Prove that G is bipartite if and only if $\alpha(H) = \beta'(H)$ for every subgraph H of G with $\delta(H) \geq 1$.

7. Let T be a tree of order $n \geq 2$. Show that $\alpha(T) \geq \lceil \frac{n}{2} \rceil$ and that the bound is sharp for each n.

8. Let T be a tree of order $n \geq 3$ in which $d(x, y)$ is even for any pair $\{x, y\}$ of end-vertices. Show that T has a *unique* maximum independent set.

9. For each of the following graphs G, find the values of $\alpha(G), \alpha'(G), \beta(G)$ and $\beta'(G)$:

   (i) the path P_n, $n \geq 2$,

(ii) the cycle C_n, $n \geq 3$,

(iii) the complete graph K_n, $n \geq 2$,

(iv) the complete bipartite graph $K(p,q)$, $p \geq q \geq 1$,

(v) the wheel W_n, $n \geq 4$,

(vi) the Petersen graph, and

(vii) the n-cube graph Q_n, $n \geq 1$.

10. For each $k \geq 3$, construct a non-bipartite graph H such that

$$\alpha'(H) = k = \beta(H).$$

11. Evaluate the values of $(\alpha, \alpha', \beta, \beta')$ for the Cartesian product $P_m \,\square\, P_n$, where $m \geq n \geq 2$.

12. Let G be a graph of order $n \geq 1$. Show that

$$\alpha(G) \geq \frac{n}{1 + \Delta(G)}.$$

For each n, construct two graphs G of order n for which the equality holds.

13. Let G and H be two graphs.

(i) Express $\alpha(G + H)$ in terms of $\alpha(G)$ and $\alpha(H)$.

(ii) What can be said about $\beta(G + H)$?

14. (a) Let G be a bipartite graph. Show that

$$\alpha'(G) \geq \frac{e(G)}{\Delta(G)}.$$

(b) For each positive integer k, construct a bipartite graph H such that

$$\alpha'(H) = k = \frac{e(H)}{\Delta(H)}.$$

15. Let G be a bipartite graph with bipartition (X, Y). Assume that $|X| = |Y| = p \geq 2$ and $e(G) \geq (k-1)p + 1$, where k is a positive integer. Show that $\alpha'(G) \geq k$.

16. Let G be a bipartite graph with bipartition (X, Y). For any v-cover Q of G, write $Q_x = Q \cap X$ and $Q_y = Q \cap Y$. Suppose that A and B are two minimum v-covers of G. Show that $A_x \cup B_x \cup (A_y \cap B_y)$ and $(A_x \cap B_x) \cup A_y \cup B_y$ are also minimum v-covers of G.

17. Prove Theorem 7.6.6 and Corollary 7.6.7.

18. Let G be a graph of order $n \geq 2$. Prove that

$$\left\lceil \frac{n}{1 + \Delta(G)} \right\rceil \leq \alpha'(G) \leq \left\lfloor \frac{n}{2} \right\rfloor.$$

For each $n \geq 2$, construct, respectively, a graph G of order n such that

(i) the left-hand equality holds;

(ii) the right-hand equality holds.

19. Let G be a graph of order $n \geq 2$. Show that

$$\alpha(G) + \alpha'(G) \leq n \leq \alpha(G) + 2\alpha'(G)$$

and that the bounds are sharp.

20. Let G be a graph of order $n \geq 2$ and with $\delta(G) \geq 1$. A matching in G is said to be **maximal** if it is not properly contained in a matching in G (thus every maximum matching is maximal). An e-cover of G is said to be **minimal** if any of its proper subsets is no longer an e-cover of G (thus every minimum e-cover is minimal). Suppose M is a maximal matching in G and W a minimal e-cover of G.

(a) Show that M is a maximum matching in G if and only if M is contained in a minimum e-cover of G.

(b) Show that W is a minimum e-cover of G if and only if W contains a maximum matching in G.

7.7 Problem Set VII

1. Let G be a bipartite graph with bipartition (X, Y). Suppose that $d(x) \geq 4$ for all $x \in X$ and $d(y) \leq 10$ for all $y \in Y$. Prove that $\alpha'(G) \geq \frac{2}{5}|X|$.

2. Let G be an $(n, n-1)$-graph such that $e(H) \leq v(H) - 1$ for every subgraph H of G. Let $v \in V(G)$ and B be the bipartite graph with bipartition $X = V(G) - \{v\}$ and $Y = E(G)$ such that $x \in X$ and $y \in Y$ are adjacent if and only if the vertex x is incident with the edge y in G.

(a) Show that B has a perfect matching.

(b) Using the matching obtained in (a), show that for every vertex $u \in X$, there is a $u - v$ path in G.

3. Let G be a bipartite graph with bipartition (X, Y) that satisfies Hall's condition, namely, $|S| \leq |N(S)|$ for all $S \subseteq X$. Suppose that $a \in X$ and $u, v \in Y$ such that $au, av \in E(G)$. Let $G_u = G - au$ and $G_v = G - av$. Show that either G_u or G_v also satisfies Hall's condition.

4. Let G be any 3-regular Hamiltonian graph. Show that its edge set $E(G)$ can be partitioned into three perfect matchings.

5. Let G be a connected graph of even order n, where $n \geq 4$, satisfying the condition that every edge in G is contained in a perfect matching of G. Show that G is 2-connected.

6. Let G be the following graph:

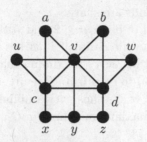

Fig. 7.7.1

(a) Does G have a perfect matching?

(b) Find four maximum matchings in G.

(c) Is there any maximum matching in G that contains the edge cd?

(d) Find four maximal matchings (for definition, see Problem 7.6.20) that are not maximum.

(e) Find in G

   (i) a maximum independent set,

   (ii) a minimum v-cover, and

   (iii) a minimum e-cover.

(f) Find the values of $\alpha(G), \alpha'(G), \beta(G)$ and $\beta'(G)$.

7. (a) Let L be a maximal matching and M a maximum matching in a graph G. Show that $|M| \leq 2|L|$.

(b) For each positive integer k, construct a connected graph G in which there exist a maximal matching L with $|L| = k$ and a maximum matching M with $|M| = 2k$.

8. Let G be a connected $(8, m)$-graph, and let

$$m^* = \min\{m \mid G \text{ contains a perfect matching}\}.$$

(i) Determine the value of m^*.

(ii) Construct a connected $(8, m^* - 1)$-graph which has no perfect matching.

9. Let G be a 3-regular connected graph which has at most two bridges and S, a subset of V. Assume that $o(G-S) = q$ and $G_1, G_2, \cdots, G_q$ are the odd components of $G - S$. For each $i = 1, 2, \cdots, q$, let k_i denote the number of edges joining S and G_i. Show that

   (a) k_i is odd for each $i = 1, 2, \cdots, q$;
   (b) $|S|$ is even if and only if q is even;
   (c) $q \le |S|$;
   (d) G contains a perfect matching.

   Is the result still valid if G contains more than two bridges?

   (Petersen (1891))

10. (A generalization of Problem VII.9) Let G be a k-regular graph of even order with $\kappa'(G) \ge k - 1$. Show that G contains a perfect matching.

    (Bäbler (1938))

11. Let G be a graph with vertex set V. A proper subset A of V is called an **anitfactor set** of G if $o(G - A) > |A|$. An antifactor set A of G is said to be **minimal** if every proper subset of A is no longer an antifactor set of G.

    (I) Assume that $v(G)$ is even and that G contains no perfect matchings. By Theorem 7.5.8 (Tutte's theorem), G contains an antifactor set. Let S be a minimal antifactor set of G with $|S| = p$ and $o(G - S) = q$.

    (a) Show that

       (i) $q > p + 2$;
       (ii) every vertex in S is adjacent to vertices in at least $q-p+1$ distinct odd components of $G - S$.

    (b) A vertex v in G is called a **claw center** if $N(v)$ contains x, y and z such that $\{x, y, z\}$ is an independent set in G (note that $[\{v, x, y, z\}] \cong K(1,3)$ in G). Show that every vertex in S is a claw center of G.

    (II) Let $W(G)$ denote the set of claw centers of G.

    (a) Show that G has a perfect matching if and only if $o(G - B) \le |B|$, for all $B \subseteq W(G)$.

    (b) Assume that G is connected and $v(G)$ is even. Show that G has a perfect matching if $W(G) = \emptyset$.

    (III) Let G be a k-connected graph of even order which contains no $K(1, k+1)$ as an induced subgraph, where $k \ge 2$. Show that G

has a perfect matching.

(IV) Let G be a 3-regular connected graph that has no perfect matchings. Let S be a minimal antifactor set of G. Show that

(a) S is an independent set;

(b) $G - S$ has exactly $|S| + 2$ odd components and no even components;

(c) if $|S| \neq 1$, then $|S|$ contains at least three cut-vertices of G;

(d) $|S| \neq 2$.

(Sumner (1976))

12. Let A be a set of natural numbers. A spanning subgraph H of a graph G is called an A-**factor** of G if $d_H(v) \in A$ for each $v \in V(H)$. (Thus, a perfect matching of G is a $\{1\}$-factor of G.) Let T be a tree of even order, and let k be a positive integer. Show that

    (i) T has a $\{1, 3, \cdots, 2k-1\}$-factor if and only if $o(T-v) \leq 2k-1$ for each vertex v in T;

    (ii) if T has a $\{1, 3, \cdots, 2k-1\}$-factor, then such a factor is unique.

(Amahashi (1985))

13. Let G be a graph. A matching M in G is called a **defect-r matching** if there are exactly r vertices in G which are not M-saturated. Assume that $v(G) + r$ is even, where $0 \leq r \leq v(G)$. Show that G has a defect-r matching if and only if $o(G - S) \leq |S| + r$ for all $S \subset V(G)$.

    (Hint: Let $G^* = G + O_r$. Then G has a defect-r matching if and only if G^* has a perfect matching.)

14. Let G be a connected graph of even order. Show that the square G^2 of G has a perfect matching. (You may wish to apply the result of Problem III.6.)

15. Let G be a graph of order n with $\delta(G) \geq 1$. Show that

    (a) $\beta'(G) \leq \beta(G) \cdot \Delta(G)$;

    (b) $\beta(G) + \beta'(G) \geq n$ and that the bound is sharp for each n.

16. Let G be a connected graph of order $n \geq 2$. Prove that

$$\lceil \frac{n}{2} \rceil \leq \beta'(G) \leq \lfloor \frac{n \cdot \Delta(G)}{1 + \Delta(G)} \rfloor.$$

For each $n \geq 2$, construct, respectively, a connected graph G of order n such that

    (i) the left-hand equality holds;

    (ii) the right-hand equality holds.

17. Let G be a graph and A be a maximum independent set in G. Suppose that B is an independent set in G such that $B \cap A = \emptyset$.

    (i) Show that there is a complete matching from B to A.
    (ii) An independent set in G is said to be **maximal** if it is not properly contained in any independent set in G. Does the conclusion in (i) still hold if A is only a maximal independent set in G?

18. Let G be a graph and A, an independent set in G satisfying the following condition: there exists a complete matching from B to A for any independent set B in G such that $B \cap A = \emptyset$. Show that A is a maximum independent set in G.

19. Let T be a tree of order at least two. Show that the following statements are equivalent:

    (a) T has a perfect matching;
    (b) T has two disjoint maximum independent sets.

20. Let G be a bipartite graph. Show that

    (a) $\alpha(G) \cdot \beta(G) \geq e(G)$;
    (b) the above equality holds if and only if G is a complete bipartite graph.

    (Andrásfai (1962))

21. Let G be a graph with $V = \{v_1, v_2, \cdots, v_n\}$. Show that

$$\alpha(G) \geq \sum_{i=1}^{n} \frac{1}{1 + d(v_i)}.$$

    (Caro (1979) and Wei (1981))

22. Let G be an (n, m)-graph. Applying the result in Problem VII.21, or otherwise, show that

$$\alpha(G) \geq \frac{n^2}{2m + n}.$$

Note. Let G be a connected (n, m)-graph. Some other lower bounds for $\alpha(G)$ in terms of n and m are given below.

(1) Hansen (1975) (see also Favaron et al. (1991)):

$$\alpha(G) \geq \lceil \frac{2n - \frac{2m}{\lceil \frac{2m}{n} \rceil}}{\lceil \frac{2m}{n} \rceil + 1} \rceil;$$

(2) Koh (1977):

$$\alpha(G) \geq \lceil \frac{n^3 + 3n + 1}{2(2m + n)} \rceil;$$

(3) Harant and Schiermeyer (2001):

$$\alpha(G) \geq \frac{1}{2}\lceil (2m+n+1) - \sqrt{(2m+n+1)^2 - 4n^2}\rceil;$$

(4) Kettani (2007, preprint):

$$\alpha(G) \geq \frac{1}{8}\left((2m+n+2) - \sqrt{(2m+n+2)^2 - 16n^2} \right).$$

23. Let G be a connected graph of order $n \geq 2$. Prove that

$$\alpha(G) \leq n - \frac{1}{2}(cv(G)+1),$$

where $cv(G)$ denotes the number of cut-vertices in G. Show also that the bound is sharp.

(Larson and Pepper (2014))

24. (a) Draw the diagram of the Cartesian product $K(1,3) \square P_5$ and evaluate its independence number.
    (b) Let G and H be graphs. Show that $\alpha(G \square H) \geq \alpha(G) \cdot \alpha(H)$.
    (c) Can the equality hold for some non-trivial G and H?

25. Let G be a graph.

    (i) Show that $\alpha(G) - 1 \leq \alpha(G - v) \leq \alpha(G)$ for any vertex v in G.

    (ii) A vertex z in G is said to be α-**critical vertex** if $\alpha(G - z) = \alpha(G) - 1$. Let H be the graph shown in Fig. 7.7.2. Find $\alpha(H)$ and all α-critical vertices in H.

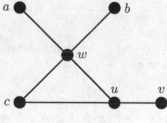

Fig. 7.7.2

    (iii) Show that a vertex z in G is α-critical if and only if z is contained in every maximum independent set in G.

    (iv) Is every isolated vertex α-critical? Is every end-vertex α-critical?

    (v) For each $n \geq 3$, construct a graph of order n in which no vertex is α-critical.

(vi) For each $n \geq 3$, construct a graph of order n in which every vertex is α-critical.

(vii) For each $n \geq 3$, construct a graph of order n in which every vertex, except one, is α-critical.

26. Let G be a graph.

(i) Show that $\alpha(G) \leq \alpha(G - e) \leq \alpha(G) + 1$ for any edge e in G.

(ii) An edge e in G is said to be α-**critical** if $\alpha(G-e) = \alpha(G)+1$. Let H be the graph shown in Fig. 7.7.3. Find $\alpha(H)$ and all α-critical edges in H.

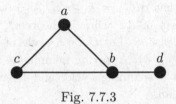

Fig. 7.7.3

(iii) Show that an edge $e = xy$ in G is α-critical if and only if both x and y are contained in any maximum independent set in $G - e$.

(iv) If the edge $e = xy$ is α-critical in G, show that the vertices x and y are not α-critical.

(v) If w is an α-critical vertex, show that w is not incident with any α-critical edge.

(vi) Assume that G contains no α-critical vertices. Show that for any independent set A in G, there is a complete matching from A to $V(G)\backslash A$.

27. A non-trivial connected graph G is said to be α-**critical** if $\alpha(G) < \alpha(G-e)$ for each edge e in G (that is, every edge in G is α-critical). Let G be an α-critical graph of order at least three, and let u and v be two distinct vertices in G. Show that

(i) there exists a maximum independent set in G which contains either u or v, but not both;

(ii) if u and v are adjacent, then there exists a maximum independent set S in G such that $S \cap \{u, v\} = \emptyset$.

28. Let G be a non-trivial graph.

(a) Show that $\beta(G) - 1 \leq \beta(G - x) \leq \beta(G)$ for any vertex x in G.

(b) A vertex x in G is said to be β-**critical** if $\beta(G-x) = \beta(G)-1$. Let H be the graph shown in Fig. 7.7.2. Find $\beta(H)$ and all β-critical vertices in H.

(c) Let w be a vertex in G. Show that w is β-critical if and only if w is contained in some minimum v-cover of G.

29. Let G be a graph of order $n \geq 3$ satisfying the condition that $\alpha(G) \leq \kappa(G)$.

   (i) Assume that $\alpha(G) = 1$. Explain why G is Hamiltonian.

   (ii) Assume that $\alpha(G) = k \geq 2$. As $\kappa(G) \geq \alpha(G) = k \geq 2$, by Theorem 6.4.12 (Dirac's theorem), G contains a cycle of order at least k. Let C be a cycle of maximum order in G (thus $v(C) \geq k$). Applying Lemma 6.4.13, or otherwise, show that C is a spanning cycle, and hence conclude that G is Hamiltonian.

   (Chvátal and Erdös (1972))

30. Let G be a graph of order $n \geq 3$. Two sufficient conditions for G to be Hamiltonian are given below.

   Ore's condition (Theorem 5.3.6): $d(u) + d(v) \geq n$ for any two non-adjacent vertices u, v in G;

   Chvátal and Erdös' condition (see Problem VII.29): $\alpha(G) \leq \kappa(G)$.

   Show that Ore's condition implies Chvátal and Erdös' condition. Is the converse true?

31. Let G be a k-regular graph of order $2k + 1$, where $k \geq 2$. Show that if $\kappa(G) = k$, then $\alpha(G) \leq k$. Deduce that G is Hamiltonian.

Chapter 8

Vertex-colorings and Planar Graphs

8.1 The Four-color Problem

Figure 8.1.1(a) shows a map with a number of regions. One wishes to color the regions in such a way that adjacent regions (i.e., regions sharing some common boundary) are colored by different colors. What is the minimum number of colors needed?

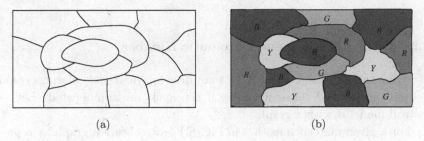

(a)   (b)

Fig. 8.1.1

Figure 8.1.1(b) shows a way of coloring the above map using four colors.

Can all maps be colored with at most four colors? Many people believed that the answer is in the affirmative, but no one could prove it for a long time. This is known as the **four-color problem**.

The history of the four-color problem started in 1852 with an English man named Francis Guthrie (1831–1899), who made this conjecture while coloring the regions on a map of England. He told his younger brother Frederick about it, and Frederick, in turn mentioned it to his teacher Augustus De Morgan (1806–1871). De Morgan was then a very well-known professor of mathematics at what is now University College, London, and quite often, he spoke of it to other mathematicians, which certainly helped spread the problem. The problem received further attention after it was formally asked by another prominent English mathematician Arthur Cayley (1821–1895) at a meeting of the London Mathematical Society on June 13, 1878.

Amongst many mathematical problems, perhaps the four-color problem was one of the most famous ones. Many researchers had claimed that they had proved it, but their 'proofs' always contained some fatal flaw somewhere.

It was not until 40 years ago that the problem was eventually settled. In the summer of 1976, Kenneth Appel and Wolfgang Haken, two professors at the University of Illinois, USA, announced that they had solved the problem affirmatively. Their approach was to divide the problem into nearly 2,000 cases and then to write computer programs to analyze the various cases. This approach required more than 1,200 hours of computer calculations. Twenty years later, a new and simpler proof (but still computer-assisted) was provided by Robertson, Sanders, Seymour and Thomas (1997). For more information about the history and development of the problem, the reader is referred to Wilson (2002).

8.2   Vertex-colorings and Chromatic Number

Studying the four-color problem by map drawings and real colorings could be quite messy and time-consuming if the maps are complicated. Let us instead model it with a graph.

For a given map such as that in Fig. 8.1.1, we obtain a graph by representing each region by a vertex, and joining two vertices by an edge if and only if the corresponding regions are adjacent. The graph corresponding to the map of Fig. 8.1.1 is shown in Fig. 8.2.1.

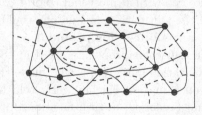

Fig. 8.2.1

Now, 'coloring regions' in a map means 'coloring vertices' in its corresponding graph. Instead of painting the regions by real colors such as blue, green, red, yellow, etc., let us make it more simple by coloring the vertices with colors $1, 2, 3, 4$ and so on. Thus the coloring of vertices of the graph in Fig. 8.2.1 corresponding to the coloring of the regions in Fig. 8.1.1(b) is now shown in Fig. 8.2.2. Note that adjacent vertices in the graph are colored by different colors.

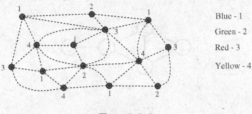

Blue - 1

Green - 2

Red - 3

Yellow - 4

Fig. 8.2.2

The above discussion motivates us to introduce the following notion of 'vertex-coloring' for a general graph.

Let G be a graph and k a positive integer. A k-**coloring** of G is a way of coloring the vertices in G by **at most** k colors in such a way that adjacent vertices are colored by different colors.

Mathematically, a k-coloring of G can be regarded as a mapping

$$\theta : V(G) \to \{1, 2, ..., k\}$$

(not necessarily onto) such that $\theta(u) \neq \theta(v)$ if the vertices u and v are adjacent in G.

Example 8.2.1. Consider the graph G of Fig. 8.2.3.

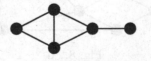

Fig. 8.2.3

A 5-coloring, a 4-coloring and a 3-coloring of G are, respectively, shown in (a), (b) and (c) of Fig. 8.2.4.

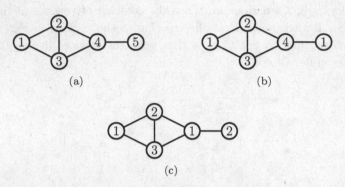

Fig. 8.2.4

Remark 8.2.2. By definition, a p-coloring of G is also a q-coloring of G if $p \leq q$. Thus, it is absolutely correct to say that the three colorings in Fig. 8.2.4 are 5-colorings of G.

Question 8.2.3.   *Let G be the graph of Fig. 8.2.5.*

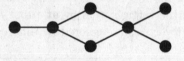

Fig. 8.2.5

Does there exist a 2-coloring of G?

Question 8.2.4.   *Let G be the graph of Fig. 8.2.6.*

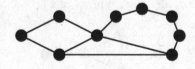

Fig. 8.2.6

(i) Find a 3-coloring of G.

(ii) Does there exist a 2-coloring of G? Why?

Example 8.2.5. For a graph G, let k be the smallest integer such that G admits a k-coloring.

  (i) What is the value of k if G is the graph in Fig. 8.2.5?
  (ii) What is the value of k if G is the graph in Fig. 8.2.6?
 (iii) What is the value of k if G is the graph in Fig. 8.2.3?

  (i) In Question 8.2.3, you should be able to provide a 2-coloring of G of Fig. 8.2.5; and it is clear that G admits no 1-colorings (since G contains an edge). Thus $k = 2$.

  (ii) In Question 8.2.4, you should be able to provide a 3-coloring of G of Fig. 8.2.6; and it can be checked that G admits no 2-colorings (since G contains a triangle). Thus $k = 3$.

 (iii) In Fig. 8.2.4(c), we provided a 3-coloring of G of Fig. 8.2.3; and as G contains a triangle, G admits no 2-colorings. Thus $k = 3$.

The above discussion on determining such a smallest 'k' motivates us to introduce the following 'key' notion on vertex-colorings.

Let G be a graph. The **chromatic number** of G, denoted by $\chi(G)$, is the minimum value of k such that G admits a k-coloring. That is, $\chi(G)$ is the smallest number of colors needed to color the vertices of G in such a way that adjacent vertices are colored by different colors. We also say that G is k-**chromatic** if $\chi(G) = k$.

Thus, if G is the graph of Fig. 8.2.3, then $\chi(G) = 3$; if G is the graph of Fig. 8.2.5, then $\chi(G) = 2$; and if G is the graph of Fig. 8.2.6, then $\chi(G) = 3$.

Question 8.2.6. *Find $\chi(G)$ for each graph G in Fig. 8.2.7.*

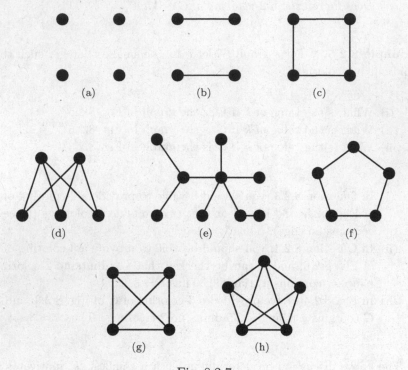

Fig. 8.2.7

Question 8.2.7. *Let G be a graph of order n. What is the least possible value of $\chi(G)$? What is the largest possible value of $\chi(G)$?*

Question 8.2.8. *If H is a subgraph of G, what is the relation between $\chi(H)$ and $\chi(G)$?*

Remark 8.2.9.

(1) There is another convenient way to express the fact that there exists a k-coloring of a graph. A graph G is said to be k-**colorable** if G admits a k-coloring. Thus, the chromatic number $\chi(G)$ is the smallest k such that G is k-colorable. Using this terminology, the four-color problem (theorem) states that if G is the graph obtained from a map of regions as shown in Fig. 8.2.1, then G is 4-colorable, that is, $\chi(G) \leq 4$.

(2) Let G be a graph with $\chi(G) = k$ and a k-coloring θ. For each

$i = 1, 2, ..., k$, let
$$V_i = \{v \in V(G) \mid \theta(v) = i\}.$$
We call V_i the ith **color class** with respect to θ. Note that each V_i is non-empty, every two different color classes are disjoint and
$$V(G) = V_1 \cup V_2 \cup ... \cup V_k.$$
Recall that a set S of vertices in G is said to be **independent** if no two vertices in S are adjacent in G. Clearly, each V_i is independent in G. Thus, equivalently, the chromatic number $\chi(G)$ of G can well be defined as the **minimum** number k such that $V(G)$ is partitioned into k pairwise disjoint independent sets in G.

Exercise for Section 8.2

1. Let p and q be integers such that $1 \le p \le q$. Explain by definition why a p-coloring of a graph is also a q-coloring of the graph.
2. Prove that if H is a subgraph of G, then $\chi(H) \le \chi(G)$.
3. Construct two graphs H and G such that H is a proper subgraph of G but $\chi(H) = \chi(G)$.
4. Construct two connected graphs H and G such that H is a spanning subgraph of G and $\chi(H) = \chi(G) - 1$.
5. For each of the following graphs, find its chromatic number.

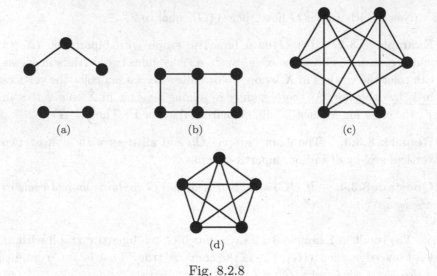

(a)   (b)   (c)

(d)

Fig. 8.2.8

8.3   Computation of Chromatic Number

Throughout this section, unless otherwise stated, let G be a graph of order n. From Question 8.2.6 and Question 8.2.7, you have probably come to the conclusion that $1 \leq \chi(G) \leq n$.

Which graphs G are such that $\chi(G) = 1$? Which graphs G have their $\chi(G)$ equal to its order n?

Example 8.3.1.   For $n \geq 1$,

   (i) $\chi(O_n) = 1$;
   (ii) $\chi(K_n) = n$.

Whenever there is an edge in G, its two ends are adjacent and must be colored by different colors. Thus, if $G \ncong O_n$ for $n \geq 2$ (that is, $e(G) \geq 1$), then $\chi(G) \geq 2$. We have:

Observation 1: Let G be a graph of order n. Then $\chi(G) = 1$ if and only if $G \cong O_n$.

Whenever there are two vertices in G which are not adjacent, we may color these two vertices by the same color, and so we use at most $n - 1$ colors to color G. Thus, if $G \ncong K_n$ for $n \geq 2$, then $\chi(G) \leq n - 1$. We have:

Observation 2: Let G be a graph of order $n \geq 2$. Then $\chi(G) = n$ if and only if $G \cong K_n$.

Now, which graphs G have their $\chi(G)$ equal to 2?

Example 8.3.2.   Let G be a bipartite graph with bipartition (X, Y) which contains at least one edge. Since no edge joins two vertices in X, we can color the vertices in X by one color. Likewise, we can color the vertices in Y by one color. As there is an edge joining a vertex in X to a vertex in Y, the color for X must be different from that for Y. Thus, $\chi(G) = 2$.

Remark 8.3.3.   The empty graphs O_n and all trees with at least two vertices are, by definition, bipartite graphs.

Question 8.3.4.   *If $\chi(G) = 3$, is it true that G contains an odd cycle as a subgraph?*

The result in Example 8.3.2 says that if G is a bipartite graph with at least one edge, then $\chi(G) = 2$. Is the converse true? That is, if G contains at least one edge and $\chi(G) = 2$, must G be bipartite?

Since bipartite graphs and odd cycles are closely related by Theorem 3.1.6, to answer this question, we first evaluate $\chi(C_n)$.

Example 8.3.5. For $n \geq 3$,

$$\chi(C_n) = \begin{cases} 2 \text{ if } n \text{ is even}; \\ 3 \text{ if } n \text{ is odd.} \end{cases}$$

In the discussion of Question 8.2.8, we learn that if H is a subgraph of a graph G, then $\chi(H) \leq \chi(G)$. Now, if G is not bipartite, then by Theorem 3.1.6, G contains an odd cycle C_{2k+1} as a subgraph, and so $\chi(G) \geq \chi(C_{2k+1}) = 3$, by Example 8.3.5. It thus follows that if $\chi(G) = 2$, then G must be bipartite; which says that the converse of the result in Example 8.3.2 is true. We now re-state this important observation as follows:

Observation 3: Let G be a graph with at least one edge. Then $\chi(G) = 2$ if and only if G is bipartite.

Question 8.3.6. *For $2 \leq n \leq 5$, construct all graphs G of order n such that $\chi(G) = n - 1$.*

We have determined all graphs G with $\chi(G)$ equal to 1 or 2. The problem of characterizing the graphs G with $\chi(G) = 3$ remains unsettled. However, from our discussion before Observation 3, we do have:

Observation 4: Let G be a graph which contains an odd cycle as a subgraph. Then $\chi(G) \geq 3$.

Likewise, we have:

Observation 5: Let G be a graph and let p be any positive integer such that G contains a K_p as a subgraph. Then $\chi(G) \geq p$.

To get a sharper lower bound in Observation 5, of course, we look for the largest such p. A set Q of vertices in G is called a **clique** if $[Q]$ is a complete subgraph of G. The **clique number** of G, denoted by $\omega(G)$, is defined as

$$\omega(G) = \max\{|Q| \mid Q \text{ is a clique in } G\}.$$

Thus we have

$$\chi(G) \geq \omega(G).$$

Example 8.3.7. Determine the chromatic number of the graph G in Fig. 8.3.1(a).

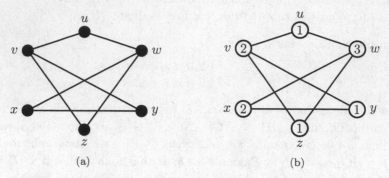

Fig. 8.3.1

Since G contains a C_5: $uvyxwu$, by Observation 4, $\chi(G) \geq 3$. As a 3-coloring of G is given in Fig. 8.3.1(b), we conclude that $\chi(G) = 3$.

Example 8.3.8. Consider the graph G of Fig. 8.2.2. There are many K_3's in G. Is there any K_4 in G? Yes, there is one (try to find it!). Is there any K_5 in G? Definitely no! Thus, $\omega(G) = 4$, and so $\chi(G) \geq \omega(G) = 4$. As a 4-coloring of G is shown in Fig. 8.2.2, we thus conclude that $\chi(G) = 4$.

Question 8.3.9.   *Consider the graph G in Fig. 8.3.2.*

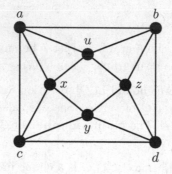

Fig. 8.3.2

Since G contains a triangle, $\chi(G) \geq 3$. What is the value of $\chi(G)$?

Exercise for Section 8.3

1. Given $p \geq q \geq 3$ and $k \geq 1$, let $B(k)$ be a graph obtained by adding k new edges to the complete bipartite graph $K(p, q)$. Evaluate $\chi(B(1))$ and find all possible values of $\chi(B(k))$ for each $k \in \{2, 3, 4\}$.

2. Let G be the graph given in Fig. 8.3.3. Explain why $\chi(G) \geq 3$. Then provide a 3-coloring for G, thereby proving that $\chi(G) = 3$.

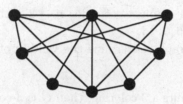

Fig. 8.3.3

3. Let G be the graph given in Fig. 8.3.4. Explain why $\chi(G) \geq 4$. Then provide a 4-coloring for G, thereby proving that $\chi(G) = 4$.

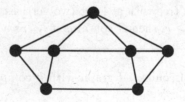

Fig. 8.3.4

4. Let G be the graph given in Fig. 8.3.5. Explain why $\chi(G) \geq 4$. Then provide a 4-coloring for G, thereby proving that $\chi(G) = 4$.

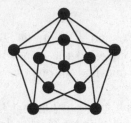

Fig. 8.3.5

5. Determine all graphs G of order $n \geq 2$ with $\chi(G) = n - 1$. (See Question 8.3.6.).

6. Let G be a graph. Determine the validity of each of the following statements.

    (i) If G admits a 3-coloring, then G is 3-colorable.

    (ii) If G is 3-colorable, then G is 5-colorable.

    (iii) If G is 3-colorable, then $\chi(G) \geq 3$.

    (iv) If G is 3-colorable, then $\chi(G) \leq 3$.

    (v) If G is 3-colorable, then G contains an odd cycle.

    (vi) If G contains an odd cycle, then G is 3-colorable.

    (vii) If G admits no 3-colorings, then $\chi(G) \geq 3$.

    (viii) If G admits no 3-colorings, then $\chi(G) = 2$.

    (ix) If G admits no 3-colorings, then $\chi(G) \leq 2$.

    (x) If $\chi(G) = 3$, then G contains a triangle.

    (xi) If $\chi(G) = 3$, then G contains an odd cycle.

    (xii) If G is a tree with at least two vertices, then $\chi(G) = 2$.

    (xiii) If $\chi(G) \geq r$, then G contains a K_r as a subgraph (i.e., $\omega(G) \geq r$).

7. Let G be a disconnected graph with k components $G_1, G_2, ..., G_k$. Show that

$$\chi(G) = \max\{\chi(G_1), \chi(G_2), ..., \chi(G_k)\}.$$

8. Let G_1 and G_2 be two connected graphs and let G be the graph obtained from G_1 and G_2 by identifying a vertex in G_1 with a vertex in G_2 as shown Fig. 8.3.6.

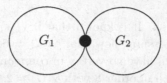

Fig. 8.3.6: G

Show that $\chi(G) = \max\{\chi(G_1), \chi(G_2)\}$.

9. Let G be a graph. Show that

   (i) $\chi(G) - 1 \le \chi(G - v) \le \chi(G)$ for each vertex v in G.
   (ii) $\chi(G) - 1 \le \chi(G - e) \le \chi(G)$ for each edge e in G.

10. For each integer $n \ge 2$, construct a graph G of order n such that $\chi(G - v) = \chi(G) - 1$ for each vertex v in G.

11. For each integer $n \ge 4$, construct a graph G of order n such that $\chi(G - v) = \chi(G)$ for each vertex v in G.

12. For each integer $n \ge 2$, construct a graph G of order n such that $\chi(G - e) = \chi(G) - 1$ for each edge e in G.

13. For each integer $n \ge 3$, construct a graph G of order n such that $\chi(G - e) = \chi(G)$ for each edge e in G.

14. Let G be the graph shown in Fig. 8.3.7.

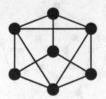

Fig. 8.3.7

   (i) Find $\chi(G)$.
   (ii) Verify that $\chi(G - e) = \chi(G) - 1$ for each edge e in G.

15. Construct a graph G such that $\chi(G) = 3$ and G contains no triangles.

16. Construct a graph G such that $\chi(G) = 4$ and G contains no triangles.

17. Let G be a graph which is not bipartite. Assume that there is a vertex in G which is contained in every odd cycle in G. Show that

$\chi(G) = 3$.

18. Let G be a graph. It is known that if $\chi(G) = 3$, then G contains an odd cycle. Assume now that $\chi(G) = 6$. Does G contain two odd cycles which have no vertex in common? Why?

19. Let G be a graph of order 8 with $\chi(G) = 2$. Show that $e(G) \leq 16$. Construct one such G with $e(G) = 16$.

20. Let G be a graph of order 7 with $\chi(G) = 3$. Show that $e(G) \leq 16$. Construct one such G with $e(G) = 16$.

21. Let G be a graph of order 6 with $\chi(G) = 4$. Show that $e(G) \leq 13$. Contruct one such G with $e(G) = 13$.

22. Determine the chromatic number of each of the following graphs:

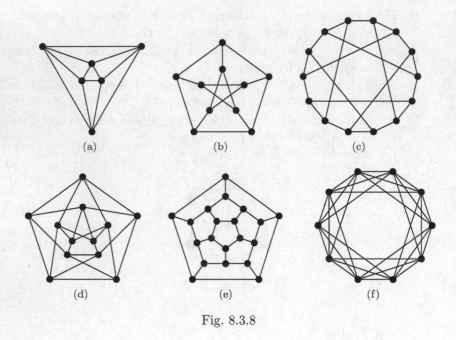

Fig. 8.3.8

23. Let $G + H$ denote the join of the graphs G and H (see Problem 2.5.6).

   (i) Find $\chi(C_4), \chi(C_5)$ and $\chi(C_4 + C_5)$.
   (ii) In general, how are $\chi(G)$, $\chi(H)$ and $\chi(G+H)$ related? Prove your result.

24. The wheel W_n of order n is defined as $W_n = C_{n-1} + K_1$.

(i) Draw W_6 and W_7.

(ii) Find a 3-coloring for W_7.

(iii) Find a 4-coloring for W_6.

(iv) Show that $\chi(W_n) = 3$ for odd $n \geq 5$.

(v) Show that $\chi(W_n) = 4$ for even $n \geq 4$.

25. Let G be a k-chromatic graph, where $k \geq 1$. Show that
$$k(k-1) \leq 2e(G).$$

26. Let G be a graph of order n.

(i) Show that $\chi(G)\alpha(G) \geq n$.

(ii) Construct a connected graph H such that $v(H) = 12$, $\chi(H) = 4$ and $\alpha(H) = 3$.

(iii) Show that $\chi(G) + \alpha(G) \leq n + 1$.

(iv) Construct a connected graph H such that $v(H) = 11$, $\chi(H) = 5$ and $\alpha(H) = 7$.

27. Let G be a graph which is not bipartite. Assume that G contains an independent set S such that $V(C) \cap S$ is non-empty for every odd cycle C in G. Show that $\chi(G) = 3$.

8.4 Greedy Coloring Algorithm

The problem of evaluating the chromatic number of a graph in general is very difficult; and until now, there is no 'efficient' way to compute its **exact value**. However, there is an 'efficient' heuristic algorithm, called the **greedy coloring algorithm**, which enables us to color a graph and obtain an approximation to its chromatic number.

Algorithm 8.4.1. (Greedy coloring algorithm) Let G be a graph of order n and list its vertices as $v_1, v_2, ..., v_n$.

Step 1: Color v_1 by color '1'. Set $i := 1$.

Step 2: If $i = n$, stop and output coloring of graph. Otherwise, consider v_{i+1}. Find the smallest color number which is not used to color the neighbors of v_{i+1}. Color v_{i+1} using this color number. Increase i by 1 and return to Step 2.

Example 8.4.2. Let G be the graph shown in Fig. 8.4.1(a). Fix two different orderings of its vertices as shown in Fig. 8.4.1(b) and (c). We shall apply Algorithm 8.4.1 to color G in two ways following the respective ordering.

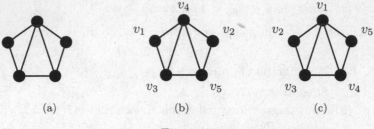

Fig. 8.4.1

Following the ordering shown in Fig. 8.4.1(b), the algorithm is applied step by step as shown in the sequence Fig. 8.4.2(a) to (e).

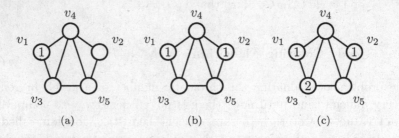

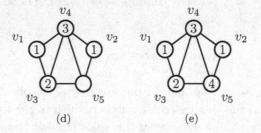

Fig. 8.4.2

The number of colors produced in this case is '4'.

Now, following the ordering shown in Fig. 8.4.1(c), the algorithm is applied step by step as shown in the sequence Fig. 8.4.3(a) to (e).

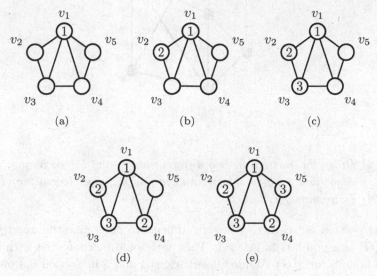

Fig. 8.4.3

The number of colors produced in this case is '3'.

Remark 8.4.3.

(1) The above example tells us that the number of colors produced by applying Algorithm 8.4.1 depends on the ordering of the vertices.

(2) The number of colors produced by applying Algorithm 8.4.1 provides only an upper bound for the chromatic number of the graph. Thus, following the ordering in Fig. 8.4.1(b), $\chi(G) \leq 4$; while following the ordering in Fig. 8.4.1(c), $\chi(G) \leq 3$.

(3) As G contains a triangle, $\chi(G) \geq 3$. Thus, combining it with the result above, we see that $\chi(G) = 3$.

Question 8.4.4. *Let G be the graph shown in Fig. 8.4.4.*

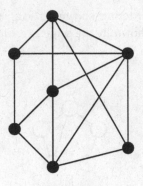

Fig. 8.4.4

(i) *Order the vertices in two different ways, and for each way, apply Algorithm 8.4.1 to color G and to find an upper bound for $\chi(G)$.*

(ii) *Determine $\chi(G)$.*

As pointed out earlier, it is very difficult to determine the exact value of $\chi(G)$ of a graph G in general. Thus we should be contented with some good bounds for $\chi(G)$. While Observations 4 and 5 in Section 8.3 provide two lower bounds for $\chi(G)$, the greedy coloring algorithm presented in this section does provide an upper bound for $\chi(G)$ as shown in Example 8.4.2 and Question 8.4.4.

Question 8.4.5.   *Let us revisit Question 8.4.4 and re-consider the coloring of G of Fig. 8.4.4.*

(i) *Find $\Delta(G)$.*

(ii) *While you applied Algorithm 8.4.1 to color v_i by a color number in Question 8.4.4, was this color number always in $\{1, 2, ..., \Delta(G)+1\}$? Why?*

(iii) *Did you need to use more than '$\Delta(G) + 1$' colors to color G when you applied Algorithm 8.4.1?*

Recall the procedure of applying Algorithm 8.4.1 to color the vertices $v_1, v_2, ..., v_n$. In each step of coloring v_i, we first look at its neighbors, then find out the smallest color number which is not used to color its neighbors, and finally color v_i by this 'missing' color. How many neighbors does v_i have? There are '$d(v_i)$' that many, which is certainly not more than '$\Delta(G)$'. Clearly, the 'worst' case is when the neighbors of v_i are colored by $\Delta(G)$ different colors, but even then, there is still one color available in $\{1, 2, ..., \Delta(G) + 1\}$ to color v_i. Thus, when applying Algorithm 8.4.1 to color G, we never have to use more than $\Delta(G) + 1$ colors. That is, the number of colors produced by applying Algorithm 8.4.1 is never more than '$\Delta(G) + 1$'. Now, by the definition of $\chi(G)$, we have:

Theorem 8.4.6.   *For any G, $\chi(G) \leq \Delta(G) + 1$.*

$\square$

The number of colors produced by applying Algorithm 8.4.1 depends on the ordering of the vertices. We shall now recall a special ordering of vertices introduced in Example 1.5.4, and see that by applying Algorithm 8.4.1, using this ordering on a large family of graphs, the above bound can be improved.

Theorem 8.4.7.   *If G is a non-regular connected graph, then $\chi(G) \leq \Delta(G)$.*

Proof. Assume that G is of order n and let w be a vertex in V such that $d(w) < \Delta(G)$. Label the n vertices $v_1, v_2, \cdots, v_{n-1}, v_n(= w)$ so that each vertex other than w has a higher labeled neighbor in G. This can be achieved (see Problem 1.5.9) by arranging the vertices in $V \backslash \{w\}$ such that

$$d(w, v_{n-1}) \leq d(w, v_{n-2}) \leq \cdots \leq d(w, v_2) \leq d(w, v_1).$$

Thus, when applying Algorithm 8.4.1 at v_i, $i \neq n$, at least one of v_i's neighbors has not been colored. Hence one of the colors in $\{1, 2, \cdots, \Delta(G)\}$ is available to color v_i. As for the last vertex $v_n(= w)$, this is also the case, since $d(w) < \Delta(G)$.

Let k be the number of colors produced by applying Algorithm 8.4.1. Then $\chi(G) \leq k \leq \Delta(G)$.

$\square$

Exercise for Section 8.4

1. Let G be the graph shown in Fig. 8.4.5 with two different ways of ordering its vertices. Apply Algorithm 8.4.1 to color G and find the number of colors produced in each case.

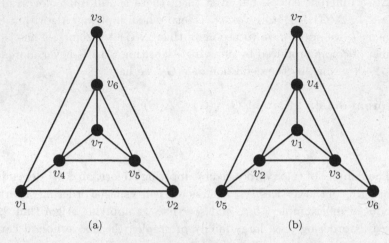

Fig. 8.4.5

2. In each of the following graphs, the vertices are labeled as $v_1, v_2, ..., v_n$.

   (i) Find the number of colors produced by applying Algorithm 8.4.1 on each graph according to the ordering of vertices $v_1, v_2, ..., v_n$.

   (ii) Find the chromatic number of each graph.

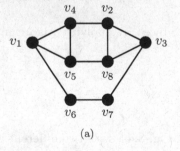

(a)

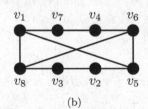

(b)

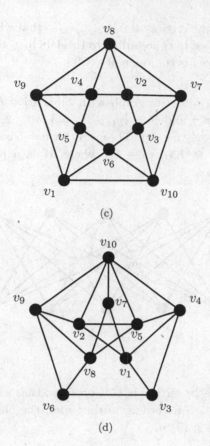

(c)

(d)

Fig. 8.4.6

3. Let G be the graph shown in Fig. 8.4.7.

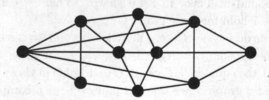

Fig. 8.4.7

(i) Find $\chi(G)$.

(ii) Label the vertices as $v_1, v_2, ..., v_n$ so that when Algorithm 8.4.1 is applied to G according to this labeling, the number of colors produced is the value of $\chi(G)$.

4. Let G be a bipartite graph with bipartition (X, Y) where $X = \{x_1, x_2, ..., x_n\}$ and $Y = \{y_1, y_2, ..., y_n\}$, $n \geq 2$, and x_i is adjacent to y_j if and only if $i \neq j$. The graph for $n = 5$ is shown in Fig. 8.4.8. (Indeed, $G \cong K(n, n) - M$, where M is a perfect matching of $K(n, n)$.)

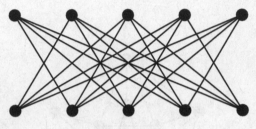

Fig. 8.4.8

Arrange the $2n$ vertices in G in order, so that when Algorithm 8.4.1 is applied to G according to the order, the number of colors produced is (i) 2; (ii) n.

8.5 Brooks' Theorem

Is the upper bound in Theorem 8.4.6 sharp? That is, can the equality $\chi(G) = \Delta(G) + 1$ hold for some graph G?

Yes! Take the odd cycle $G = C_{2k+1}$. Clearly, $\chi(G) = 3$ while $\Delta(G) = 2$. Take the complete graph $G = K_n$. Clearly, $\chi(G) = n$ while $\Delta(G) = n - 1$. And we see that the equality $\chi(G) = \Delta(G) + 1$ holds in these two instances.

Can you find a graph other than an odd cycle or a complete graph for which the equality holds?

Question 8.5.1. *Let G be the disconnected graph with two components as shown in Fig. 8.5.1.*

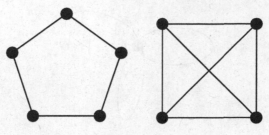

Fig. 8.5.1

(i) Find $\Delta(G)$ *and* $\chi(G)$.

(ii) Is $\chi(G)$ *equal to* $\Delta(G) + 1$?

Can you find a **connected** graph other than an odd cycle or a complete graph for which the equality holds?

In this section, we shall answer this question by presenting a celebrated result, known as Brooks' theorem on coloring. The idea of the proof given below is essentially due to Lovász (1975).

Recall that a non-trivial connected graph that has no cut-vertices is called a **block** (see Section 6.4). Every graph of order at least three is a block if and only if it is 2-connected. A **block** of a graph G is a subgraph of G that is a block and is maximal with respect to this property. An **end-block** of G is a block containing exactly one cut-vertex of G. It is known that every connected graph with at least one cut-vertex contains at least two end-blocks (see Problem 6.4.15).

Theorem 8.5.2. (Brooks (1941)) *If* G *is a connected graph, which is neither an odd cycle nor a complete graph, then* $\chi(G) \leq \Delta(G)$.

Proof. Write $\Delta = \Delta(G)$. If $\Delta = 1$, then $G = K_2$. If $\Delta = 2$, then G is a path or a cycle, and the result follows. We now assume that $\Delta \geq 3$.

If G is not regular, then by Theorem 8.4.7, the inequality holds. We may now assume that G is Δ-regular.

Case 1. G has a cut-vertex (say w, see Fig. 8.5.2).

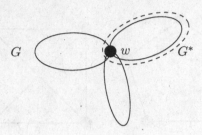

Fig. 8.5.2

Let G^* be a subgraph of G induced by the union of $\{w\}$ and a component of $G - w$. Since the degree of w in G^* is strictly less than Δ, each G^* is a non-regular connected graph and so $\chi(G^*) \leq \Delta(G^*) \; (= \Delta(G))$ by Theorem 8.4.7. By permuting the names of colors of such G^*s, these colorings can be re-defined in such a way that the colors assigned for w are the same. This gives rise to a Δ-coloring of G and thus $\chi(G) \leq \Delta(G)$.

Case 2. G is 2-connected.

Claim. G contains three vertices u, v_1 and v_2 as shown in Fig. 8.5.3 (v_1 and v_2 are not adjacent) such that $G - \{v_1, v_2\}$ is connected.

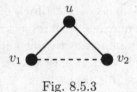

Fig. 8.5.3

Let z be a vertex in G. Assume that $\kappa(G-z) \geq 2$. Since G is regular and not complete, there exists a vertex x in G with $d(z,x) = 2$ (see Problem I.3). Let zux be a path of length two and let $z = v_1$ and $x = v_2$. As $\kappa(G - v_1) \geq 2$, $G - \{v_1, v_2\}$ is connected.

If $\kappa(G - z) = 1$, let B_1 and B_2 be two end-blocks of $G - z$. Since G has no cut-vertices, z has a neighbor v_1 in B_1 and a neighbor v_2 in B_2. Note that (1) v_1 and v_2 are non-adjacent and (2) $G - \{v_1, v_2\}$ is connected as $\Delta \geq 3$ (see Fig. 8.5.4).

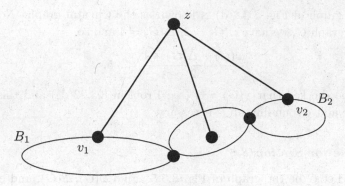

Fig. 8.5.4

This establishes the above claim.

Now, since $G - \{v_1, v_2\}$ is connected, its $n - 2$ vertices can be labeled $v_3, v_4, \cdots, v_{n-1}, v_n(= u)$ so that each vertex other than u has a higher-labeled neighbor. By applying Algorithm 8.4.1 to this vertex ordering, we see that v_1 and v_2 are colored '1'; each v_i $(i = 3, 4, \cdots, n-1)$ has at most $\Delta - 1$ lower-indexed neighbors; two neighbors of v_n, namely, v_1 and v_2, are colored by the same color. Thus, Algorithm 8.4.1 uses at most Δ colors and so $\chi(G) \leq \Delta(G)$.

□

Question 8.5.3. *In each of the following cases, is $\chi(G) = \Delta(G)$?*

    (i) G is a path;
    (ii) G is an even cycle;
    (iii) G is obtained from K_n, $n \geq 3$, by deleting an edge;
    (iv) G is the Petersen graph;
    (v) $G = K(p, q)$.

Remark 8.5.4. For any graph G, we have

$$\omega(G) \leq \chi(G) \leq \Delta(G) + 1.$$

Reed (1997) conjectured that the rounded up of the average of the above lower and upper bounds of $\chi(G)$ is also an upper bound of $\chi(G)$; that is:

$$\chi(G) \leq \lceil \frac{\omega(G) + \Delta(G) + 1}{2} \rceil.$$

The above conjecture is still open though some partial results have been found.

The graph of Fig. 8.3.8(d) is known as the Chvátal graph. Note that for this graph G, we have $\omega(G) = 2$, $\Delta(G) = 4$ and so

$$\frac{\omega(G) + \Delta(G) + 1}{2} = 3.5.$$

It can be checked that $\chi(G) = 4$ (see Problem 8.3.22(d)) and this shows that the above rounding up is necessary.

Exercise for Section 8.5

1. Let G be the graph of Fig. 8.5.5. Find $\omega(G), \Delta(G)$, and evaluate $\chi(G)$ by Theorem 8.5.2.

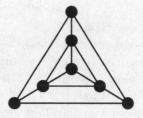

Fig. 8.5.5

2. Let G be a connected graph which is not complete. Show that if G contains K_r as a subgraph, where $r = \Delta(G) \geq 3$, then $\chi(G) = r$.
3. Does there exist a graph G satisfying the following conditions:
   (i) $\chi(G) = 7$, and
   (ii) the degree sequence of G is $(6, 6, 6, 6, 6, 6, 5, 5, 5, 5, 4, 4, 3, 3, 3, 3)$?
4. Let G be a cubic connected graph. What are the possible values of $\chi(G)$? Classify G according to the values of $\chi(G)$.
5. Let G be a regular and connected graph of order n. Show that $\chi(G) + \chi(\overline{G}) = n + 1$ if and only if $G \cong K_n$ or $G \cong C_5$.

8.6   Critical Graphs

Let G be a graph, and H a subgraph of G. Then either $\chi(H) = \chi(G)$ or $\chi(H) < \chi(G)$. In this section, we shall introduce a family of graphs G which are 'minimal' with respect to $\chi(G)$ in the sense that $\chi(H) < \chi(G)$ for any proper subgraph H of G.

More precisely, G is said to be **critical** if $\chi(H) < \chi(G)$ for any proper subgraph H of G. If G is critical and $\chi(G) = k$, then G is said to be **k-critical**.

The concept of a critical graph was first introduced and studied by Dirac (1952b). It is noted that, quite often, problems for k-chromatic graphs could be reduced to those for k-critical ones. For instance, the notion of criticality could be employed to prove Theorem 8.5.2, Corollary 8.6.7 etc.

Amongst all k-chromatic graphs, k-critical ones are the cream, and we naturally expect that they have special properties. Some of them will be presented in this section.

Example 8.6.1.

(i) G is 1-critical if and only if $G \cong O_1$.
(ii) G is 2-critical if and only if $G \cong K_2$.
(iii) G is 3-critical if and only if G is an odd cycle.
(iv) K_n is n-critical for all $n \geq 1$.
(v) The following graph is 4-critical. (See Problem 8.3.14.) No characterizations of 4-critical graphs have been found yet.

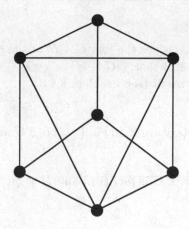

Fig. 8.6.1

Observation 8.6.2.   (See Problem 8.6.1.)

(i) Every critical graph is connected.
(ii) Every k-chromatic graph G contains a subgraph H which is k-critical.

(iii) If G is critical, then

    (a) $\chi(G - v) = \chi(G) - 1$ for each v in $V(G)$;
    (b) $\chi(G - e) = \chi(G) - 1$ for each e in $E(G)$.

(iv) Let G be a connected graph. Then G is critical if and only if $\chi(G - e) = \chi(G) - 1$ for each edge e in G.

Discussion 8.6.3.

    (i) Is it true that a connected graph G is critical if and only if $\chi(G - v) = \chi(G) - 1$ for each vertex v in G? (See Problem 8.6.2.)
    (ii) By Observation 8.6.2(i), critical graphs must be connected. Can a critical graph contain a cut-vertex?

It is not so easy to construct critical graphs other than odd cycles and complete graphs. We provide in what follows two general methods to construct larger critical graphs from smaller ones (see also Problems 8.6.3 and 8.6.4).

Construction I Consider the join $G + H$ of the graphs G and H. We learned in Problem 8.3.23 that $\chi(G + H) = \chi(G) + \chi(H)$. For criticality, we have the following result (see Problem 8.6.3):

The join $G + H$ is critical if and only if both G and H are critical.

Thus, in particular, if G is p-critical and H is q-critical, then $G + H$ is a $(p + q)$-critical graph.

Construction II (Hajós (1961)) Let G and H be two non-trivial graphs which have a vertex, say w, in common. Let wg be an edge in G and wh be an edge in H. Form a new graph, denoted by $G \wedge H$, as shown in Fig. 8.6.2:

$$G \wedge H = (G - wg) \cup (H - wh) + gh.$$

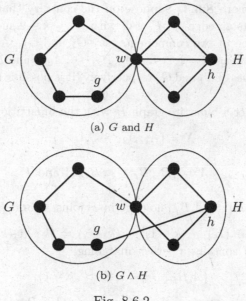

(a) G and H

(b) $G \wedge H$

Fig. 8.6.2

We have the following result (see Problem 8.6.4).

If both G and H are k-critical, then $G \wedge H$ is also k-critical.

We shall now present two results on critical graphs. The first one provides a sharp upper bound for the chromatic number of a critical graph in terms of its edge-connectivity.

Theorem 8.6.4. *If G is a k-critical graph, where $k \geq 1$, then $\chi(G) = k \leq 1 + \kappa'(G)$.*

To prove the above theorem, we shall make use of the following result (a special case of the result in Problem 7.6.15).

Lemma 8.6.5. *Let H be a bipartite graph with bipartition (X, Y), where $|X| = |Y| = p \geq 1$. If $e(H) > p(p-1)$, then $\alpha'(H) \geq p$, i.e., H contains a perfect matching.*

$\square$

Proof of Theorem 8.6.4. Suppose on the contrary that $\kappa'(G) \leq k - 2$. Then there exists an edge-cut F of G with $|F| \leq k - 2$ such that $G - F$ is the disjoint union of two components, say G_1 and G_2. As G is k-critical, $\chi(G_i) \leq k-1$, for each $i = 1, 2$. Let $A_1, A_2, \cdots, A_{k-1}$ and $B_1, B_2, \cdots, B_{k-1}$ be the color classes of G_1 and G_2 respectively (it is possible that some classes may be empty).

Now construct a bipartite graph H with the bipartition

$$X = \{A_1, A_2, \cdots, A_{k-1}\},$$

$$Y = \{B_1, B_2, \cdots, B_{k-1}\} \text{ and}$$

$$E(H) = \{A_i B_j \mid \text{no edge in } F \text{ joins } A_i \text{ and } B_j\}.$$

As $|F| \leq k - 2$, $e(H) > (k - 1)^2 - (k - 1) = (k - 1)(k - 2)$. Thus, by Lemma 8.6.5, H contains a perfect matching

$$\{A_i B_{\phi(i)} \mid i = 1, 2, \cdots, k - 1\},$$

where ϕ is a permutation of $\{1, 2, \cdots, k - 1\}$. But then $A_1 \cup B_{\phi(1)}$, $A_2 \cup B_{\phi(2)}, \cdots, A_{k-1} \cup B_{\phi(k-1)}$ form color classes of G, which implies that $\chi(G) \leq k - 1$, a contradiction.

$\square$

By Theorem 6.3.1, $\kappa'(G) \leq \delta(G)$. We thus have:

Corollary 8.6.6.   If G is a k-critical graph, where $k \geq 1$, then $\chi(G) = k \leq 1 + \delta(G)$.

$\square$

Corollary 8.6.7.   Let G be a k-chromatic graph, where $k \geq 1$. Then G contains a set S of at least k vertices such that $d(x) \geq k - 1$ for each x in S.

Proof. By Observation 8.6.2(ii), let H be a k-critical subgraph of G. Let $S = V(H)$. Clearly, $|S| \geq k$, and for each x in S, $d(x) \geq d_H(x) \geq \delta(H) \geq k - 1$, by Corollary 8.6.6.

$\square$

In Discussion 8.6.3(ii), it is not hard to argue that a critical graph contains no cut-vertex. Recall that a clique of a graph G is a subset S of

$V(G)$ such that $[S]$ is a complete subgraph of G. More generally, we have the following:

Theorem 8.6.8. *Let G be a connected graph. If G is critical, then G contains no set S of vertices such that S is both a cut and a clique.*

Proof. Suppose G contains a cut S which is also a clique. Assume that the components of $G - S$ have vertex sets $U_1, \cdots, U_p$, where $p \geq 2$, and let $G_i = [U_i \cup S]$ for each $i = 1, 2, \cdots, p$.

Let $k = \chi(G)$. As G is critical, each G_i has a $(k-1)$-coloring θ_i. Since S is a clique, $\theta_i(u) \neq \theta_i(v)$ for any two distinct vertices u, v in S and for each $i = 1, 2, \cdots, p$. Thus we may assume that for each u in S, $\theta_j(u) = \theta_1(u)$ for all $j = 1, 2, \cdots, p$. Now, define a labeling $\theta : V(G) \to \{1, 2, \cdots, k - 1\}$ as follows:

$$\theta(v) = \begin{cases} \theta_1(v) & \text{if } v \in S; \\ \theta_i(v) & \text{if } v \in U_i, \ i = 1, 2, \cdots, p. \end{cases}$$

It can be shown that θ is a $(k-1)$-coloring of G, which, however, contradicts the fact that $\chi(G) = k$. $\qquad\square$

Discussion 8.6.9.

(i) In each of the following graphs, can you find a set S of vertices such that S is both a cut and a clique?

(ii) Which graph is critical? Can it be obtained via Hajós construction?

(iii) Is the converse of Theorem 8.6.8 true?

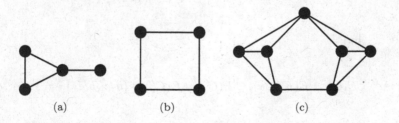

(a)   (b)   (c)

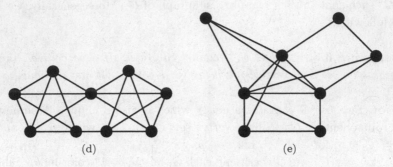

(d)   (e)

Fig. 8.6.3

Exercise for Section 8.6

1. Prove the four statements in Observation 8.6.2.
2. Consider the graph G in Problem 8.3.3. Is it true that $\chi(G - v) = \chi(G) - 1$ for each vertex v in G? Is G critical?
3. (i) Let G and H be any two graphs. Show that the join $G + H$ is critical if and only if both G and H are critical.
   (ii) Construct a non-complete k-critical graph, for $k = 5, 6$.
4. (i) Let G and H be any two graphs. Show that if both G and H are k-critical, then the graph $G \wedge H$ is also k-critical.
   (ii) Construct a 5-critical graph of order n, for $n = 5, 9, 13$.
5. Suppose $q \geq 3$ is an odd number. Let $K(q, q)$ be a complete bipartite graph with partite sets $\{a_1, a_2, \cdots, a_q\}$ and $\{b_1, b_2, \cdots, b_q\}$, and let C_q and C'_q be two disjoint cycles of order q disjoint from $K(q, q)$ with $V(C_q) = \{c_1, c_2, \cdots, c_q\}$ and $V(C'_q) = \{d_1, d_2, \cdots, d_q\}$. A graph G is defined by

$$V(G) = V(K(q,q)) \cup V(C_q) \cup V(C'_q)$$

and

$$E(G) = E(K(q,q)) \cup E(C_q) \cup E(C'_q) \cup \{a_i c_i, b_i d_i \mid i = 1, \cdots, q\}.$$

The graph G when $q = 3$ is shown in Fig. 8.6.4.

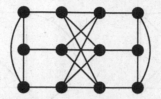

Fig. 8.6.4

Find the value of $\chi(G)$ and show that G is critical.

6. Give a direct proof of Corollary 8.6.6 without applying Theorem 8.6.4.

7. Construct examples to show that the bound in Theorem 8.6.4 is sharp.

8. Is it true that if G is a k-critical graph, where $k \geq 1$, then $\chi(G) = k \leq 1 + \kappa(G)$?

8.7 Planar Graphs and the Four-color Theorem

We now return to Section 8.1 and Section 8.2 and reconsider the four-color problem introduced therein. Observe that the graph of Fig. 8.2.2 constructed correspondingly from the map of Fig. 8.2.1 has the following feature: no two edges cross each other. A graph G is said to be **planar** if it can be drawn in the plane in such a way that no two edges intersect except at a vertex. Such a drawing, if it exists, is called a **plane drawing** of G. G is said to be **non-planar** if no plane drawing of G exists. A planar graph with a plane drawing is called a **plane** graph.

Example 8.7.1.

(i) The complete graph K_4 is planar and its plane drawing is shown in Fig. 8.7.1.

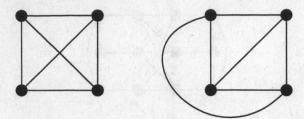

Fig. 8.7.1

(ii) The complete graph K_5 with an edge deleted is planar and its plane drawing is shown in Fig. 8.7.2.

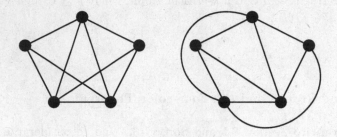

Fig. 8.7.2

(iii) The complete bipartite graph $K(3,3)$ with an edge deleted is planar and its plane drawing is shown in Fig. 8.7.3.

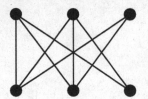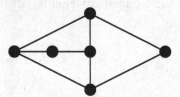

Fig. 8.7.3

Question 8.7.2. *Do there exist plane drawings for the complete graph* K_5 *and the complete bipartite graph* $K(3,3)$?

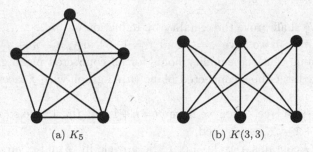

(a) K_5 (b) $K(3,3)$

Fig. 8.7.4

Let G be a planar graph. Then any plane drawing of G divides the plane into regions, called **faces**. The face with infinite extent is called the **infinite face**.

Example 8.7.3. The plane graph shown in Fig. 8.7.5 is of order 8 and size 15; including the infinite face, there are 9 faces in its plane drawing.

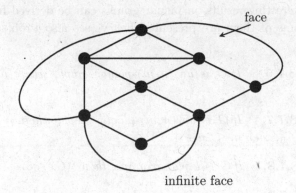

face

infinite face

Fig. 8.7.5

There is a very interesting relationship linking the order, the size and the number of faces in any plane drawing of a planar graph. This was observed by Euler around 1750 and is stated below.

Theorem 8.7.4. (Euler's Formula) *Let G be a connected plane graph of order n and size m. Let f denote the number of faces in its plane drawing. Then $n - m + f = 2$.*

Proof. We shall prove the equality by induction on 'm'.

When $m = 0$, we have $n = 1$ and $f = 1$. Clearly, $n - m + f = 2$ in this case. Assume that the equality holds for all connected plane graphs with $m = k - 1$. Let G be a connected plane (n, k)-graph with f faces. We shall show that $n - k + f = 2$.

Case 1: G is a tree. Then we have $n = k + 1$ and $f = 1$, and so $n - k + f = k + 1 - k + 1 = 2$, as required.

Case 2: G contains a cycle. Let C be a cycle in G and e, an edge on C. Consider $G - e$. Clearly, $G - e$ is a connected plane $(n, k - 1)$-graph with $f - 1$ faces. By the induction hypothesis, $n - (k - 1) + (f - 1) = 2$; that is $n - k + f = 2$. The proof is thus complete.

$\square$

Remark 8.7.5. There are more than 20 proofs available for Theorem 8.7.4. See, for instance,

$\qquad$ http://www.ics.uci.edu/~eppstein/junkyard/euler/.

Many interesting results on planar graphs can be derived from Euler's formula. Some of them are presented below (see also Problems 8.7.4 to 8.7.8).

Corollary 8.7.6. *If G is an (n, m)-planar graph, where $n \geq 3$, then $m \leq 3n - 6$.*

Corollary 8.7.7. *If G is an (n, m)-planar graph, where $n \geq 3$, which contains no C_3, then $m \leq 2n - 4$.*

Corollary 8.7.8. *If G is a planar graph, then $\delta(G) \leq 5$.*

Corollary 8.7.9. *The complete graph K_n, where $n \geq 1$, is planar if and only if $n \leq 4$.*

Corollary 8.7.10. *The complete bipartite graph $K(3, 3)$ is non-planar.*

A graph G is called a **subdivision** of a graph H if G can be obtained from H by replacing some of its edges by paths of finite lengths. Figure 8.7.6(b) shows a subdivision of the wheel W_7 shown in Fig. 8.7.6(a).

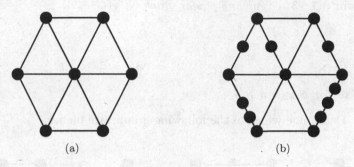

Fig. 8.7.6

Question 8.7.11. *Verify that the following two graphs are subdivisions of K_5 and $K(3,3)$ respectively.*

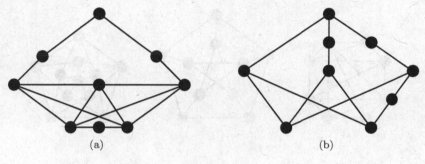

Fig. 8.7.7

The two smallest non-planar graphs, namely K_5 and $K(3,3)$, play vital roles in characterizing planar graphs. This can be seen from the following classic result found by Kuratowski (1930).

Theorem 8.7.12. *A graph G is planar if and only if G contains no subgraph isomorphic to a subdivision of K_5 or $K(3,3)$.*

$\square$

To end this section, we now state below an equivalent statement of the four-color theorem using graph terminology.

Theorem 8.7.13.   *For any planar graph G, $\chi(G) \leq 4$.*

$\square$

Exercise for Section 8.7

   1. Determine which of the following graphs are planar.

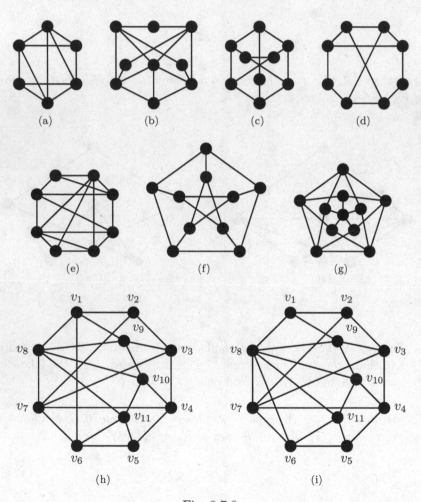

Fig. 8.7.8

2. Determine all the values of n such that $\overline{C}_n$ is planar. Justify your answer.

3. Let $V(C_{2k}) = \{v_1, v_2, ..., v_{2k}\}$, where $k \geq 2$. Let H_k be the graph obtained from C_{2k} by joining v_i to v_{i+k} for each $i = 1, 2, ..., k$. For which values of k is H_k planar? Justify your answer.

4. Let G be a connected plane graph of order $n \geq 3$ and size m. Let f denote the number of faces in its plane drawing. Show that

   (i) $2m \geq 3f$;

   (ii) $m \leq 3n - 6$;

   (iii) $m \leq 2n - 4$ if G contains no triangle;

   (iv) $\delta(G) \leq 5$;

   (v) G contains at least 12 vertices of degree 5 if $\delta(G) = 5$;

   (vi) $\delta(G) \leq 4$ if $n \leq 11$;

   (vii) if $n \geq 4$, then $|\{v \in V(G) \mid d(v) \leq 5\}| \geq 4$.

5. Construct a 5-regular planar graph of order 12.

6. Let G be a 3-regular plane graph in which every vertex lies on one face of length 4, one face of length 6 and one face of length 8. Determine the number of faces of G. Justify your answer.

7. Show that the complete graph K_n, where $n \geq 1$, is planar if and only if $n \leq 4$.

8. Show that the complete bipartite graph $K(p, q)$, where $1 \leq p \leq q$, is planar if and only if $p \leq 2$.

9. Show that if G is a planar graph of order $n \geq 11$, then $\overline{G}$ is non-planar.

10. Prove, without using Theorem 8.7.13, that $\chi(G) \leq 5$ for any planar graph G.

11. Which of the following graphs are planar?

    (i) $P_m \square C_n$, where $m \geq 1$ and $n \geq 3$;

    (ii) $C_m \square C_n$, where $m \geq n \geq 3$.

8.8 Applications

There are a number of applications of vertex coloring. These include the allocation of variables to hardware registers during program execution by a computer, and the scheduling of traffic lights. In this section, we shall explain in more detail two other applications namely, scheduling an exam timetable and the storage of combustible chemicals.

Exam Timetable

There are ten students A to J in a postgraduate mathematics program. At the end of the semester, the students are to take exams for the subjects they are enrolled in. A, B and I took Probability; A, D, I and J took Graph Theory; C, D, F and H took Henstock Integration; E and F took Lie Groups; B, E and G took Non-Linear Dynmical Systems; D and J took Topology.

Of course, the exams must be scheduled in such a way that no student has to take two subjects at the same time. In addition, for efficiency, the University would like to minimize the number of time slots. Can we schedule an exam timetable with a minimum number of time slots?

To do this, we will draw a graph to model the situation so that we may assign time slots to the subjects directly on the graph in such a way that no two subjects with a common student are assigned the same time slot. Here are the questions we will consider:

(i) What will the vertices represent?
(ii) Under what condition(s) will two vertices be joined by an edge?
(iii) How can we use vertex coloring to assign time slots to the subjects?

Here are the answers:

(a) Each vertex represents a subject. Thus there will be 6 vertices: Probability (P), Graph Theory (GT), Henstock Integration (H), Lie Groups (L), Non-Linear Dynamical Systems (N), Topology (T).
(b) Two vertices (subjects) are joined by an edge if and only if there is a student taking both the subjects.
(c) A vertex coloring of the resulting graph will ensure that no two adjacent vertices will have the same color. Thus, the color represents a time slot and a viable coloring represents a viable exam time table. The chromatic number of the graph will be the minimum number of time slots required for a viable time table.

Let us solve the example by drawing the graph G, shown in Fig. 8.8.1(a), to model the situation.

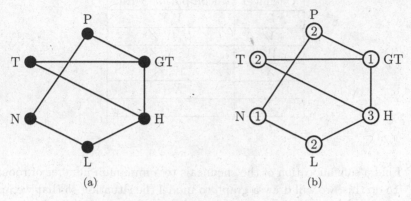

Fig. 8.8.1

As G contains a triangle, $\chi(G) \geq 3$. The 3-coloring of G in Fig. 8.8.1(b) shows that indeed $\chi(G) = 3$. Thus, 3 time slots suffice for a viable exam timetable and we may allocate 1 as, say, Monday morning, color 2 as Tuesday morning and color 3 as Wednesday morning.

Chemical Storage

A common scenario in a factory is the storage of different materials, some of which may be incompatible with others. For example, one would not store gas canisters and matches together.

The case we want to consider here is the storage of chemicals in a factory. Some chemicals when in contact with some other chemicals may result in a combustible or poisonous mixture. These incompatible chemicals ought then to be stored in different rooms. In addition, to optimize the use of space, the factory would like to minimize the number of storerooms needed. Can we allocate the chemicals to a minimum number of storerooms in such a way that chemicals which are combustible together are placed in different rooms?

Let us consider the following concrete example. There are six chemicals U to Z which are used in a factory to manufacture air conditioners. The following table shows which chemicals are incompatible, i.e., they are combustible or poisonous together.

Chemical	Incompatible with:
U	V, Y
V	U, W, Z
W	V, X, Z
X	W, Y
Y	X, U
Z	V, W

Find a safe allocation of the chemicals to a minimum number of rooms.

To do this, we will draw a graph to model the situation so that we may assign rooms to the subjects directly on the graph in such a way that no two incompatible chemicals are assigned to the same room. Here are the questions we will consider:

(i) What will the vertices represent?

(ii) Under what condition(s) will two vertices be joined by an edge?

(iii) How can we use vertex coloring to assign rooms to the chemicals?

Here are the answers:

(a) Each vertex represents a chemical. Thus there will be 6 vertices U to Z.

(b) Two vertices (chemicals) are joined by an edge if and only if they are incompatible.

(c) A vertex coloring of the resulting graph will ensure that no two adjacent vertices will have the same color. Thus, the color represents a room and a viable coloring represents a safe allocation of rooms. The chromatic number of the graph will be the minimum number of rooms required for a safe allocation.

Let us solve the example above by drawing a graph H, shown in Fig. 8.8.2, to model the situation.

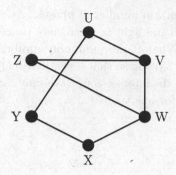

Fig. 8.8.2

As it turns out, the graph H is isomorphic to the graph G in the previous example. Thus, using the argument of the previous example, $\chi(H) = 3$. Hence, 3 rooms suffice for a safe allocation.

Exercise for Section 8.8

1. A chemist wishes to ship chemicals A, B, C, D, W, X, Y, Z using as few containers as possible. Certain chemicals cannot be shipped in the same container since they will react with each other. In particular, any two of the chemicals in each of the following 6 groups

$$\{A, B, C\}, \{A, B, D\}, \{A, B, X\}, \{C, W, Y\}, \{C, Y, Z\}$$

$$\text{and } \{D, W, Z\}$$

react with each other. Draw a graph to model these relations between the chemicals. Use this graph to find the minimum number of containers needed to ship the chemicals. Is it possible to have an allocation of the chemicals that uses the minimum number of containers and such that there are at most two chemicals in each container?

2. The following figure shows the intersection of a major road and a small road. There are 10 traffic lanes, L_1 to L_{10}, along which vehicles approach the intersection. The directions in which vehicles along each of the lanes are allowed to negotiate the intersection and go on to a prescribed exit lane are shown. A traffic light system is installed to control movement through the intersection. The system

consists of a certain number of phases. At each phase, vehicles in lanes for which the light is green may proceed safely through the intersection. What is the minimum number of phases needed for the traffic light system so that (eventually) all vehicles may proceed safely through the intersection? (We may assume that each lane is broad enough for one vehicle at a time.)

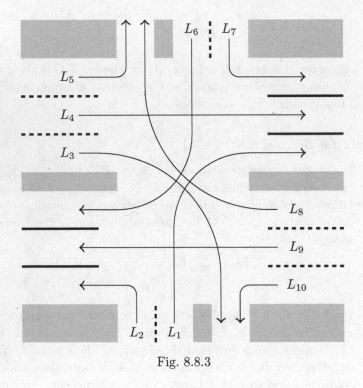

Fig. 8.8.3

3. A Student Council has 8 committees. Ten councilors A, B, C, D, E, F, G, H, I, J are appointed to be members of the committees as shown below:

Publicity:	A, B, C, D
Recreation:	A, E, F, G
Welfare:	G, H, I, J
School Liaison:	C, J
Community:	D, E
Projects:	A, C
Secretariat:	B, F, H
Finance:	G, I

If each committee is scheduled to meet for two hours each week, what is the smallest number of two-hour sessions required to schedule all 8 committee meetings so that each of these councilors is able to attend the meetings of the committees he/she is a member of?

4. A school is preparing a timetable for exams in 7 different subjects, labeled A to G. It is understood that if there is a pupil taking two of these subjects, their exams must be held in different time slots. The table below shows (by crosses) the pairs of subjects which are taken by at least one pupil in common. The school wants to find the minimum number of time slots necessary and also to allocate subjects to the time slots accordingly. Interpreting this problem as a vertex-coloring problem, find the minimum number of time slots needed and a suitable time allocation of the subjects.

	A	B	C	D	E	F	G
A		x	x	x		x	
B	x		x			x	x
C	x	x		x			x
D	x		x		x		
E				x		x	x
F	x	x			x		x
G		x	x		x	x	

5. **The Art Gallery Problem** Consider the following layout of an art gallery, which is an n-sided polygon (here $n = 16$):

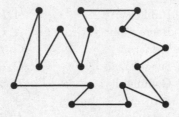

Fig. 8.8.4

A number of guards are to be placed at the vertices of the polygon in such a way that they, as a team, are able to watch the whole gallery. We call them *vertex guards*. An example of placing four vertex guards to do the job is shown in Fig. 8.8.5:

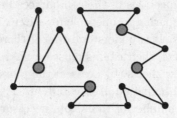

Fig. 8.8.5

In general, what is the minimum number of vertex guards needed to do the job for an n-sided polygon? This is the so-called Art Gallery Problem.

(i) Find the minimum number of vertex guards needed to do the job for the above 16-sided polygon.

(ii) Add new edges within the above polygon to join non-adjacent vertices so that the resulting graph H is a planar graph bounded by the polygon in which every interior region is a triangle. (This process is called a **triangulation** of the polygon.)

(iii) Verify that $\chi(H) = 3$.

(iv) Explain why the team of vertex guards in each of the three color classes obtained in part (iii) is able to do the job.

(v) Let V_1 be a color class obtained in part (iii) with minimum cardinality. Verify that $|V_1| \leq 5$.

(vi) Verify that the minimum number of vertex guards needed to do the job for the following 15-sided polygon is 5.

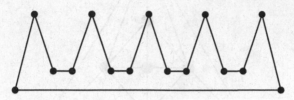

Fig. 8.8.6

Note: The Art Gallery Problem was first asked by Klee in 1973 (see Honsberger (1976)) who further conjectured that for any n-polygon, $\lceil \frac{n}{3} \rceil$ vertex guards are sufficient and sometimes necessary. The conjecture was proved to be true by Chvátal (1975). A simpler and elegant proof of this result using 3-coloring argument as shown above was given by Fisk (1978). The example in part (vi) and its natural extensions show that $\lceil \frac{n}{3} \rceil$ vertex guards are sometimes necessary. Generalizations, extensions and variations of this problem are available in literature.

8.9   Problem Set VIII

1. Let G be a $(10, 40)$-graph.

   (i) What is the maximum possible value of $\chi(G)$?
   (ii) What is the minimum possible value of $\chi(G)$?

2. For each of the following two graphs G, evaluate $\chi(G)$ and show a $\chi(G)$-coloring. Are they critical?

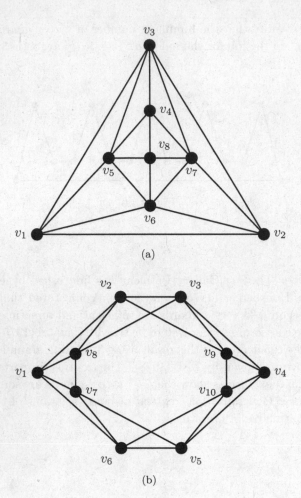

(a)

(b)

Fig. 8.9.1

3. Let G be a connected graph.

   (i) Show that $\chi(G) \leq \beta(G) + 1$, where $\beta(G)$ is the v-covering number of G.

   (ii) For each $k \geq 2$, construct a non-complete connected graph G of order $2k$ such that $\beta(G) = k$ and the equality in (i) holds.

4. Let $G \,\square\, H$ denote the Cartesian product of the graphs G and H.

   (i) Show that $\chi(G \,\square\, H) = \max\{\chi(G), \chi(H)\}$.

   (ii) Is $G \,\square\, H$ critical if both G and H are critical?

5. Let G be a graph of order n and H, an induced bipartite subgraph of order h in G.

   (i) Show that $\chi(G) \leq n - h + 2$.

   (ii) Deduce that $\chi(G) \leq n - \mathrm{diam}(G) + 1$.

6. Let G be an (n, m)-graph.

   (i) Show that $\chi(G) \geq \frac{n^2}{n^2 - 2m}$.

   (ii) Deduce that if G is k-regular, then $\chi(G) \geq \frac{n}{n-k}$.

7. Construct a planar graph G of order n and provide a labeling for its n vertices so that when Algorithm 8.4.1 is applied on G according to this labeling, the number of colors produced is at least 5. What is the least possible value of n?

8. For each $k \geq 1$, construct a tree T and provide a labeling for its vertices so that when Algorithm 8.4.1 is applied on T according to this labeling, the number of colors produced is k.

9. Let G be a graph of order n with non-increasing degree sequence $(d_1, d_2, \cdots, d_n)$. Show that

$$\chi(G) \leq \max\{\min\{i, 1 + d_i\} \mid 1 \leq i \leq n\}.$$

(Welsh and Powell (1967))

10. Let G be a non-empty graph. Show that $\chi(G) \leq \lceil \sqrt{2e(G)} \rceil$.

11. (i) Apply Theorem 8.5.2 to prove the following result:

      If H is a k-critical graph of order n, where $n > k \geq 4$, and H is not complete, then $e(H) \geq \frac{1}{2}(1 + n(k - 1))$.   (*)

   (ii) Assume that the statement (*) holds. Apply (*) to prove Theorem 8.5.2.

12. Show that for any graph G,

$$\chi(G) \leq 1 + \max\{\delta(H) \mid H \text{ is an induced subgraph of } G\}.$$

13. Let G be a k-critical graph, where $k \geq 1$.

   (i) Show that for any two distinct vertices u and v in G, $N(u)$ is never a subset of $N(v)$.

   (ii) Is it possible that $v(G) = k + 1$?

   (iii) For every vertex w in G, show that G admits a k-coloring for which the singleton $\{w\}$ forms a color class.

14. Let G be a graph of order n. Show that

   (i) $2\sqrt{n} \leq \chi(G) + \chi(\overline{G}) \leq n + 1$;

   (ii) $n \leq \chi(G) \cdot \chi(\overline{G}) \leq \left(\frac{n+1}{2}\right)^2$. (Nordhaus and Gaddum (1956))

Note: For any parameter, say ψ, inequalities of the form (i) and (ii) for $\psi(G) + \psi(\overline{G})$ and $\psi(G) \cdot \psi(\overline{G})$ in terms of n are called the **Nordhaus-Gaddum-type** results for ψ.

15. Let $k, \overline{k}$ and n be three positive integers such that $k + \overline{k}$ and $k \cdot \overline{k}$ satisfy

    (i) $2\sqrt{n} \leq k + \overline{k} \leq n + 1$;
    (ii) $n \leq k \cdot \overline{k} \leq \left(\frac{n+1}{2}\right)^2$.

    Show that there exists a graph G of order n such that $\chi(G) = k$ and $\chi(\overline{G}) = \overline{k}$.   (Stewart (1969))

16. Let p and q be two integers with $p \geq 2$ and $q \geq 2$. Let G be a graph of order $1 + (p-1)(q-1)$. Assume that G does not contain K_p as a subgraph. Show that $\chi(\overline{G}) \geq q$.

17. (i) Let G be a graph such that $\chi(G) = k$ and $\omega(G) \leq k - 1$, where $k \geq 3$. Show that $v(G) \geq k + 2$.
    (ii) For each integer $k \geq 3$, construct a graph H such that $v(H) = k + 2$, $\chi(H) = k$ and $\omega(H) = k - 1$.

18. **(Mycielski construction (1955))** Let G be a graph with $V(G) = \{v_1, v_2, \cdots, v_n\}$. Let $M(G)$ denote the graph defined by

$$V(M(G)) = V(G) \cup \{u_1, u_2, \cdots, u_n, w\},$$
$$E(M(G)) = E(G) \cup \{u_i x \mid x \in N_G(v_i), i = 1, 2, \cdots, n\}$$
$$\cup \{u_i w \mid i = 1, 2, \cdots, n\}.$$

For instance, for the graph G shown in Fig. 8.9.2(a), $M(G)$ is shown in Fig. 8.9.2(c).

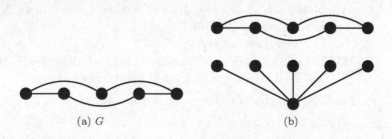

(a) G (b)

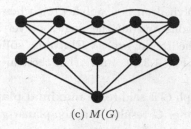

(c) $M(G)$

Fig. 8.9.2

Show that

   (i) G is k-chromatic if and only if $M(G)$ is $(k+1)$-chromatic.
   (ii) G is k-critical if and only if $M(G)$ is $(k+1)$-critical.

Note: We have pointed out in Section 8.3 that for any graph G, $\chi(G) \geq \omega(G)$, where $\omega(G)$ is the clique number of G. Thus $\chi(G) \geq 3$ if G contains a triangle. The bound $\chi(G) \geq \omega(G)$ can be very tight, but it can be arbitrarily bad as well. If we begin with $G = K_2$, and for any $k \geq 1$, apply Mycielski construction inductively k times, we would obtain a critical graph G with $\chi(G) = k$ but $\omega(G) = 2$.

19. Let G be a P_4-free graph (see Problems II.24 and II.25). Show that $\chi(G) = \omega(G)$.

(Seinische (1974))

20. Let G be a 2-connected (n, m)-graph with $n = k+6$, $m = \binom{k}{2} + 7$ and $\chi(G) = k$, where $k \geq 4$.

   (i) Show that $\omega(G) = k$.
   (ii) Determine the structure of G.

(You may wish to apply the result in Problem VIII.11.)

(Koh and Goh (1990))

21. Let G be a graph of order $n \geq 5$ with $\omega(G) = n-3$ and $\chi(G) = n-2$. Prove that $G \cong C_5 + K_{n-5}$.

22. Let G be a connected plane graph having n vertices, m edges and f faces, where $f \geq 2$. The **girth** of G, denoted by g, is the order of a smallest cycle in G. Show that

   (i) $2m \geq gf$;
   (ii) $m(g-2) \leq g(n-2)$.
   (iii) Apply (ii) to show that the Petersen graph is non-planar.

23. A regular polyhedron is a polyhedron whose faces are congruent regular polygons and all the angles at vertices are equal. Five regular polyhedra (known as **Platonic solids**) and their names are shown in Fig. 1.3.10. Prove that these are the only five regular polyhedra.

24. A planar graph G is said to be **maximal planar** if the addition of any new edge of G results in a non-planar graph. An example is given in Fig. 8.9.3.

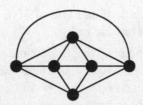

Fig. 8.9.3

Let G be a maximal planar (n, m)-graph with $n \geq 4$. Prove that

   (i) each face in G is bounded by a triangle;
   (ii) $m = 3n - 6$;
   (iii) $\delta(G) \geq 3$; and
   (iv) $3n_3 + 2n_4 + n_5 = 12 + n_7 + 2n_8 + 3n_9 + \cdots + (\Delta - 6)n_\Delta$,
      where n_i denotes the number of vertices of degree i in G, and $\Delta = \Delta(G)$.

25. A planar graph is said to be **outerplanar** if it can be drawn on a plane in such a way that all its vertices lie on the boundary of a face (usually, this face is chosen to be the infinite face). The graph $K_4 - e$ is outerplanar as shown in Fig. 8.9.4(a) and (b). The planar graphs K_4 shown in Fig. 8.9.4(c) and $K(2,3)$ shown in Fig. 8.9.4(d) are not outerplanar. Indeed, K_4 is the non-outerplanar graph of the smallest order while $K(2,3)$ is a non-outerplanar graph of the smallest size.

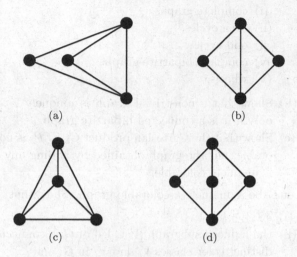

Fig. 8.9.4

Similar to Theorem 8.7.12, we have the following characterization of outerplanar graphs established by Chartrand and Harary (1967).

A graph is outerplanar if and only if it contains no subgraph isomorphic to a subdivision of K_4 or $K(2,3)$.

(i) Show that a graph G is outerplanar if and only if $G + K_1$ is planar.

(ii) Let G be an outerplanar (n,m)-graph. Show that

   (a) $m \leq 2n - 3$;

   (b) $\chi(G) \leq 3$.

26. **[Uniquely colorable graphs]** Let G be a graph with $\chi(G) = k$ and a k-coloring θ. Then, as pointed out in Remark 8.2.9(2), θ induces a partition of $V(G)$ into k color classes $\{V_1, V_2, \cdots, V_k\}$. We say that G is **uniquely k-colorable** if every k-coloring of G induces the same partition. For instance, the graph C_4 is uniquely 2-colorable while C_5 is not uniquely 3-colorable. A graph G is said to be **uniquely colorable** if it is uniquely k-colorable, where $k = \chi(G)$.

   (a) Determine if the following graphs are uniquely colorable:

    (i) complete graphs,

   (ii) even cycles,

  (iii) odd cycles,

  (iv) complete bipartite graphs,

   (v) wheels.

(b) Show that a non-trivial graph is uniquely 2-colorable if and only if it is a connected bipartite graph.

(c) Show that the Cartesian product $C_3 \ \square \ P_2$ is not uniquely colorable; but the graph obtained by adding any new edge to it is uniquely colorable.

27. Let G be a uniquely k-colorable graph. Show that

   (i) $\delta(G) \geq k - 1$;

  (ii) the induced subgraph $[V_i \cup V_j]$ of G is connected for any two distinct color classes V_i and V_j in G; and

  (iii) G is connected.

(Cartwright and Harary (1968))

28. Given a positive integer q, the class of q-**trees** is defined recursively as follows: any complete graph K_q is a q-tree, and any q-tree of order $n + 1$ is a graph obtained from a q-tree G of order n, where $n \geq q$, by adding a new vertex and joining it to each vertex of a K_q in G. Thus, the class of 1-trees is exactly the family of trees. Let G be a q-tree with $v(G) > q$. Show that

   (i) $\chi(G) = q + 1$;

  (ii) G is uniquely colorable.

29. (i) Let G be a uniquely k-colorable (n, m)-graph. Show that

$$m \geq (k - 1)n - \binom{k}{2}.$$

(ii) Construct such graphs G for which the above equality holds.

(Truszczyński (1981) and Xu (1990))

Chapter 9

Domination

9.1 Introduction

In international chess, the most dominant piece is the queen. It can move any number of squares along a row, column or diagonal from the square it is placed (see Fig. 9.1.1). Still, a single queen cannot reach every other square on the 8×8 chessboard from where it sits by a single move. In 1862, Carl Friedrich de Jaenisch asked, 'What is the minimum number of queens to be placed on an 8×8 chessboard so as to ensure that every square (not occupied by a queen) is attacked by at least one queen?'

Fig. 9.1.1

As an approach to solve this problem, a 'Queen Graph' can be drawn.

This is a graph with 64 vertices (representing the 64 squares on the chessboard) where two vertices are adjacent if and only if the corresponding squares can mutually be reached by a single move of a queen. Figure 9.1.2 shows 4 selected vertices and the adjacency among them.

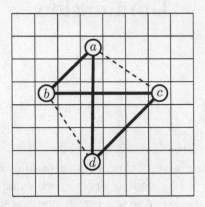

Fig. 9.1.2: Vertex a is adjacent to vertices b and d but not c

Figure 9.1.3 shows two selections of 5 vertices in a Queen Graph such that all other vertices are adjacent to at least one of the 5 selected vertices.

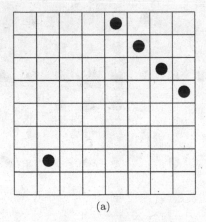

(a)

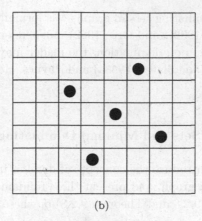

(b)

Fig. 9.1.3

Placing 5 queens on the corresponding squares of the selected vertices will give an upper bound for de Jaenisch's problem. The set of 5 vertices in the Queen Graph is said to be a *dominating set* to convey the idea that all other vertices are 'dominated' by at least one of the 5 vertices.

The same idea of dominating sets in graphs can be used to solve real-world problems. For example, suppose transmitting stations are to be built in some cities in a country so that every city without a transmitting station can receive messages from some adjacent city with one. The problem for selecting the cities for the transmitting stations is that of finding a dominating set in the graph corresponding to the network of cities in the country (see Fig. 9.1.4).

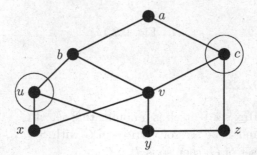

Fig. 9.1.4

The study of dominating sets in graphs was formally initiated by Berge in 1958 (see Berge (1962)) and Ore (1962) independently. For more comprehensive information on domination, the reader may refer to the books: Haynes, Hedetniemi and Slater (1998a) and Haynes, Hedetniemi and Slater (1998b).

9.2 Dominating Sets and Minimal Dominating Sets

In this section, we introduce the concepts of dominating sets and minimal dominating sets of a graph, and present three fundamental results (Theorem 9.2.3, Theorem 9.2.6 and Theorem 9.2.8) on these sets that are due to Ore (1962).

Let $G = (V, E)$ be an (n, m)-graph with vertex set V and edge set E. A set S of vertices in G is called a **dominating set** of G if each vertex in $V \backslash S$ is adjacent to (i.e., dominated by) a vertex in S.

Example 9.2.1. Consider the graph G of Fig. 9.2.1. While the sets $\{u, v, w\}$ and $\{t, y, v\}$ are dominating sets of G, the set $\{t, u, w\}$ is, however, not.

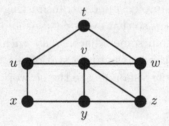

Fig. 9.2.1

Observation 9.2.2.

(1) The vertex set V itself is a dominating set of G.
(2) Let S and T be sets of vertices in G with $S \subseteq T$. If S is a dominating set of G, then so is T.
(3) Let $v \in V$. Then $\{v\}$ is a dominating set of G if and only if $d(v) = n - 1$.
(4) Let H be a spanning subgraph of G. If S is a dominating set of H,

then S is a dominating set of G.

For any graph G containing no isolated vertices, we can always find two dominating sets of G which are disjoint. Indeed, we have:

Theorem 9.2.3. *Let G be a graph with $\delta(G) \geq 1$. Then G possesses a set S of vertices such that both S and $V \backslash S$ are dominating sets of G.*

Proof. We may assume that G is connected. Then G contains a spanning tree, say T. Fix a vertex in T, say w. Let $S = \{x \in V \mid d(x, w)$ is even in $T\}$. Then $V \backslash S = \{x \in V \mid d(x, w)$ is odd in $T\}$ ($\neq \emptyset$). It can be checked readily that both S and $V \backslash S$ are dominating sets of G. $\square$

As indicated in Observation 9.2.2, finding 'large' dominating sets is not interesting. We thus aim at searching for 'small' such sets. A dominating set S of G is said to be **minimal** if no proper subset of S is a dominating set of G.

Example 9.2.4. For the graph G of Fig. 9.2.1, the dominating set $\{u, v, w\}$ is not minimal as it contains a proper subset (i.e., $\{u, v\}$), which is also a dominating set of G. Now, this $\{u, v\}$, together with $\{t, x, z\}$ are, among many others, examples of minimal dominating sets of G.

Observation 9.2.5.

(1) It is clear by definition that every dominating set S of G, if not minimal, would contain

    (i) a vertex v such that $S \backslash \{v\}$ is a dominating set of G and
    (ii) a proper subset that is a minimal dominating set of G.

(2) It follows from (1) that every graph always contains a minimal dominating set.

Minimal dominating sets could be candidates for S that plays in Theorem 9.2.3 as shown in Theorem 9.2.6.

Theorem 9.2.6. *Let G be a graph with $\delta(G) \geq 1$. If S is a minimal dominating set of G, then $V \backslash S$ is also a dominating set of G.*

Proof. Let S be a minimal dominating set of G. Suppose on the contrary that $V \backslash S$ is not a dominating set of G. Then, by definition, there exists a vertex w in S which is not adjacent to any vertex in $V \backslash S$. As $\delta(G) \geq 1$, w is adjacent to a vertex in S. But then this implies that $S \backslash \{w\}$ is a dominating set of G, contradicting the minimality of S.

$\square$

Among the dominating sets of a graph, the minimal ones are, of course, more interesting and more desirable. To end this section, we identify the characteristics of vertices in any minimal dominating set.

Let $G = (V, E)$ be a graph. For $v \in V$, write $N[v] = N(v) \cup \{v\}$.

Fix $S \subseteq V$ and $w \in S$. A vertex v in V (v could be w) is called a **private neighbor** of w (with respect to S) in G if $N[v] \cap S = \{w\}$.

Let us try to understand this definition better.

(1) Suppose $v \in S$. Then the equality $N[v] \cap S = \{w\}$ means that $v = w$ and that w is not adjacent to any vertex in S, that is, $v(= w)$ is an isolated vertex in the induced subgraph $[S]$ of G.

(2) Suppose $v \notin S$. Then the equality $N[v] \cap S = \{w\}$ means that w is the only vertex in S that has v as its neighbor, that is, v is adjacent to w, but not adjacent to any other vertex in S.

Write

$$pn[w, S] = \{v \in V \mid v \text{ is a private neighbor of } w \text{ with respect to } S\}.$$

Example 9.2.7.

(i) Consider the dominating set $S_1 = \{u, v, w\}$ of the graph G of Fig. 9.2.1, which is **not minimal**. Note that $pn[u, S_1] = \{x\}$, $pn[v, S_1] = \{y\}$ and $pn[w, S_1] = \emptyset$.

(ii) Consider the dominating set $S_2 = \{t, x, z\}$ of the graph G of Fig. 9.2.1, which is **minimal**. Note that $pn[t, S_2] = \{t\}$, $pn[x, S_2] = \{x\}$ and $pn[z, S_2] = \{v, z\}$.

As shown in the above example, it seems that the minimality of a dominating set of a graph has something to do with the existence of a private neighbor for each member in the set. Indeed, we have:

Theorem 9.2.8. *Let G be a graph and S a dominating set of G. Then S is a minimal dominating set of G if and only if $pn\,[w, S] \neq \emptyset$ for each $w \in S$.*

Proof. ($\Rightarrow$) Suppose that S is a minimal dominating set of G. Let $w \in S$. We shall show that $pn\,[w, S] \neq \emptyset$. By the minimality of S, the set $S \backslash \{w\}$ is not a dominating set of G. Thus, there exists $v \in V \backslash (S \backslash \{w\})(= (V \backslash S) \cup \{w\})$ which is not adjacent to any vertex in $S \backslash \{w\}$.

Case 1. $v = w$. Then w is not adjacent to any vertex in $S \backslash \{w\}$; that is, w is an isolated vertex in $[S]$. Thus $w \in pn\,[w, S]$, and so $pn\,[w, S] \neq \emptyset$.

Case 2. $v \in V \backslash S$. As S is a dominating set of G, v is adjacent to some vertex in S. Since v is not adjacent to any vertex in $S \backslash \{w\}$, v must be adjacent to w. Thus, w is the only vertex in S that has v as its neighbor. By definition, $v \in pn\,[w, S]$, and so $pn\,[w, S] \neq \emptyset$.

($\Leftarrow$) Assume that $pn\,[w, S] \neq \emptyset$ for each w in the dominating set S. We shall prove that S is minimal.

Suppose on the contrary that S is not minimal. Then there exists a vertex u in S such that $S \backslash \{u\}$ is a dominating set of G. It follows that u is adjacent to some vertex in $S \backslash \{u\}$, and so u is not an isolated vertex in $[S]$. Since $pn\,[u, S] \neq \emptyset$ by assumption, u has a private neighbor in $V \backslash S$, say v. By definition, v is not adjacent to any vertex in $S \backslash \{u\}$. This, however, contradicts the fact that $S \backslash \{u\}$ is a dominating set of G. Hence, S is a minimal dominating set of G, as required. $\qquad\square$

Question 9.2.9. *Let G be the graph of Fig. 9.2.2 and $S = \{a, c, g\}$.*

(i) *Is S a dominating set of G?*

(ii) *Find $pn\,[a, S]$, $pn\,[c, S]$ and $pn\,[g, S]$.*

(iii) *Is S a minimal dominating set of G?*

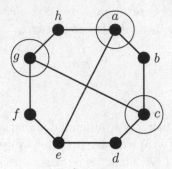

Fig. 9.2.2

Exercise for Section 9.2

1. Let G be the following graph:

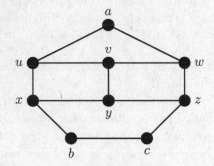

Fig. 9.2.3

   (i) Show that G contains no dominating set of size 2.

   (ii) Let $S = \{u, x, z\}$. Is S a dominating set of G? Find $pn\,[u, S]$, $pn\,[x, S]$ and $pn\,[z, S]$. Is S a minimal dominating set of G?

  (iii) Find all minimal dominating sets A of G with $|A| = 3$.

2. Let S and T be sets of vertices in a graph G with $S \subseteq T$. Show that if S is a dominating set of G, then so is T.

3. Let H be a spanning subgraph of a graph G, and $S \subseteq V(H)$.

   (i) Show that if S is a dominating set of H, then S is a dominating set of G.

(ii) Assume that S is a minimal dominating set of H. Is S a minimal dominating set of G?

4. Let S be a dominating set of a graph G. Show that if S is independent, then S is a minimal dominating set of G.

5. For any two integers p, q with $p > q > 2$, construct a connected graph H such that

$$\{|A| \mid A \text{ is a minimal dominating set in } H\} = \{2, p, q\}.$$

6. Let S be a dominating set of the following graph G $(= P_3 \square P_3)$:

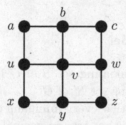

Fig. 9.2.4

(a) Show that $|S| \geq 3$.
(b) Assume that S is a minimal dominating set of G.

   (i) Show that $|S| \leq 5$.
   (ii) List all such S with $3 \leq |S| \leq 5$.

7. Let G be a graph. A set S of vertices in G is said to be **irredundant** if $pn\,[x, S] \neq \emptyset$ for every x in S. An irredundant set of G is said to be **maximal** if it is not a proper subset of any other irredundant set of G.

   (i) Find all maximal irredundant sets that contain b in the following graph:

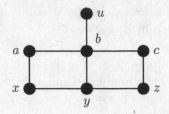

Fig. 9.2.5

(ii) Show that if S is an irredundant set in a graph G, then every non-empty subset of S is also an irredundant set of G.

(iii) Is it true that every maximal irredundant set of G is a dominating set of G?

(iv) Show that every minimal dominating set of G is a maximal irredundant set of G.

9.3 Domination Number

Let G be a graph. A dominating set S of G is said to be **minimum** if $|S| \leq |S'|$ for any dominating set S' of G.

It is clear by definition that every minimum dominating set of G is also minimal, but the converse is not true in general.

Example 9.3.1. Let us revisit Example 9.2.7(ii). The dominating set $S_2 = \{t, x, z\}$ of the graph G of Fig. 9.2.1 is indeed minimal; but not minimum as there exists in G a dominating set (for instance, $\{u, v\}$) of smaller size. It is clear that no single vertex in G can form by itself a dominating set of G. Thus, $\{u, v\}$ is a minimum dominating set of G.

Question 9.3.2. *Let G be the graph of Fig. 9.2.2 and $S = \{a, c, g\}$.*

(i) *Does G contain a dominating set of size 2?*

(ii) *Is S a minimum dominating set of G?*

(iii) *Is it possible to find a minimum dominating set A of G such that $A \cap S = \emptyset$?*

Minimal dominating sets of a graph may have different cardinalities. Those having the smallest cardinality are the minimum dominating sets of the graph. We give a name and assign a notation to this unique smallest number for a given graph.

The **domination number** of a graph G, denoted by $\gamma(G)$, is defined as $|S|$, where S is a **minimum** dominating set of G; that is,

$$\gamma(G) = \min\{|D| \mid D \text{ is a dominating set in } G\}.$$

For simplicity, we call a minimum dominating set of G a γ-**set** of G.

Example 9.3.3.

(i) For the graph G of Fig. 9.2.1, $\gamma(G) = 2$; and $\{u, v\}$, $\{u, z\}$ and $\{t, y\}$ are some γ-sets of G.

(ii) For the graph G of Fig. 9.2.2, $\gamma(G) = 3$; and $\{a, c, g\}$ and $\{a, d, f\}$ are some γ-sets of G.

Example 9.3.4.

(i) Let G be a graph of order $n \geq 2$. Trivially, $1 \leq \gamma(G) \leq n$. These two bounds are attainable. Indeed,

(1) $\gamma(G) = 1$ if and only if $\Delta(G) = n - 1$; and

(2) $\gamma(G) = n$ if and only if G is an empty graph.

(ii) For $1 \leq p \leq q$, $\gamma(K(p,q)) = 1$ if $p = 1$; and $\gamma(K(p,q)) = 2$ otherwise.

(iii) $\gamma(P_n) = \lceil \frac{n}{3} \rceil$ for all $n \geq 1$.

(iv) $\gamma(C_n) = \lceil \frac{n}{3} \rceil$ for all $n \geq 3$.

We note that the path P_n is a spanning subgraph of the cycle C_n, and observe in the above example that $\gamma(P_n) = \gamma(C_n)$ for all $n \geq 3$. On the other hand, take the complete graph K_n and a spanning subgraph, say C_n, where $n \geq 4$; then we have, from the above example again, that $\gamma(C_n) = \lceil \frac{n}{3} \rceil \geq 2 > 1 = \gamma(K_n)$.

In general, if H is a spanning subgraph of a graph G, is there any relation between $\gamma(H)$ and $\gamma(G)$? We have the following (see Observation 9.2.2(4) and Problem 9.3.1):

Observation 9.3.5. If H is a spanning subgraph of a graph G, then $\gamma(H) \geq \gamma(G)$.

$\square$

In general, the problem of determining $\gamma(G)$ for an arbitrary graph G is very difficult. In view of this, researchers have focused their attention on finding good bounds for it. Some results along this line will be presented here. We assume the trivial condition that the graphs considered contain no isolated vertices.

The upper bounds of the first type are those in terms of order n and minimum degree δ. Our first one follows immediately from Theorem 9.2.6.

Theorem 9.3.6. *Let G be a graph of order n with $\delta(G) \geq 1$. Then* $\gamma(G) \leq \lfloor \frac{n}{2} \rfloor$.

Proof. Let S be a γ-set of G. By Theorem 9.2.6, $V \backslash S$ is also a dominating set of G. Thus,

$$n = |S| + |V \backslash S| = \gamma(G) + |V \backslash S| \geq \gamma(G) + \gamma(G) = 2\gamma(G),$$

and the inequality follows.

$\square$

Remark 9.3.7.

(1) The bound for γ given in Theorem 9.3.6 is attainable. Suppse that $n \geq 2$ is given. We assume first that n is even and write $n = 2k$. Let H be any connected graph of order k with $V(H) = \{v_1, v_2, \cdots, v_k\}$, say. Construct a new graph H^* based on H as follows:

$$V(H^*) = V(H) \cup \{u_1, u_2, \cdots, u_k\} \text{ and}$$

$$E(H^*) = E(H) \cup \{u_i v_i \mid i = 1, 2, \cdots, k\}.$$

It is easy to see that H^* is a connected graph of order n and $\gamma(H^*) = k = \frac{n}{2}$ (the graph H^* is often called the **corona** of H; see Fig. 9.3.1).

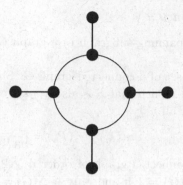

Fig. 9.3.1

If n is odd, say $n = 2k + 1$, then we add another new vertex w to H^* and join it to a vertex in H. The resulting graph is a desired one.

(2) The structure of graphs G of even order n with $\delta(G) \geq 1$ such that $\gamma(G) = \frac{n}{2}$ was determined independently by a number of researchers. The result is as follows: Let G be a graph of even order n with $\delta(G) \geq 1$. Then $\gamma(G) = \frac{n}{2}$ if and only if each component of G is either a C_4 or isomorphic to H^* for some connected graph H.

(3) The graphs G of order n with $\delta(G) \geq 1$ such that $\gamma(G) = \lfloor \frac{n}{2} \rfloor$ were characterized by Cockayne, Haynes and Hedetniemi (Extremal graphs for inequalities involving domination parameters, *preprint* (1996)).

(4) While McCuaig and Shepherd (1989) showed that if $\delta(G) \geq 2$, then $\gamma(G) \leq \frac{2n}{5}$, Reed (1996) showed that if $\delta(G) \geq 3$, then $\gamma(G) \leq \frac{3n}{8}$.

(5) Haynes, Hedetniemi and Slater (1998a) conjectured that for any graph G of order n with $\delta(G) \geq k \geq 4$,

$$\gamma(G) \leq \frac{kn}{3k - 1}.$$

This conjecture has been proved to be true by a number of researchers for different cases of 'k' as shown below:

(i) $k \geq 7$ (Caro and Roditty (1990)),
(ii) $k = 4$ (Sohn and Yuan (2009)),
(iii) $k = 5$ (Xing, Sun and Chen (2006)),
(iv) $k = 6$ (Cao, Shi, Sohn and Yuan (2008)).

Exercise for Section 9.3

1. Let H be a spanning subgraph of a graph G. Show that $\gamma(H) \geq \gamma(G)$.

2. Let S be a γ-set of a connected graph G. Show that G contains a spanning tree T such that S is also a γ-set of T.

3. Show that for all $n \geq 3$,
$$\gamma(C_n) = \gamma(P_n) = \lceil \frac{n}{3} \rceil.$$

4. Let G be a connected graph of order $n \geq 2$. Example 9.3.4(i)(1) states that $\gamma(G) = 1$ if and only if $\Delta(G) = n - 1$. What can be said about G with $\gamma(G) = 2$?

5. Let G be a graph with $\operatorname{diam}(G) = 2$. Show that

   (i) for each vertex v in G, the set $N(v)$ is a dominating set in G;

   (ii) $\gamma(G) \leq \delta(G)$.

6. Let G be a connected graph of order $n \geq 3$. Show that $\gamma(G) \leq n - \kappa(G)$.

7. Let G be an (n, m)-graph.

   (i) Prove that $\gamma(G) \geq n - m$.

   (ii) Characterize G such that $\gamma(G) = n - m$.

8. Vizing (1965) proved the following result: If G is an (n, m)-graph such that $\gamma(G) \geq 2$, then
$$2m \leq (n - \gamma(G))(n - \gamma(G) + 2).$$

Applying this result, or otherwise, show that
$$\gamma(G) \leq n + 1 - \sqrt{1 + 2m}.$$

9. Let G be a graph of order $n \geq 2$.

   (i) Show that
$$\lceil \frac{n}{1 + \Delta(G)} \rceil \leq \gamma(G) \leq n - \Delta(G).$$

   (ii) For each $n \geq 2$, construct a graph G for which the left-hand equality in (i) holds.

   (iii) For each $n \geq 2$, construct a graph G for which the right-hand equality in (i) holds.

10. Let G be a graph with $\delta(G) \geq 1$. Let M be a maximum matching in G. Show that there exists a dominating set S of G such that $|S| = |M|$. Deduce that $\gamma(G) \leq \alpha'(G)$.

11. Let G be a disconnected graph. Show that $\gamma(\overline{G}) \leq 2$.

12. Let G be a graph such that $\min\{\gamma(G), \gamma(\overline{G})\} \geq 3$. Show that G is connected. What is the value of $\mathrm{diam}(G)$?

13. Let G be a graph of order $n \geq 2$. Show that

   (i) $\gamma(G) + \gamma(\overline{G}) \leq n + 1$;

   (ii) $\gamma(G) \cdot \gamma(\overline{G}) \leq n$.

(Jaeger and Payan (1972))

9.4 Independent Domination

Much research on domination has been carried out by imposing additional conditions on a dominating set of a graph. In this section, we study one such variation, known as **independent domination**, which was introduced by Cockayne and Hedetniemi (1974).

Let us recall the two dominating sets of 5 queens shown in the Queen Graph of Fig. 9.1.3. The special feature of the second dominating set (Fig. 9.1.3(b)) is that no two of the 5 vertices are adjacent. Thus, this dominating set is also an independent set. Independence among members of the dominating set may be an important feature in some applications. For example, in the Transmitting Stations Problem, the transmitting stations should not be placed in adjacent cities so that there would be no interference should the same frequency be used.

Let G be a graph and $S \subseteq V$. Recall that S is an **independent set** of G if $[S]$ is an empty graph; that is, no two vertices in S are adjacent in G. The set S is called an **independent dominating set** of G if S is both an **independent** set and a **dominating** set of G.

Clearly, every subset of an independent set of G is also independent. An independent set of G is said to be **maximal** if it is not a proper subset of an independent set of G.

Example 9.4.1. Some of the independent sets of the graph G of Fig. 9.4.1 are shown below:

$$\{a\}, \{a, u\}, \{a, u, v\}, \{a, u, v, z\}, \{c, u\}, \{a, x, y\}, \{b, c, x, y\}.$$

Clearly, only the last four independent sets are **maximal**. Note that they are also **dominating** sets of G.

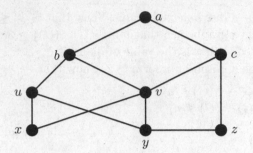

Fig. 9.4.1

Theorem 9.4.2.   (Berge (1973)) *Let S be an independent set of a graph G. Then S is a maximal independent set of G if and only if S is a dominating set of G.*

Proof. ($\Rightarrow$) Assume that S is a maximal independent set of G. We shall show that S is a dominating set of G. Suppose not; then there is a vertex, say w, in $V \backslash S$ which is not adjacent to any vertex in S. This, however, implies that $S \cup \{w\}$ is an independent set of G, contradicting the maximality of S. Thus, S is a dominating set of G, as required.

($\Leftarrow$) Assume that S is an independent dominating set of G. We shall show that S is a maximal independent set of G. Suppose not; then there is a vertex, say u, in $V \backslash S$ such that $S \cup \{u\}$ is an independent set of G. This, however, implies that u is not adjacent to any vertex in S, and so S is not a dominating set of G, a contradiction. Thus, S is a maximal independent set of G, as asserted.

□

It follows from Theorem 9.4.2 that every maximal independent set of G is a dominating set of G. Thus, we define the **independent domination number** of a graph G, denoted by $i(G)$, as follows:

$$i(G) = \min\{|S| \mid S \text{ is a maximal independent set of } G\}.$$

For simplicity, we call a maximal independent set of G with **minimum** cardinality an *i*-**set** of G. Equivalently, an *i*-**set** is an independent dominating set of G with minimum cardinality.

Example 9.4.3.

   (i) $i(O_n) = \gamma(O_n) = n$ and $i(K_n) = \gamma(K_n) = 1$.
   (ii) For the graph G of Fig. 9.2.1, $i(G) = \gamma(G) = 2$.
  (iii) For the graph G of Fig. 9.4.1, $i(G) = \gamma(G) = 2$.
  (iv) For the graph G of Fig. 9.4.2, $\gamma(G) = 3$ while $i(G) = 7$.

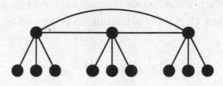

Fig. 9.4.2

Since every i-set of G is both a dominating set and an independent set of G, we have:

$$\gamma(G) \leq i(G) \leq \alpha(G).$$

As shown in Example 9.4.3, there are graphs G for which $\gamma(G) = i(G)$ and there are graphs G for which $i(G) - \gamma(G)$ can be arbitrarily large.

Call a graph G a $[\gamma, i]$-**graph** if $\gamma(G) = i(G)$. Perhaps, the first non-trivial family of $[\gamma, i]$-graphs was found by Hedetniemi and Mitchell (1977) who showed that the line graph (see Problem II.11) of any tree is such a graph. Allan and Laskar (1978) generalized it by giving in the next theorem a sufficient condition for G to be a $[\gamma, i]$-graph.

A graph is said to be **claw-free** if it doesn't contain $K(1,3)$ as an induced subgraph (see also Problem VII.11).

Theorem 9.4.4.   *Every claw-free graph is a $[\gamma, i]$-graph.*

   □

As a very special case, we see that every graph G with $\Delta(G) \leq 2$ (i.e., each component of G is either a path or a cycle) is a $[\gamma, i]$-graph.

Extending Theorem 9.4.4, Bollobás and Cockyane (1979) proved the following interesting result.

Theorem 9.4.5.   *Let G be a graph which does not contain $K(1, h + 1)$ as an induced subgraph, where $h \geq 2$. Then $i(G) \leq (h - 1)\gamma(G) - (h - 2)$.*

Proof. Let S be a γ-set of G (thus $\gamma(G) = |S|$) and A be a maximal independent set of $[S]$. Clearly, $A \neq \emptyset$. If $A = S$, then A is an i-set and $\gamma(G) = |S| = |A| = i(G)$; and the inequality holds trivially. Thus, assume that $S \backslash A \neq \emptyset$. Note that each vertex in $S \backslash A$ is adjacent to at least one vertex in A (why?). Extend A to form a maximal independent set B of G. As S is a dominating set, each vertex in $B \backslash A$ is adjacent to a vertex in $S \backslash A$. On the other hand, as G contains no $K(1, h + 1)$ as an induced subgraph, each vertex in $S \backslash A$ is adjacent to at most $h - 1$ vertices in $B \backslash A$. Thus,

$$|B \backslash A| \leq (h - 1)|S \backslash A|,$$

and we have

$$
\begin{aligned}
i(G) &\leq |B| \\
&= |A| + |B \backslash A| \\
&\leq |A| + (h - 1)|S \backslash A| \\
&= |A| + (h - 1)(|S| - |A|) \\
&= (h - 1)|S| - (h - 2)|A| \\
&\leq (h - 1)\gamma(G) - (h - 2),
\end{aligned}
$$

as was to be shown.

$\square$

Remark 9.4.6. Consider the very special case of Theorem 9.4.5 when $h = 2$. In this case, the inequality becomes $i(G) \leq \gamma(G)$ (and so $i(G) = \gamma(G)$) and Theorem 9.4.4 follows.

It is generally believed that a complete characterization of $[\gamma, i]$-graphs is perhaps impossible. This is, however, not the case for $[\gamma, i]$-trees. The first characterization of $[\gamma, i]$-trees was found in Harary and Livingston (1986), but their characterization is rather complicated. Cockayne et al. (2000) gave a simpler one, and Dorfling et al. (2006) provided a simple constructive characterization of such trees.

Exercise for Section 9.4

1. For each of the following graphs G, find the values of $\gamma(G)$, $i(G)$ and $\alpha(G)$.

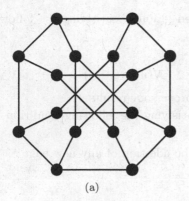
(a)

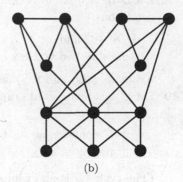
(b)

Fig. 9.4.3

2. For each of the following graphs G, find the value of $i(G)$:

    (i) P_n, $n \geq 2$;   (iii) $K(p, q)$, $p \geq q \geq 1$;

    (ii) C_n, $n \geq 3$;   (iv) the Petersen graph.

3. Let H be a spanning subgraph of a graph G. It is known that $\gamma(H) \geq \gamma(G)$. Is it true that $i(H) \geq i(G)$?

4. Let G be a bipartite graph of order $n \geq 2$ with $\delta(G) \geq 1$. Show that $i(G) \leq \frac{n}{2}$.

5. Let G be a connected bipartite graph of even order $n \geq 2$ with bipartition (X, Y). Show that $i(G) = \frac{n}{2}$ if and only if, for any $S \subseteq X$,

$$|\{v \in X \mid N(v) \subseteq N(S)\}| \geq |N(S)|.$$

(Ma and Chen (2004))

6. Let G be a graph of order $n \geq 2$ with maximum degree Δ.

    (i) Show that $\gamma(G) \leq i(G) \leq n - \Delta$.

    (ii) Suppose that $i(G) = n - \Delta$ and let w be a vertex in G with $d(w) = \Delta$. Show that $V \backslash N[w]$ is an independent set in G.

    (iii) Show that $i(G) \geq \frac{n}{1+\Delta}$, and that the bound is sharp.

7. Let p, q and r be integers such that $2 \leq p \leq q \leq r$. Construct a graph G such that $(\gamma(G), i(G), \alpha(G)) = (p, q, r)$.

8. Let G be a k-regular connected graph of order n. Show that

$$\frac{n}{k+1} \leq i(G) \leq \alpha(G) \leq \frac{n}{2}.$$

9. Let S be a γ-set of a connected graph G satisfying the following conditions:

   (i) $|S| \geq 2$ and
   (ii) for any two vertices $u, v \in S$, $N(u) \cap (V \backslash S) = N(v) \cap (V \backslash S)$.

   Evaluate $\gamma(G)$ and $i(G)$, and prove your results.

10. Let G be a connected graph satisfying the following condition

   > G contains no C_3 but the addition of any new edge produces a C_3. $(*)$

   (The cycle C_5 is an example.)

   (i) Show that $i(G) \leq \delta(G) \leq \frac{n}{2}$.
   (ii) For each $k \geq 3$, construct a connected graph H satisfying $(*)$ such that $i(H) = k = \delta(H)$.

   (Wang (2008))

11. Let G be a graph of order $n \geq 2$. Prove that $3 \leq i(G) + i(\overline{G}) \leq n+1$. Show that the above bounds are sharp for each $n \geq 2$.

9.5   Roman Domination

In his article 'Defend the Roman Empire!' (*Scientific American*, 281(6) (1999), 136-138), Ian Stewart discussed a strategy of the Emperor Constantine for defending the Roman Empire in the 4th century A.D. The empire then stretched from Britain in the north to Egypt in the south, and Iberia (Spain) in the west to Asia Minor in the east. How could such a vast empire be protected with the limited number of legions that Constantine had?

Constantine had 8 states which he had to protect and in which he could place his limited number of military legions. The rules by which he placed his legions are as follows:

- A state is secured if it is occupied by at least one legion.
- A state is unsecured if no legions are stationed there.
- An unsecured state is securable if a legion can be moved to it from an adjacent state.
- At least two legions must occupy a state before a legion can be moved out of it (that is, at least one legion must remain behind).
- The empire is safe if all unsecured states are securable.

Figure 9.5.1 shows a map of the Roman Empire at that time with the 8 states. Superimposed on the map is a graph with vertices representing the states and where two vertices are adjacent if and only if the states are considered adjacent. Under the above rules set by Constantine, what is the least number of legions needed to ensure that the empire is safe?

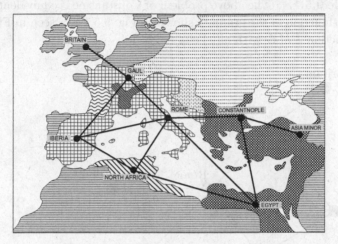

Fig. 9.5.1

Motivated by this article, Cockayne et al. (2004) introduced the following notion of Roman domination in graphs.

Let $G = (V, E)$ be a graph. A function $f : V \to \{0, 1, 2\}$ is called a **Roman dominating function** (or **RDF**) of G if every vertex v with $f(v) = 0$ is adjacent to at least one vertex w with $f(w) = 2$. The weight of f, denoted by $w(f)$, is defined as

$$w(f) = \sum \{f(v) \mid v \in V\}.$$

For $i = 0, 1, 2$, let

$$V_i = \{v \in V \mid f(v) = i\}.$$

As there is a one-one correspondence between the set of mappings $f : V \to \{0, 1, 2\}$ and the family of ordered partitions (V_0, V_1, V_2) of V, we may write $f = (V_0, V_1, V_2)$. Thus, $f = (V_0, V_1, V_2)$ is an RDF of G if every vertex in V_0 is dominated by a vertex in V_2. Note also that

$$w(f) = |V_1| + 2|V_2|.$$

The **Roman domination number** of G, denoted by $\gamma_R(G)$, is defined as

$$\gamma_R(G) = \min\{w(f) \mid f = (V_0, V_1, V_2) \text{ is an RDF of } G\}.$$

We call f a γ_R-**function** of G if f is an RDF of G and $w(f) = \gamma_R(G)$.

Example 9.5.1.   The above problem of Constantine is equivalent to finding a γ_R-function of the graph G shown in Fig. 9.5.1. As seen in Fig. 9.5.2, a γ_R-function f of G with $w(f) = 4$ is given.

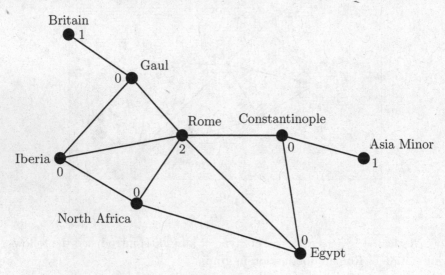

Fig. 9.5.2

Observation 9.5.2.   Let G be a graph of order $n \geq 1$.

    (i) If H is a spanning subgraph of G, then $\gamma_R(H) \geq \gamma_R(G)$.

    (ii) $\gamma_R(G) = 1$ if and only if $n = 1$.

    (iii) $\gamma_R(G) = 2$ if and only if $G = O_2$ or $\Delta(G) = n - 1$.

    (iv) $\gamma_R(P_4) = \gamma_R(C_4) = 3$, $\gamma_R(P_5) = \gamma_R(C_5) = 4$, $\gamma_R(P_6) = \gamma_R(C_6) = 4$, and in general, $\gamma_R(P_n) = \gamma_R(C_n) = \lceil \frac{2n}{3} \rceil$, for $n \geq 3$.

The following result can be found in Cockayne et al. (2004).

Theorem 9.5.3.   *For any graph G, $\gamma(G) \leq \gamma_R(G) \leq 2\gamma(G)$.*

Proof. Let $f = (V_0, V_1, V_2)$ be any γ_R-function of G. Observe that $V_1 \cup V_2$ is a dominating set of G. Thus,

$$\gamma(G) \le |V_1 \cup V_2| \le |V_1| + 2|V_2| = w(f) = \gamma_R(G).$$

On the other hand, let D be a γ-set of G. Then $f = (V \backslash D, \emptyset, D)$ is an RDF of G. Thus,

$$\gamma_R(G) \le w(f) = 2|D| = 2\gamma(G).$$

$\square$

Remark 9.5.4.

(1) In Theorem 9.5.3, the lower bound of $\gamma_R(G)$ can be achieved. Indeed, $\gamma(G) = \gamma_R(G)$ if and only if $G = O_n$.
(2) A graph G is called a **Roman** graph if $\gamma_R(G) = 2\gamma(G)$ (that is, the right-hand equality in Theorem 9.5.3 holds). It can be checked that the following graphs are Roman:

    (i) any graph G with $\Delta(G) = n - 1$;
    (ii) $P_{3k}, C_{3k}, P_{3k+2}, C_{3k+2}$;
    (iii) $K(p, q)$, where $\min\{p, q\} \ne 2$.

We end this section by stating the following results which were established by Chambers et al. (2009).

Theorem 9.5.5. *For any tree T of order $n \ge 1$, $\gamma_R(T) \le \frac{4n}{5}$.*

$\square$

By Observation 9.5.2(i), we have:

Corollary 9.5.6. *For any connected graph G of order $n \ge 1$,*

$$\gamma_R(G) \le \frac{4n}{5}.$$

$\square$

The upper bound in Theorem 9.5.5 is sharp. An example is shown in Fig. 9.5.3.

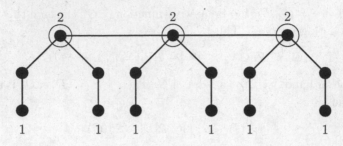

Fig. 9.5.3

Exercise for Section 9.5

1. Find the values of $\gamma(G)$, $i(G)$ and $\gamma_R(G)$ for each of the following graphs G:

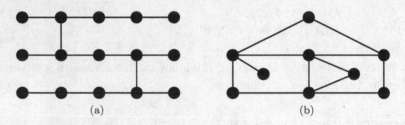

(a)   (b)

Fig. 9.5.4

2. For $n \geq 1$, let Q_n be the n-cube graph. Find the values of $\gamma(Q_n)$, $i(Q_n)$ and $\gamma_R(Q_n)$, where $1 \leq n \leq 4$.

3. Let $f = (V_0, V_1, V_2)$ be a γ_R-function for a graph G of order $n \geq 2$. Show that

    (i) $\Delta([V_1]) \leq 1$;

    (ii)   no edge in G joins V_1 and V_2;

    (iii)   each vertex in V_0 is adjacent to at most two vertices in V_1;

    (iv) V_2 is a γ-set of $[V_0 \cup V_2]$.

(For Problems 9.5.3 to 9.5.5, see Cockayne et al. (2004).)

4. Let G be a graph. Show that G is Roman if and only if G has a γ_R-function $f = (V_0, V_1, V_2)$ with $V_1 = \emptyset$.

5. Show that for any graph G of order $n \geq 2$,

$$\gamma_R(G) \geq \frac{2n}{\Delta(G) + 1}.$$

For each $n \geq 2$, construct a graph G of order n for which the above equality holds.

6. Let G be a non-empty graph of order n. Show that

   (i) there exists a γ_R-function $f = (V_0, V_1, V_2)$ of G such that $V_2 \neq \emptyset$;
   (ii) $\gamma_R(G) = \gamma(G)$ if and only if $G = O_n$;
   (iii) $\gamma_R(G) = 2$ if and only if $\Delta(G) = n - 1$;
   (iv) $\gamma_R(G) = 3$ if and only if $\Delta(G) = n - 2$, where $n \geq 3$;
   (v) if $\gamma_R(G) = 4$, then there exist two vertices u and v such that $d(u) + d(v) \geq n - 2$.

7. Show that $\gamma_R(P_n) = \gamma_R(C_n) = \lceil \frac{2n}{3} \rceil$, for $n \geq 3$.

8. Let $G = C_8 \,\square\, P_2$. Find the values of $\gamma(G)$ and $\gamma_R(G)$. Is G a Roman graph?

9. Let $f = (V_0, V_1, V_2)$ be a γ_R-function of a graph G of order $n \geq 2$. Show that

   (i) $\gamma_R(G) \geq \gamma(G) + |V_2|$;
   (ii) $\gamma_R(G) \geq 2\gamma(G) - |V_1|$.

(Yero and Rodriguez-Velazquez (2013))

9.6 Problem Set IX

1. (i) Let G be a graph of order $n \geq 2$ such that $\delta(G) \geq \frac{n}{2}$. Is it true that any two adjacent vertices in G form a minimal dominating set in G?

   (ii) Let G be a graph. Show that any two adjacent vertices in G form a minimal dominating set in G if and only if $G \cong K(p_1, p_2, \cdots, p_r)$, where $p_i \geq 2$ for each $i = 1, 2, \cdots, r$.

   (iii) Let G be a graph of order $n \geq 2$. Show that any two vertices in G form a minimal dominating set in G if and only if $n = 2r$ and G is the complete r-partite graph $K(2, 2, \cdots, 2)$.

(Jayaram (1997))

2. Let G be a graph with vertex set V and $\delta(G) \geq 1$.

   (i) Show that G contains a γ-set S such that every vertex in S has a private neighbor (with respect to S) in $V \backslash S$.

(ii) Show that there exists a matching M in G such that every vertex in S is M-saturated.

(iii) Deduce that $\gamma(G) \le \alpha'(G)$. (See Problem 9.3.10.)

3. Let G be a connected graph of order n with girth $g(G)$. Prove that

(i) if $\delta(G) \ge 2$ and $g(G) \ge 5$, then $\gamma(G) \le \lceil \frac{n}{2} - \frac{g(G)}{6} \rceil$;

(ii) if $g(G) \ge 5$, then $\gamma(G) \ge \delta(G)$;

(iii) if $g(G) \ge 6$, then $\gamma(G) \ge 2(\delta(G) - 1)$;

(iv) if $\delta(G) \ge 2$ and $g(G) \ge 7$, then $\gamma(G) \ge \Delta(G)$.

(Brigham and Dutton (1990))

Note: Löwenstein and Rautenbach (2008) proved the following interesting deep result: If G is a connected graph of order n with $\delta(G) \ge 2$ and $g(G) \ge 5$, then

$$\gamma(G) \le n \left(\frac{1}{3} + \frac{2}{3g(G)} \right).$$

4. Let G be a graph of order $n \ge 2$ without isolated vertices.

(i) It is known that $\gamma(G) \le \frac{n}{2}$. Is it true that $i(G) \le \frac{n}{2}$?

(ii) Prove that $\gamma(G) + i(G) \le n$. Show that the bound is sharp for each $n \ge 2$.

5. Construct a 3-regular graph H with $g(H) = 5$ and $\gamma(H) = \delta(H)$.

6. Show that for each $p \ge 1$ and $q \ge 1$, the graph $C_{5p} \,\square\, C_{5q}$ is Roman.

(Fu, Yang and Jiang (2009))

7. Let G be a graph of order $n \ge 2$. A Roman dominating function $f = (V_0, V_1, V_2)$ of G is called an **independent Roman dominating function** (IRDF) if $V_1 \cup V_2$ is an independent set. The **independent Roman domination number** of G, denoted by $i_R(G)$, is the minimum weight of an IRDF of G. By definition, $\gamma_R(G) \le i_R(G)$.

(a) Evaluate $i(T), i_R(T)$ and $\gamma_R(T)$ of the following tree T:

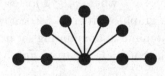

Fig. 9.6.1

(b) Show that

(i) $\gamma_R(G) = i_R(G)$ if and only if there exists a γ_R-function $f = (V_0, V_1, V_2)$ of G such that V_2 is independent;

(ii) $\gamma_R(P_n) = i_R(P_n) = \gamma_R(C_n) = i_R(C_n)$;

(iii) $\gamma_R(P_n \,\square\, P_2) = i_R(P_n \square P_2)$;

(iv) $i_R(K(p,q)) = \min\{p, q\} + 1$.

(c) Show that

(i) $i(G) \le i_R(G) \le 2i(G)$;

(ii) $i_R(G) = i(G)$ if and only if G is an empty graph;

(iii) $i_R(G) = i(G) + 1$ if and only if G has a vertex of degree $n - i(G)$.

(Adabi, Targhi, Rad and Moradi (2012))

8. Show that for all $n \ge 1$,

(i) $\gamma(P_2 \,\square\, P_n) = \lceil \frac{n+1}{2} \rceil$;

(ii) $\gamma(P_3 \,\square\, P_n) = \lceil \frac{3n+1}{4} \rceil$.

Note: For the complete evaluation of $\gamma(P_m \,\square\, P_n)$, see Goncalves et al. (2011).

9. Evaluate $\gamma(P_5 \,\square\, P_7)$ and $i(P_5 \,\square\, P_7)$.

10. [**Vizing's Conjecture**] Vizing (1968) made the following conjecture: For any two graphs G and H,

$$\gamma(G \,\square\, H) \ge \gamma(G)\gamma(H). \qquad (*)$$

(i) Let $G = H = P_4$. Verify that the equality in $(*)$ holds.

(ii) Let $G = P_7$ and $H = P_5$. Verify that the inequality $(*)$ holds.

(iii) Let G be a graph with $\gamma(G) = 1$. Show that $(*)$ holds for any graph H.

(iv) A graph G is said to satisfy Vizing's conjecture if $(*)$ holds for any graph H. Let G be a graph satisfying Vizing's conjecture. Suppose that F is a spanning subgraph of G with $\gamma(F) = \gamma(G)$. Show that F also satisfies Vizing's conjecture.

(Hartnell and Rall (1991))

(v) Let G be a graph satisfying Vizing's conjecture. Suppose that x is a vertex in G such that $\gamma(G - x) < \gamma(G)$. Show that $G - x$ also satisfies Vizing's conjecture.

(Rall)

Note: For a survey on Vizing's conjecture, see Brešar et al. (2012).

11. Let G be the following graph:

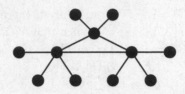

Fig. 9.6.2

(i) Verify that $i(G) = 6$.

(ii) Let $H = \overline{G}$. Show that $i(H) = \gamma(H) = 2$ and $i(G \,\Box\, H) = 11$.

Note: While Vizing's conjecture remains open, the above example, which can be found in Brešar et al. (2012), says that there exist graphs G and H such that $\gamma(G \,\Box\, H) \le i(G \,\Box\, H) < i(G)\gamma(H) = i(G)i(H)$. See also Nowakowski and Rall (1996).

12. Let G and H be any two graphs. Show that

$$\gamma_R(G \,\Box\, H) \le \min\{v(G) \cdot \gamma_R(H), v(H) \cdot \gamma_R(G)\}.$$

Construct two non-trivial graphs G and H for which the above equality holds.

(Yero and Rodriguez-Velazquez (2013))

13. Let G be a connected graph of order $n \ge 2$. Show that

$$\left\lceil \frac{1}{2}(\operatorname{diam}(G) + 2) \right\rceil \le \gamma_R(G) \le n - \left\lfloor \frac{1}{3}(1 + \operatorname{diam}(G)) \right\rfloor.$$

(Mobaraky and Sheikholeslami (2008))

14. Let G be a connected graph of order $n \ge 2$ with vertex set V. Two vertices x and y in V are called **false twin** vertices if $N(x) = N(y)$.

(i) Show that there exists in G a γ-set which contains no pair of false twin vertices.

(ii) Let S be a γ-set of G that contains no pair of false twin vertices. Show that $V \setminus S$ is a resolving set for G. (See Problem II.27.)

(iii) Deduce from (ii) that $\gamma(G) \le n - \dim(G)$.

(iv) Prove that the equality in (iii) holds if and only if $G \cong K_n$ or $G \cong K(p, q)$ for some $p \ge q \ge 2$.

(Bagheri et al. (2011))

15. Following Problem VIII.18, let $M(G)$ denote the graph obtained from a non-trivial graph G by Mycielski construction. Show that

(a) if G is non-empty, then $\omega(M(G)) = \omega(G)$;

(b) if G is complete, then $\text{diam}(M(G)) = 2$;

(c) if G is non-complete and $\delta(G) \geq 1$, then $\text{diam}(M(G)) = \min\{4, \text{diam}(G)\}$;

(d) if G is Hamiltonian, then so is $M(G)$;

(e) if $\delta(G) \geq 1$, then $\kappa(M(G)) \geq \kappa(G) + 1$;

(f) $\alpha(M(G)) \geq \max\{2\alpha(G), v(G)\}$;

(g) $\beta(M(G)) \leq 2\beta(G) + 1$;

(h) if G has no perfect matching, then $\alpha'(M(G)) \geq 2\alpha'(G) + 1$;

(i) $\gamma(M(G)) = \gamma(G) + 1$;

(j) $i(M(G)) = i(G) + 1$;

(k) $\gamma_R(M(G)) \leq \gamma_R(G) + 2$. (Note that Kazemi (2012) showed that $\gamma_R(G) + 1 \leq \gamma_R(M(G))$.)

16. Let G be a graph and v a vertex in G.

(a) Show that $\gamma(G) - 1 \leq \gamma(G - v)$.

(b) Is it true that $\gamma(G - v) \leq \gamma(G)$ in general?

(c) For each $p \geq 1$, construct a graph H such that

$$\gamma(H - w) \geq \gamma(H) + p,$$

for some vertex w in H.

(d) Show that if $\gamma(G - v) < \gamma(G)$, then $\gamma(G - v) = \gamma(G) - 1$.

(e) G is said to be $(\gamma = k)$-**critical** if $\gamma(G) = k$ and $\gamma(G - w) = k - 1$ for each vertex w in G. It is obvious that G is $(\gamma = 1)$-critical if and only if $G = K_1$. Show that G is $(\gamma = 2)$-critical if and only if $G = K_{2r} - M$, for some $r \geq 1$, where M is a perfect matching in K_{2r}.

(f) Construct a $(\gamma = 3)$-critical graph.

(See Brigham et al. (1988))

17. Let G be a graph.

(a) Show that $\gamma(G) - 1 \leq \gamma(G + e) \leq \gamma(G)$ for each edge e in $\overline{G}$.

(b) G is said to be $(\gamma = k)$-**edge-critical** if $\gamma(G) = k$ and $\gamma(G + e) = k - 1$, for each edge e in $\overline{G}$. It is obvious that G is $(\gamma = 1)$-edge-critical if and only if G is a complete graph. Show that G is $(\gamma = 2)$-edge-critical if and only if $\overline{G}$ is the disjoint union of stars.

(c) Construct a $(\gamma = 3)$-edge-critical graph.

(See Sumner and Blitch (1983))

18. Let G be a graph.

(a) Show that $\gamma_R(G) - 1 \leq \gamma_R(G+e) \leq \gamma_R(G)$ for each edge e in $\overline{G}$.

(b) Following Hansberg et al. (2013), we say that G is $(\gamma_R = k)$-**edge-critical** if $\gamma_R(G) = k$ and $\gamma_R(G+e) = k-1$, for each edge e in $\overline{G}$. Show that

    (i) no paths are γ_R-edge-critical;

    (ii) the cycle C_n is γ_R-edge-critical if and only if $n = 3, 4$ and 5.

19. Let G be a graph of order $n \geq 2$. It is obvious that G is $(\gamma_R = 2)$-edge-critical if and only if $G = K_n$. Assume that $n \geq 3$. Show that G is $(\gamma_R = 3)$-edge-critical if and only if $\Delta(G) = n-2$ and for every pair of non-adjacent vertices u and v in G, either $d(u) = \Delta(G)$ or $d(v) = \Delta(G)$.

(Chellali et al. (2012))

20. (a) Let G be a graph of order $n \geq 4$. Show that G is $(\gamma_R = 4)$-edge-critical if and only if $\Delta(G) = n-3$ and for every pair of non-adjacent vertices u and v, either $d(u) = \Delta(G)$ or $d(v) = \Delta(G)$.

(b) Let M_1 and M_2 be two disjoint perfect matchings of K_n, where $n \geq 4$ and n is even. Show that

    (i) $K_n - M_1$ is a $(\gamma_R = 3)$-edge-critical graph; and

    (ii) $K_n - (M_1 \cup M_2)$ is a $(\gamma_R = 4)$-edge-critical graph.

21. (a) Let G be a graph, and v a vertex in G. Show that

    (i) $\gamma_R(G) - 1 \leq \gamma_R(G - v)$;

    (ii) $\gamma_R(G-v) < \gamma_R(G)$ if and only if there exists a γ_R-function g of G such that $g(v) = 1$.

(b) A vertex in G is called a **supporting vertex** if it is adjacent to an end-vertex. Let S be the set of all supporting vertices in G. We say that G is $(\gamma_R = k)$-**critical** if $\gamma_R(G) = k$ and $\gamma_R(G - w) = k-1$, for each vertex w in $V \setminus S$. It is obvious that G is $(\gamma_R = 2)$-critical if and only if $G = K_2$. Assume that G is of order $n \geq 6$. Show that G is $(\gamma_R = 3)$-critical if and only if $G \cong K_{2r} - M$, where $n = 2r$ and M is a perfect matching of K_{2r}. (See Problem IX.16(e).)

(Ang (2013))

Chapter 10

Digraphs and Tournaments

10.1 Digraphs

The multigraph G of Fig. 10.1.1 models the street network of the business section of a town.

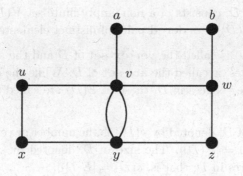

Fig. 10.1.1

As the traffic in this section has increased significantly, the town council has decided to convert the current two-way traffic system in this section to a one-way system, and one possible solution is depicted in Fig. 10.1.2, where the arrow on an edge indicates the proposed direction of the corresponding street.

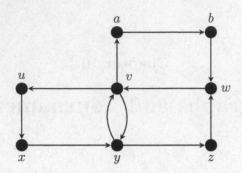

Fig. 10.1.2

The diagram shown in Fig. 10.1.2 is called a **directed graph** or, in short, a **digraph**. Like a multigraph, a digraph has a set of vertices; but a digraph differs from a multigraph in that any edge which links two vertices has a 'direction'.

A **digraph**, D, consists of a non-empty finite set $V(D)$ together with a set $E(D)$ of **ordered** pairs of distinct elements of $V(D)$.

The set $V(D)$ is called the **vertex set** of D and the set $E(D)$ of 'directed edges' is called the **arc set** of D. While the elements in $V(D)$ are the vertices in D, those in $E(D)$ are called the **arcs** in D.

The **order** of D, denoted by $v(D)$, is the number of vertices in D; that is, $v(D) = |V(D)|$. The **size** of D, denoted by $e(D)$, is the number of arcs in D; that is, $e(D) = |E(D)|$.

Example 10.1.1.   Let D be the digraph shown in Fig. 10.1.2. Then
$$V(D) = \{a, b, u, v, w, x, y, z\}$$
and
$$E(D) = \{(a,b), (b,w), (u,x), (v,a), (v,u), (v,y), (w,v), (x,y), (y,v),$$
$$(y,z), (z,w)\}.$$
Also, the order of D is 8 (that is, $v(D) = 8$) and the size of D is 11 (that is, $e(D) = 11$).

In Example 10.1.1, the **ordered** pair of vertices (a, b) represents the arc:

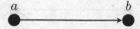

In this case,

    (1) we may write ab or $a \to b$ for (a, b);

    (2) we may say that

        (i) a is adjacent to b,

        (ii) b is adjacent from a,

       (iii) a dominates b,

       (iv) b is dominated by a,

       (v) the arc ab is incident from a,

      (vi) the arc ab is incident to b.

Remark 10.1.2.

    (1) As **ordered** pairs, $(p, q) \neq (q, p)$ in general. In the digraph D of Fig. 10.1.2, we have the arc (a, b), but no (b, a); we have both (v, y) and (y, v), but they represent different arcs.

    (2) Consider the diagram depicted in Fig. 10.1.3.

Fig. 10.1.3

There are two arcs incident from x to y. We call them parallel arcs. Also, there is an arc incident from y to y. We call it a loop. Throughout this chapter, we shall assume the following: *All digraphs contain neither parallel arcs nor loops*, unless otherwise stated.

Exercise for Section 10.1

1. Let D be the digraph shown in Fig. 10.1.4.

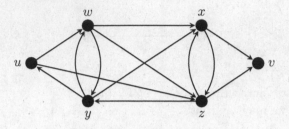

Fig. 10.1.4

Find

(i) $V(D)$ and $E(D)$;

(ii) $v(D)$ and $e(D)$;

(iii) all vertices adjacent from w;

(iv) all vertices adjacent to y;

(v)   all vertices dominated by x;

(vi) all vertices that dominate z;

(vii) all arcs incident from u;

(viii) all arcs incident to z.

2. Let D be the digraph defined as follows:

$$V(D) = \{a, b, u, v, w, x, y, z\}$$

and

$$E(D) = \{aw, ay, bx, ux, vz, wu, wx, wy, xz, ya, yv, yx, yz, zb, zy\}.$$

(i)   Draw the diagram of D.

(ii) Find $v(D)$ and $e(D)$.

(iii) Which vertices are adjacent from a?

(iv) Which vertices are adjacent to y?

(v)   Which vertices are dominated by w?

(vi) Which vertices dominate x?

(vii) Which arcs are incident to z?

(viii) Which arcs are incident from y?

3. Let D be the digraph defined as follows:

$$V(D) = \{1, 2, 3, 4, 5, 6\}$$

and $(i, j) \in E(D)$, where i, j are in $V(D)$, if and only if $i > j$.

(i) Draw the diagram of D.

(ii) Find $e(D)$.

(iii) Which vertices are adjacent to '2'?

(iv) Which vertices are adjacent from '2'?

4. Two table tennis teams A and B, each consisting of 3 players as shown below:

$$A = \{x, y, z\} \text{ and } B = \{u, v, w\},$$

had a friendly match between their players in singles. Each player in the team must play each player in the other exactly once with no ties allowed. At the end of the match, it was reported that

(1) x won all the matches;

(2) y was defeated only by w;

(3) team A defeated team B by just one match.

(i) Construct a digraph D to model the situation where $V(D)$ is the set of all players, and $a \to b$ in D if player a defeated player b.

(ii) Find $e(D)$.

(iii) Did z win any game?

(iv) Which player in team B won the largest number of matches?

10.2 Basic Concepts

In this section, we shall introduce a number of basic concepts in digraphs which are parallel to their counterparts in multigraphs.

The in-degree and out-degree of a vertex

The degree of a vertex in a multigraph is defined as the number of edges incident with the vertex. Due to the presence of **directions** of the arcs, we have two different types of degrees for vertices in a digraph.

Let D be a digraph, and v a vertex in D.

The **in-degree** of v, denoted by $d^-(v)$, is the number of arcs incident to v in D.

The **out-degree** of v, denoted by $d^+(v)$, is the number of arcs incident from v in D.

A vertex in D is called a **source** if its in-degree is zero.

A vertex in D is called a **sink** if its out-degree is zero.

Question 10.2.1. *Let D be the digraph of Fig. 10.1.2. Complete the following table:*

Vertex	In-degree	Out-degree
a		
b		
u		
v		
w		
x		
y		
z		
Total sum		

Does D contain any source or sink?

Can you see any relationship between $e(D)$ and the two total sums you found in the above table?

Indeed, as an arc in a digraph contributes exactly one towards the total sum of the in-degrees and exactly one towards the total sum of the out-degrees, we immediately have the following result for digraphs, which is analogous to Euler's handshaking lemma in multigraphs.

Result (1) Let D be a digraph. Then

$$\sum_{v \in V(D)} d^-(v) = e(D) = \sum_{v \in V(D)} d^+(v).$$

Isomorphic digraphs

Just like the situation for graphs, given two digraphs, we may wish to know whether they are the 'same'. This leads to the following concept.

Let D_1 and D_2 be two digraphs. We say that D_1 is **isomorphic** to D_2, and we write $D_1 \cong D_2$, if there exists a one-one and onto mapping $f : V(D_1) \to V(D_2)$ such that $(u, v) \in E(D_1)$ if and only if $(f(u), f(v)) \in E(D_2)$, where u, v are vertices in D_1 (that is, the dominance relation is preserved under f). Such a one-one and onto mapping f is called an **isomorphism** from D_1 to D_2.

Example 10.2.2. Let D_1 and D_2 be the two digraphs shown in Fig. 10.2.1.

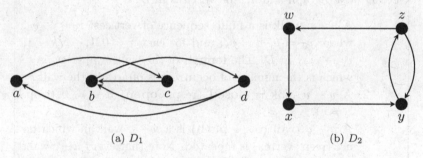

(a) D_1 (b) D_2

Fig. 10.2.1

We claim that $D_1 \cong D_2$. Indeed, define the mapping

$$f : V(D_1) \to V(D_2)$$

as follows:

$$f(a) = w, f(b) = y, f(c) = x \text{ and } f(d) = z.$$

Clearly, f is both one-one and onto. Moreover, it can be checked that f preserves the dominance relation (for instance, $a \to c$ and $f(a) \to f(c)$, $d \to b$ and $f(d) \to f(b)$, etc.). Thus, we conclude that $D_1 \cong D_2$.

Question 10.2.3.   *There are three digraphs of the same order and same size shown in Fig. 10.2.2. Which two are isomorphic? Justify your answer.*

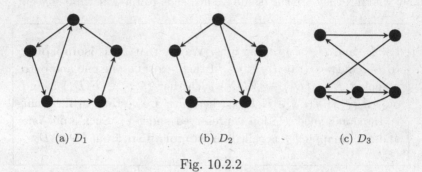

(a) D_1 (b) D_2 (c) D_3

Fig. 10.2.2

Connectedness

Let D be a digraph and u, v be vertices in D.

1. A $u - v$ **walk** is a finite sequence of vertices: $v_0 v_1 \cdots v_k$, where $v_0 = u$, $v = v_k$, and for each $i = 0, 1, \cdots, k - 1$, $v_i \to v_{i+1}$ in D. The **length** of the walk $v_0 v_1 \cdots v_k$ is k, which is the number of occurrences of arcs in the walk.

2. A $u - v$ walk is **closed** (resp. **open**) if $u = v$ (resp. $u \neq v$).

3. A $u - v$ **trail** (resp. **path**) is a $u - v$ walk in which no arc (resp. vertex) is repeated. Note that every $u - v$ walk contains a $u - v$ path; and every $u - v$ path is a $u - v$ trail, but not conversely.

4. A **circuit** is a closed trail.

5. A **cycle** is a closed walk: $v_1 v_2 \cdots v_k v_1$, $k \geq 2$, where all $v_1, v_2, \cdots, v_k$ are distinct. Such a cycle is also called a k-**cycle**, and is denoted by C_k.

6. A vertex v is **reachable from** a vertex u if there is a $u - v$ path in D.

7. If v is reachable from u, the **distance from** u **to** v, $d(u, v)$, is the smallest length of all $u - v$ paths in D. We define $d(u, v) = \infty$ if v is not reachable from u.

8. The **underlying graph** (or **underlying multigraph**) of D, denoted by $G(D)$, is the graph (or multigraph) obtained from D by disregarding the direction of each arc in D.
9. The digraph D is **connected** if $G(D)$ is connected.
10. The digraph D is **strongly connected**, or simply **strong**, if every two vertices in D are mutually reachable.

Example 10.2.4. We show again the digraph D of Fig. 10.1.2 below for convenience.

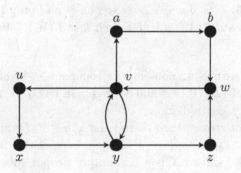

Fig. 10.2.3

(i) Disregarding the directions of arcs in D, we obtain its underlying multigraph $G(D)$, which is the multigraph of Fig. 10.1.1. As $G(D)$ is connected, the digraph D is thus connected.

(ii) There are only two $a - y$ paths in D, namely, $abwvuxy$ and $abwvy$. The former one is of length 6 while the latter of length 4. By definition, $d(a, y) = 4$.

(iii) The closed walk $abwvuxyva$ is a circuit but not a cycle.

(iv) In D, we can find a 2-cycle, namely, vyv but no 3-cycle. A number of 4-cycles are available; one of them is $abwva$. The 6-cycle $xyzwvux$ is a largest cycle.

(v) It can be checked that every two vertices in D are mutually reachable. Thus D is a strong digraph. Note that $d(v, y) = d(y, v) = 1; d(v, a) = 1$ but $d(a, v) = 3$; while $d(w, u) = 2, d(u, w) = 4$.

We see from (v) that the distance function 'd' in digraphs is not symmetric.

Strong digraphs are important models for applications. The following result provides a simple necessary and sufficient condition for a digraph to be strong.

For $W \subseteq V(D)$, let

$$R(W) = \{v \in V(D) \mid v \text{ is reachable from some vertex in } W\}.$$

Note that a vertex is always reachable from itself and therefore $W \subseteq R(W)$.

Theorem 10.2.5. *A digraph D is strong if and only if there exists no non-empty proper subset W of $V(D)$ such that $R(W) \subseteq W$.*

Proof. ($\Rightarrow$) If there exists a non-empty proper subset W of $V(D)$ such that $R(W) \subseteq W$, then $V(D) \backslash W \neq \emptyset$ and no vertex in $V(D) \backslash W$ is reachable from any vertex of W, a contradiction.

($\Leftarrow$) If D is not strong, then there exist $x, y \in V(D)$ such that y is not reachable from x. Let $X = \{v \in V(D) \mid v \text{ is reachable from } x\}$. Clearly, $x \in X$ and $y \notin X$, and so X is a non-empty proper subset of $V(D)$. We now claim that $R(X) \subseteq X$. Let $u \in R(X)$. Then u is reachable from some v in X. By definition, v is reachable from x. Thus u is also reachable from x. This implies that $u \in X$, and we have $R(X) \subseteq X$, as claimed. $\square$

Eccentricity and diameter

The notion of the eccentricity of a vertex in a graph was introduced in Section 1.5. It can be similarly defined for vertices in digraphs.

Let D be a digraph and v be a vertex in D. The **eccentricity** of v, denoted by $e(v)$, is defined as

$$e(v) = \max\{d(v, x) \mid x \in V(D)\}.$$

Note that v can reach any other vertex in D with a path of length at most $e(v)$.

Example 10.2.6. For the digraph D in Example 10.2.4 (see Fig. 10.2.3), we have:

$$e(z) = \max\{d(z,a), d(z,b), d(z,u), d(z,v), d(z,w), d(z,x), d(z,y), d(z,z)\}$$
$$= \max\{3, 4, 3, 2, 1, 4, 3, 0\} = 4.$$

It can be verified that the eccentricities of the eight vertices in D are shown in the following table:

Vertex	a	b	u	v	w	x	y	z
Eccentricity	5	4	5	3	3	4	3	4

Note: Let v be a vertex in a digraph D. If there exists a vertex, say x, in D which is not reachable from v, then $d(v,x) = \infty$. In this case, $e(v)$ is defined to be '∞'.

Similar to the notion of the diameter $\text{diam}(G)$ of a graph G, the **diameter** of a digraph D, denoted by $\text{diam}(D)$, is defined by

$$\text{diam}(D) = \max\{e(x) \mid x \in V(D)\}.$$

Thus, if D is the digraph in Example 10.2.4, then we observe from the table in Example 10.2.6 that $\text{diam}(D) = 5$.

Note that given any digraph D, we also have:

$$\text{diam}(D) = \max\{d(x,y) \mid x,y \in V(D)\}.$$

For the digraph D in Example 10.2.4,

$$\text{diam}(D) = 5 = d(a,x) = d(a,z) = d(u,b) = \max\{d(p,q) \mid p,q \in V(D)\}.$$

Note: Let D be a digraph. Clearly, $\text{diam}(D)$ is a finite number if and only if D is strong.

The World Wide Web can be modeled as a digraph. A webpage is represented by a vertex and there is an arc from a vertex to another vertex if there is a hyperlink from the first vertex to the other vertex. There are loops as well in this definition as some webpages have links to various sections within themselves. Of interest to researchers is the diameter of the World Wide Web as this would indicate the maximum number of links needed from any given webpage to another. Also of interest are vertices of high degree. Vertices of high out-degree have been called 'hubs'. These are usually the search engines and pages of bookmarks. Vertices of high in-degree have been called 'authorities', i.e., many webpages have links to them thus showing the importance of their content. In fact, some search engines work on this property and rank a webpage according to its in-degree, the in-degrees of the webpages adjacent to it, the in-degrees of the webpages adjacent to them, and so on.

Subdigraph

Let D be a digraph. A digraph H is called a **subdigraph** of D if $V(H) \subseteq V(D)$ and $E(H) \subseteq E(D)$.

Every digraph is a subdigraph of itself. A subdigraph H of D is said to be **proper** if $V(H) \subset V(D)$ or $E(H) \subset E(D)$.

A subdigraph H of D is called a **spanning subdigraph** if $V(H) = V(D)$.

Let $A \subseteq V(D)$. The **subdigraph of D induced by** A, denoted by $[A]$, is the digraph whose vertex set is A and whose arc set consists of all arcs xy in D whenever x, y are in A. An induced subdigraph of D is a subdigraph induced by some subset A of $V(D)$.

Example 10.2.7. Consider the four digraphs shown in Fig. 10.2.4.

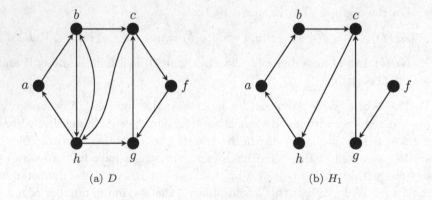

(a) D (b) H_1

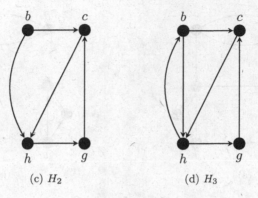

(c) H_2 (d) H_3

Fig. 10.2.4

We observe that

(i) H_1, H_2 and H_3 are all proper subdigraphs of D;

(ii) H_1 is a spanning subdigraph of D, H_2 and H_3 are not spanning subdigraphs of D;

(iii) H_3 is the subdigraph induced by $\{b, c, g, h\}$ (i.e., $H_3 = [\{b, c, g, h\}]$), H_1 and H_2 are not induced subdigraphs of D.

Strong component

Let D be a digraph. Define a binary relation '$\equiv$' on $V(D)$ as follows: For $x, y \in V(D)$, write $x \equiv y$ if and only if x and y are mutually reachable in D.

It is easy to see that the binary relation '$\equiv$' is reflexive (i.e., $x \equiv x$ for each $x \in V(D)$), symmetric (i.e., for x and y in $V(D)$, $x \equiv y$ implies $y \equiv x$) and transitive (i.e., for x, y, z in $V(D)$, $x \equiv y$ and $y \equiv z$ imply $x \equiv z$). Thus '$\equiv$' is an equivalence relation on $V(D)$, and hence it partitions $V(D)$ into disjoint equivalence classes, say, $V_1, V_2, \cdots, V_k$. For each $i = 1, 2, \cdots, k$, the subdigraph $[V_i]$ of D induced by V_i is called a **strong component** of D.

Example 10.2.8. The digraph in Example 10.2.5(a) has three strong components as shown in 10.2.5(b).

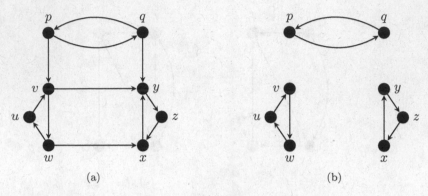

Fig. 10.2.5

Remark 10.2.9.   Instead of employing the notion of equivalence relation, a strong component of a digraph D can equivalently be defined as a subdigraph S of D such that

(1) S is itself a strong digraph and
(2) S is not a proper subdigraph of any strong subdigraph of D.

Question 10.2.10.

(i) *How many strong components does D have if D is a strong digraph?*
(ii) *The following digraph contains no cycles. How many strong components does it have?*

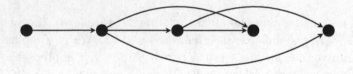

Fig. 10.2.6

Condensation

Based on the notion of strong components, we shall now introduce a new digraph obtained from a give one, called its **condensation**.

Let D be a digraph with k strong components, $S_1, S_2, \cdots, S_k$. The **condensation** of D is the digraph D^* with k vertices, denoted by $s_1, s_2, \cdots s_k$, say, such that for $i, j \in \{1, 2, \cdots, k\}$ with $i \neq j$, $s_i \to s_j$ if and only if $u \to v$ for some u in S_i and v in S_j.

Example 10.2.11. For the digraph in Fig. 10.2.7(a), its condensation is shown in Fig. 10.2.7(b).

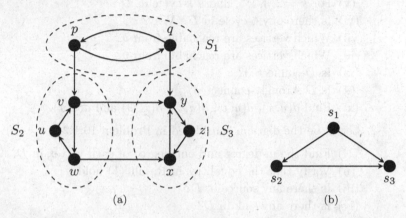

(a)   (b)

Fig. 10.2.7

Question 10.2.12.

(i) *What is the condensation of a strong digraph?*

(ii) *What is the condensation of the digraph shown in Fig. 10.2.6?*

(iii) *Does there exist a cycle in the condensation of a digraph?*

While every strong digraph always contains a cycle, there are digraphs that contain no cycles.

A digraph is said to be **acyclic** if it contains no cycles.

As shown in Example 10.2.11, we notice that the condensation is acyclic. More generally, the answer to Question 10.2.12(iii) is 'no' as stated in the following result (see Problem 10.2.22).

Theorem 10.2.13. *The condensation of any digraph is acyclic.*

$\square$

Exercise for Section 10.2

1. Let D be the digraph considered in Problem 10.1.1.

   (i) Find the in-degree and out-degree of each vertex in D.
   (ii) Verify that the equalities in Result (1) hold.
   (iii) Is there any source in D?
   (iv) Is there any sink in D?
   (v) For $k = 2, 3, 4, 5$, find a k-cycle in D.
   (vi) Is there any 6-cycle in D? Why?
   (vii) Which vertices are reachable from u?
   (viii) Which vertices are reachable from v?
   (ix) Is D connected?
   (x) Is D strongly connected?
   (xi) Find $d(u,x)$, $d(u,z)$, $d(u,v)$, $d(x,w)$ and $d(v,x)$.

2. Let D be the digraph considered in Problem 10.1.2.

   (i) Find the in-degree and out-degree of each vertex in D.
   (ii) Verify that the equalities in Result (1) hold.
   (iii) Is there any source in D?
   (iv) Is there any sink in D?
   (v) Find a 6-cycle in D.
   (vi) Is there any 7-cycle in D?
   (vii) Is there any 8-cycle in D?
   (viii) Which vertices are reachable from u?
   (ix) Is D connected?
   (x) Is D strongly connected?
   (xi) Find $d(a,u)$, $d(u,a)$, $d(a,b)$ and $d(b,a)$.
   (xii) Find two vertices in D such that the distance from one of them to the other is 5.
   (xiii) Find two vertices in D such that the distance from one of them to the other is 6.

3. Let D be the digraph considered in Problem 10.1.3.

   (i) Find $e(D)$.
   (ii) Find the in-degree and out-degree of each vertex in D.
   (iii) Verify that the equalities in Result (1) hold.
   (iv) Is there any source in D?
   (v) Is there any sink in D?
   (vi) Are there any two vertices in D which have the same out-degree?

(vii) Is there any cycle in D?

(viii) Is there any path of length 5 in D?

(ix) What is $G(D)$?

(x) Is D strong?

4. Are the following two digraphs isomorphic? If 'yes', find an isomorphism from one of them to the other.

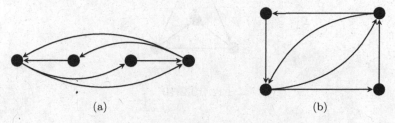

Fig. 10.2.8

5. Are the following two digraphs isomorphic? Justify your answer.

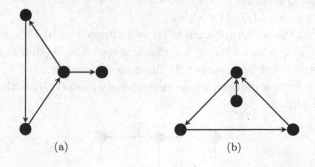

Fig. 10.2.9

6. Among the following three digraphs, which two are isomorphic? Justify your answer.

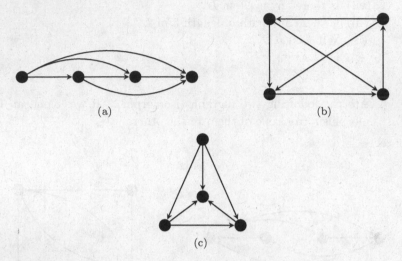

(a)   (b)

(c)

Fig. 10.2.10

7. Let D be a digraph. Prove that every $u - v$ walk in D contains a $u - v$ path, where u and v are two vertices in D.

8. Let D be a digraph and $k = \min\{d^+(v) \mid v \in V(D)\}$. Show that D contains a path of length at least k. Is the result also true if $k = \min\{d^-(v) \mid v \in V(D)\}$?

9. Let D be a digraph with $k = \min\{d^+(v) \mid v \in V(D)\} \geq 1$. Show that D contains an r-cycle, where $r \geq k+1$. Is the result also true if $k = \min\{d^-(v) \mid v \in V(D)\}$?

10. Let D be a digraph with $v(D) \geq 2$. Prove that if D is strong, then $d^-(v) \geq 1$ and $d^+(v) \geq 1$ for each vertex v in D (that is D contains neither sink nor source). Is the converse true?

11. Let D be a digraph whose underlying graph $G(D)$ is the following graph:

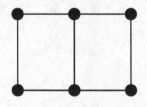

Fig. 10.2.11

Assume that D contains neither sink nor source. Show that D contains a 4-cycle.

12. Let D be a digraph whose underlying graph $G(D)$ is $K(4,4)$. Prove that if D contains an 8-cycle, then D contains a 4-cycle.

13. Let D be a digraph. Prove that if D is acyclic, then D contains a sink and a source. Is the converse true?

14. Let D be a digraph. Prove that D contains no cycles if and only if every walk in D is a path.

15. Let D be a digraph with n vertices labeled $v_1, v_2, \cdots, v_n$. The **adjacency matrix** of D, denoted by $A(D)$, is the $n \times n$ matrix in which a_{ij}, the entry in row i and column j, is 1 if there is an arc from vertex v_i to vertex v_j, and 0 otherwise. We may sometimes write $A(D) = (a_{ij})$.

   (i) Draw the digraph D which has the following adjacency matrix $A(D)$:

   $$A(D) = \begin{pmatrix} 0 & 1 & 0 & 1 \\ 0 & 0 & 1 & 1 \\ 0 & 1 & 0 & 0 \\ 0 & 1 & 1 & 0 \end{pmatrix}.$$

   (ii) Find the adjacency matrix $A(D)$ of the following digraph D.

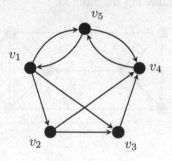

Fig. 10.2.12

16. Let D be a digraph of order $n \geq 2$. Show that D contains no cycles if and only if the vertices in D can be named as $v_1, v_2, \cdots, v_n$ such that $A(D)$ is upper triangular. (A square matrix (a_{ij}) is called an **upper triangular matrix** if $a_{ij} = 0$ for all i, j with $i > j$.)

17. Let D be a digraph.

  (i) Show that if every two vertices in D are contained in a $(k+1)$-cycle, where $k \geq 1$, then diam$(D) \leq k$.

  (ii) Is the converse of (i) true?

18. Consider the following digraph D:

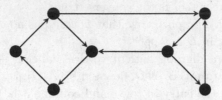

Fig. 10.2.13

  (i) Find the eccentricity of each vertex in D and diam(D). Is D strong?

  (ii) Add one new arc to join two vertices in D so that the resulting digraph has its diameter equal to diam$(D) - 1$.

  (iii) What is the least number of new arcs needed to be added to join vertices in D so that the resulting digraph has its diameter equal to diam$(D) - 2$?

19. Consider the following digraph D:

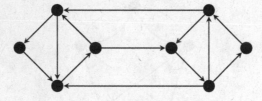

Fig. 10.2.14

  (i) Find the eccentricity of each vertex in D and diam(D). Is D strong?

  (ii) Remove an arc from D so that the resulting spanning subdigraph has its diameter equal to diam$(D) + 1$.

  (iii) What is the least number of arcs needed to be removed from D so that the resulting spanning subdigraph has its diameter equal to diam$(D) + 2$?

(iv) What is the least number of arcs needed to be removed from D so that the resulting spanning subdigraph is not strong?

20. Consider the following digraph D:

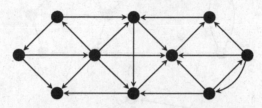

Fig. 10.2.15

(i) Is D strong?
(ii) Find all strong components of D.
(iii) Construct the condensation of D.

21. Let D^* be the condensation of a digraph D. What can be said of D^* in each of the following cases:

(i) D contains an n-cycle, where $n = V(D)$;
(ii) D is obtained by linking two disjoint cycles with an arc;
(iii) D is acyclic;
(iv) D is the condensation of some digraph.

22. Prove that the condensation of any digraph is acyclic.
23. For each of the following digraphs, find its strong components and construct its condensation.

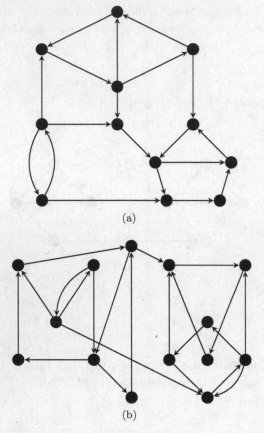

Fig. 10.2.16

24. Let D be a digraph of order $n \geq 2$ such that

$$d^{+}(u) + d^{-}(v) \geq n \qquad\qquad (*)$$

for any two distinct vertices u, v with u not adjacent to v in D. Show that

$$D \text{ is strong.} \qquad\qquad (**)$$

Can the condition $(*)$ be weakened to establish $(**)$? Can the conclusion $(**)$ be strengthened?

25. The **converse** of a digraph D, denoted by $\overrightarrow{D}$, is the digraph obtained by reversing each arc in D. Show that

(i) D is strong if and only if $\overrightarrow{D}$ is strong.

(ii) $\operatorname{diam}(D) = \operatorname{diam}(\vec{D})$.

(iii) D is acyclic if and only if $\vec{D}$ is acyclic.

Is there any relation between the condensations of D and $\vec{D}$?

26. [**Eulerian digraphs**] Let D be a connected digraph in which parallel arcs and loops are permitted. We call D an **Eulerian** (resp., **semi-Eulerian**) digraph if D contains a circuit which includes every arc in D (resp., an open trail which includes every arc in D). Show that

    (i) D is Eulerian if and only if $d^+(v) = d^-(v)$ for each vertex v in D;

    (ii) D is semi-Eulerian if and only if there exist u, w in $V(D)$ such that $d^+(u) = d^-(u) + 1$, $d^-(w) = d^+(w) + 1$, and $d^+(v) = d^-(v)$ for any other vertex v in D.

27. [**Hamiltonian digraphs**] A digraph is said to be **Hamiltonian** if it contains a spanning cycle. Let D be a digraph of order $n \geq 2$. For a vertex v in D, write $d(v) = d^+(v) + d^-(v)$. Three known sufficient conditions for a **strong** D to be Hamiltonian are stated below.

Ghouila-Houri's condition (1960): $d(v) \geq n$ for each v in $V(D)$.

Woodall's condition (1972): $d^+(u) + d^-(v) \geq n$ for every two vertices u, v with u not adajcent to v in D.

Meyniel's condition (1973): $d(u) + d(v) \geq 2n - 1$ for every two non-adjacent vertices u, v in D.

    (i) Show that Ghouila-Houri's condition implies Meyniel's condition.

    (ii) Show that Woodall's condition implies Meyniel's condition.

    (iii) Is Meyniel's condition necessary for D to be Hamiltonian?

    (iv) Construct a non-Hamiltonian digraph D such that $d(u) + d(v) = 2n - 2$ for every two non-adjacent vertices u, v in D.

28. Construct a strong digraph that has no spanning path.

10.3   Tournaments

In this section, we shall introduce and study a very special type of digraphs, called tournaments.

A **tournament** is a digraph in which every two vertices are joined by exactly one arc.

Equivalently, a **tournament** is a digraph obtained by assigning a direction to each edge of a complete graph.

A tournament of order 5 is shown in Fig. 10.3.1.

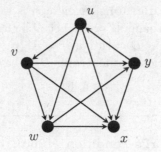

Fig. 10.3.1

Such a mathematical model is called a tournament since it can be used to show the possible outcomes of a **round-robin tournament**. In a round-robin tournament, there is a set of players (or teams) where any two players (or teams) engage in a game that cannot end in a tie, and every player (or team) must play each other once and exactly once.

The tournament of Fig. 10.3.1 shows the outcomes of a 5-player tennis (singles) round-robin tournament. Here, for any two players, say a and b, we denote by $a \to b$ if a defeats b. The five players in the tournament are: u, v, w, x and y, and the results are u defeats v, w and x, but is defeated by y, etc.

In a round-robin tournament, the **score** obtained by a player is the number of players he/she defeats, that is, the out-degree of the vertex representing the player in the tournament. Thus, in the tournament of Fig. 10.3.1, the scores of the players u, v, w, x and y are, respectively, $3, 3, 2, 0, 2$.

Question 10.3.1. *Five singles table tennis players a, b, c, d, e were involved in a round-robin tournament. It was reported that a defeated all other players; d was defeated only by a; and e defeated only two players. Were there any two players having the same score?*

Thoughout this chapter, for a positive integer n, we shall denote by T_n a tournament of order n.

Example 10.3.2. For $n = 1, 2, 3, 4$, all non-isomorphic T_n are shown in Fig. 10.3.2.

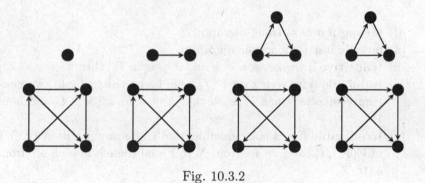

Fig. 10.3.2

Question 10.3.3.

(i) *Construct a tournament of order 5 in which every vertex has the same score.*

(ii) *Construct a tournament of order 5 in which no two vertices have the same score.*

Some useful quantitative relations in tournaments are stated below.

Result (2)

(i) $e(T_n) = \binom{n}{2}$;

(ii) $d^+(v) + d^-(v) = n - 1$ for each vertex v in T_n; and

(iii) $\sum_{v \in V(T_n)} d^+(v) = \binom{n}{2} = \sum_{v \in V(T_n)} d^-(v)$.

Some special families of tournaments are introduced below.

A tournament T_n is

- (i) **strong** if it is a strong digraph;
- (ii) **acyclic** if it is an acyclic digraph;
- (iii) **transitive** if whenever $u \to v$ and $v \to w$ in T_n, then $u \to w$ in T_n;
- (iv) **reducible** if its vertex set $V(T_n)$ can be partitioned into two non-empty subsets U and W such that $u \to w$ for all u in U and w in W;
- (v) **irreducible** if it is not reducible, that is, for any partition $\{X, Y\}$ of $V(T_n)$, there is an arc from X to Y and there is also an arc from Y to X;
- (vi) **regular** if every vertex has the same score, that is, n is odd and $d^+(v) = \frac{n-1}{2}$ for each v in $V(T_n)$ (see Problem 10.3.10);
- (vii) **near-regular** if n is even and

$$\max_{u,v \in V(T_n)} |d^+(u) - d^+(v)| = 1$$

(see Problem 10.3.11).

Question 10.3.4.   *Five tournaments are shown in Fig. 10.3.3.*

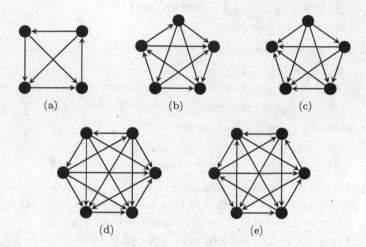

Fig. 10.3.3

Determine, for each digraph, whether it is strong, acyclic, transitive, reducible, irreducible, regular, or near-regular.

Some relations that exist amongst these families of tournaments are stated below (see Problem 10.3.12).

Theorem 10.3.5. *Let T be a tournament with $v(T) \geq 3$.*

(i) *T is transitive if and only if T is acyclic;*
(ii) *If T is transitive, then T is reducible;*
(iii) *T is strong if and only if T is irreducible;*
(iv) *If T is regular, then T is strong;*
(v) *If T is near-regular, then T is strong.*

☐

Exercise for Section 10.3

In this Exercise, we denote by T a tournament and by T_n a tournament of order $n \geq 2$.

1. Let u, v be two mutually reachable vertices in T. Prove that

(i) $d(u,v) \neq 1$ if and only if $d(v,u) = 1$.

(ii) $d(u,v) \neq d(v,u)$.

2. Let u, v be two vertices in T with $d(u,v) = k \geq 2$. Prove that

   (i) $d^+(v) \geq k - 1$;

   (ii) v is contained in an r-cycle for each $r = 3, 4, \cdots, k+1$;

   (iii) u and v are contained in a common $(k+1)$-cycle.

3. Show that, up to isomorphism,

   (i) there is only one strong tournament of order 3;

   (ii) there is only one strong tournament of order 4.

4. Let T_n be a strong tournament such that, for each arc in T_n, the reversing of the direction of this arc also results in a strong tournament. Show that $n \geq 5$ and construct one such T_n.

5. Prove the results in Result (2). (See page 418.)

6. Show that

$$\sum_{v \in V(T)} d^+(v)^2 = \sum_{v \in V(T)} d^-(v)^2.$$

7. Let $V(T) = \{v_1, v_2, \cdots, v_n\}$. Show that, for any $k = 1, 2, \cdots, n$,

$$\sum_{i=1}^{k} d^+(v_i) \geq \binom{k}{2}.$$

8. Suppose in a round-robin tournament, team A has the maximum score. Let p denote the number of teams defeated by A, and q the number of teams not defeated by A. Which of the following situations are possible?

   (i) $p > q$;

   (ii) $p = q$;

   (iii) $p < q$.

9. Assume that T_n is strong and $n \geq 4$. Show that T_n contains no more than two vertices of out-degree at least $n - 2$. Construct a strong T_n with exactly two vertices of out-degree $n - 2$.

10. Assume that T_n is regular, that is, every vertex has the same score. Show that n must be odd and $d^+(v) = \frac{n-1}{2}$ for each v in $V(T_n)$. Construct one regular T_7.

11. Assume that

$$\max_{u,v \in V(T_n)} |d^+(u) - d^+(v)| = 1.$$

(i) Show that n must be even.

(ii) Express the maximum score and minimum score of vertices in terms of n.

(iii) Construct one such tournament T_8.

12. Prove Theorem 10.3.5.

13. Construct a T such that

(i) T is reducible but not transitive;

(ii) T is strong but neither regular nor near-regular.

14. Let T be of order $n \geq 5$ and let S be a strong component of T.

(i) Is it possible that $|S| = 1$ for some T?

(ii) Is it possible that $|S| = 2$ for some T?

(iii) Is it possible that $|S| = n$ for some T?

15. Let T be of order 5. Given that T^* is a condensation of T, what are the possible values of $v(T^*)$? For each of your answers, construct one such T.

16. Let T be of order $n \geq 2$ and u, v be two vertices in T.

(i) Assume that u and v are in different strong components of T. Show that $u \to v$ if and only if $d^+(u) > d^+(v)$.

(ii) Assume that $d^+(u) = d^+(v)$. Show that u and v are in the same strong component of T.

17. Show that the condensation of any tournament is a transitive tournament.

18. Let T be a strong tournament of order $n \geq 3$.

(i) What is the maximum possible value of diam(T)?

(ii) Construct a tournament T of order n such that diam(T) attains the maximum value obtained in (i).

10.4 Paths and Cycles in Tournaments

Tournaments are very special digraphs, and so they are expected to have some properties that general digraphs do not have. In this section, we shall focus on properties regarding the existence of long paths and cycles.

Consider the tournament of Fig. 10.3.1. We observe that the five players in the round-robin tournament can be arranged in a row so that one defeats the player on his/her immediate right. For instance,

$$u \to v \to w \to y \to x, \quad y \to u \to v \to w \to x, \text{ etc.}$$

Such an arrangement involves all 'vertices' in a single 'path'. We call it a spanning path or Hamiltonian path of the tournament.

A path in a digraph D is called a **Hamiltonian path** if the path contains all vertices in D.

Not every connected digraph D always contains a Hamiltonian path even if D is strong (see Problem 10.2.28). However, such a path does exist in any tournament.

Theorem 10.4.1.   *Every tournament contains a Hamiltonian path.*

Proof. Let T be a tournament of order n, and let

$$P : x_1 \to x_2 \to \cdots \to x_{k-1} \to x_k$$

be a **longest** path (of length $k-1$) in T. If $k = n$, then P is a Hamiltonian path in T, and we are through.

Suppose $k < n$. We shall derive a contradiction. Let v be a vertex in T but not in P. As T is a tournament, either $v \to x_1$ or $x_1 \to v$. Since P is a longest path, $x_1 \to v$, otherwise $vx_1x_2\cdots x_k$ will be a path longer than P. Again, either $v \to x_2$ or $x_2 \to v$. As P is longest, $x_2 \to v$, otherwise $x_1vx_2\cdots x_k$ will be a path longer than P.

Proceeding in this manner successively, we arrive at the following, namely, $x_k \to v$. But then

$$x_1 \to x_2 \to \cdots \to x_{k-1} \to x_k \to v$$

is a path (of length k) in T longer than P, a contradiction.

$\square$

Remark 10.4.2.   Theorem 10.4.1 is indeed a very special case of a much deeper result due to Rédei (1934) which states that the number of Hamiltonian paths in any tournament is always odd.

Let D be a digraph. A cycle in D is called a **Hamiltonian** cycle if it contains all vertices in D; and D is said to be **Hamiltonian** if it contains a Hamiltonian cycle. (See Problem 10.2.27.)

It is clear that every Hamiltonian digraph is strong. While it is also obvious that not every strong digraph is Hamiltonian, Camion (1959) proved that a tournament is strong if and only if it is Hamiltonian. Harary and Moser (1966) improved Camion's result by showing that every strong tournament of order $n \geq 3$ contains a k-cycle for all $k = 3, 4, \cdots, n$. This latter result was further strengthened by Moon (1966) as shown in our next theorem.

For a vertex v in a digraph D, we define the **in-set**, $I(v)$, and **out-set**, $O(v)$, as follows:

$$I(v) = \{x \in V(D) \mid x \to v\}, \quad O(v) = \{x \in V(D) \mid v \to x\}.$$

Theorem 10.4.3. *Let T_n be a strong tournament, where $n \geq 3$. Then every vertex in T_n is contained in a k-cycle for all $k = 3, 4, \cdots, n$.*

Proof. Let v be a vertex in T. Our aim is to show that for each $k = 3, 4, \cdots, n$, v is contained in some k-cycle. The proof is by induction on k.

We begin with $k = 3$, and show that v is contained in a 3-cycle. Consider the in-set $I(v)$ and the out-set $O(v)$ of v. As T is strong, it follows that both $I(v)$ and $O(v)$ are non-empty, and there exist w in $I(v)$ and u in $O(v)$ such that $u \to w$ as shown in Fig. 10.4.1, as otherwise there would be no path from a vertex in $O(v)$ to a vertex in $I(v)$.

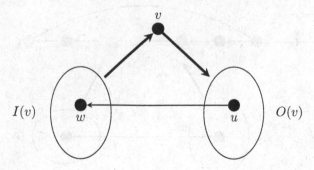

Fig. 10.4.1

Clearly, $vuwv$ is a 3-cycle containing v.

Assume now that v is contained in a k-cycle $C : vv_1 \cdots v_{k-1}v$, where $3 \leq k \leq n-1$. We shall show that there is a $(k+1)$-cycle containing v.

Let A be the set of vertices x in $V(T)\backslash V(C)$ such that x dominates all vertices in C, and B the set of vertices y in $V(T)\backslash V(C)$ such that y is dominated by all vertices in C, and let $R = V(T)\backslash(V(C)\cup A\cup B)$ as shown in Fig. 10.4.2.

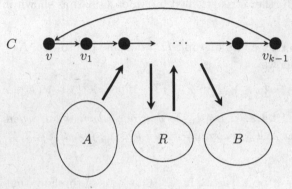

Fig. 10.4.2

Case 1. R is empty. As T is strong and $k \leq n-1$, it follows that both A and B are non-empty, and there exist w in A and u in B such that $u \to w$ as shown in Fig. 10.4.3.

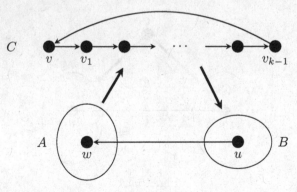

Fig. 10.4.3

In this case, $vv_1 \cdots v_{k-2}uwv$ is a $(k+1)$-cycle containing v.

Case 2. R is non-empty. Let z be a vertex in R. By the definition of R, z dominates some vertex in C and is dominated by some vertex in C as well. It follows that there exist two consecutive vertices, say s and t, in C such that $s \to z$ and $z \to t$ as shown in Fig. 10.4.4 (why?).

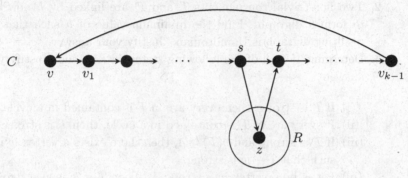

Fig. 10.4.4

In this case, v is contained in the $(k+1)$-cycle as shown in Fig. 10.4.5.

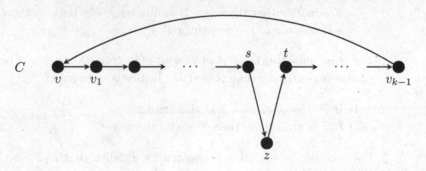

Fig. 10.4.5

This completes the proof of the theorem.

□

Exercise for Section 10.4

In this Exercise, we denote by T a tournament and by T_n a tournament of order $n \geq 2$.

1. Assume that $n \geq 5$. Is it true that $T_n - e$ always contains a Hamiltonian path, where e is an arc in T_n?

2. Two non-trivial tournaments T and T' are linked by h new arcs to form a digraph. Find the minimum value of h such that the resulting digraph is Hamiltonian. Justify your answer.

3. Determine if each of the following statements is true. Justify your answers.

    (i) If T is strong, then every arc in T is contained in a cycle.
    (ii) If every arc in T is contained in a cycle, then T is strong.
    (iii) If T is strong and $v(T) \geq 4$, then there exists a vertex w in T such that $T - w$ is strong.
    (iv) Let w be a vertex in a strong T. Then $T - w$ is also strong.
    (v) If T is strong, then every arc in T is contained in a Hamiltonian cycle.
    (vi) If $T - w$ is strong for every vertex w in T, where $v(T) \geq 4$, then T is strong.
    (vii) Suppose that T is strong. Then for any two distinct vertices u, v in T, either there is a Hamiltonian $u - v$ path or there is a Hamiltonian $v - u$ path in T.

4. Let S be a subtournament of T with $v(S) \geq 3$. Determine if each of the following statements is true. Justify your answers.

    (i) If T is strong, then S is also strong.
    (ii) If T is transitive, then S is also transitive.

5. For each odd n, $n = 2k + 1$, construct a T_n such that $E(T)$ can be decomposed into k arc-disjoint Hamiltonian cycles.

6.   (i) Assume that T_n is strong. Show that it contains at least $n - 2$ cyclic triples (i.e., 3-cycles).
    (ii) For each $n \geq 3$, construct a strong T_n which has exactly $n - 2$ cyclic triples.

10.5   The Score Sequence of a Tournament

In the context of tournaments, quite often, the out-degree of a vertex v is called the **score** of v. Thus, let T be a tournament of order n, and v a vertex in T. The **score** of v, denoted by $s(v)$, is defined as $s(v) = d^+(v)$. Assume that the n vertices in T are named $v_1, v_2, \cdots, v_n$ such that $s(v_1) \leq s(v_2) \leq \cdots \leq s(v_n)$. We call the sequence $(s(v_1), s(v_2), \cdots, s(v_n))$ the **score sequence** of T in non-decreasing order. Thus, $(1, 1, 2, 3, 4, 4)$ is the score sequence of the tournament of Fig. 10.5.1.

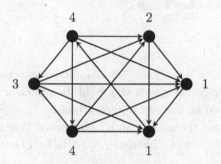

Fig. 10.5.1

Note: When we talk about the degree sequence
$$(d(v_1), d(v_2), \cdots, d(v_n))$$
of a graph, the sequence is arranged in *non-increasing* order (see Section 2.5). The order of any score sequence of a tournament is always *non-decreasing*.

A non-decreasing sequence $(s_1, s_2, \cdots, s_n)$ of non-negative integers is said to be **representable** if it is the score sequence of some tournament T. If this is the case, we also say that T **represents** the sequence or the sequence is **represented by** T.

A natural and basic problem is to determine whether a given non-decreasing sequence $(s_1, s_2, \cdots, s_n)$ is representable, that is, the score sequence of a tournament.

By Problem 10.3.7 and Result (2)(iii) in Section 10.3, we have the following two necessary conditions for $(s_1, s_2, \cdots, s_n)$ to be representable:

(i) For any k with $1 \leq k \leq n$, $\sum(s_i \mid 1 \leq i \leq k) \geq \binom{k}{2}$ and

(ii) $\sum(s_i \mid 1 \le i \le n) = \binom{n}{2}.$

Landau (1953) showed that they together are also sufficient.

Theorem 10.5.1.   *The non-decreasing sequence $(s_1, s_2, \cdots, s_n)$ of non-negative integers is representable if and only if for each $k = 1, 2, \cdots, n$,*

$$\sum(s_i \mid 1 \le i \le k) \ge \binom{k}{2},$$

with equality holding when $k = n$.

$\square$

Landau's Theorem is so 'great' that many researchers have thought it worthwhile to attempt their own proof (of its sufficiency). Until now, the theorem has received at least 10 different proofs. In what follows, we shall introduce by example the idea of an elegant proof, due to Bang and Sharp (1979), which makes use of Theorem 7.4.3 (Hall's Theorem on SDR).

Thus, suppose we are given a non-decreasing sequence $(s_1, s_2, \cdots, s_n)$ of non-negative integers satisfying the condition.

Take, for example, the sequence $(s_1, s_2, \cdots, s_5) = (1, 2, 2, 2, 3)$ with $n = 5$ that satisfies the condition. The objective here is to introduce a general method to construct a tournament of order 5 that represents the sequence $(1, 2, 2, 2, 3)$.

Step 1: For $i = 1, 2, \cdots, n$, let A_i be an arbitrary set with $|A_i| = s_i$, and the A_i's are pairwise disjoint. For our instance, let $A_1 = \{a\}$, $A_2 = \{b, c\}$, $A_3 = \{d, e\}$, $A_4 = \{f, g\}$ and $A_5 = \{p, q, r\}$.

Step 2: Form $A_i \cup A_j$, for all $1 \le i < j \le n$ (there are $\binom{n}{2}$ such unions). In our case, we have:

$$
\begin{aligned}
S_1 &= A_1 \cup A_2 = \{a, b, c\}, & S_2 &= A_1 \cup A_3 = \{a, d, e\}, \\
S_3 &= A_1 \cup A_4 = \{a, f, g\}, & S_4 &= A_1 \cup A_5 = \{a, p, q, r\}, \\
S_5 &= A_2 \cup A_3 = \{b, c, d, e\}, & S_6 &= A_2 \cup A_4 = \{b, c, f, g\}, \\
S_7 &= A_2 \cup A_5 = \{b, c, p, q, r\}, & S_8 &= A_3 \cup A_4 = \{d, e, f, g\}, \\
S_9 &= A_3 \cup A_5 = \{d, e, p, q, r\}, & S_{10} &= A_4 \cup A_5 = \{f, g, p, q, r\}.
\end{aligned}
$$

Step 3: Given that the sequence $(s_1, s_2, \cdots, s_n)$ satisfies the condition, Bang and Sharp showed that the family $(S_1, S_2, \cdots, S_m)$, where $m = \binom{n}{2}$, satisfies the inequality:

$$\left| \bigcup_{i \in I} S_i \right| \geq |I|,$$

for any subset I of $\{1, 2, \cdots, m\}$. Thus, by Theorem 7.4.3, the family $(S_1, S_2, \cdots, S_m)$ has an SDR.

In our case, we have, for instance,

$$S_1 - b, S_2 - a, S_3 - f, S_4 - p, S_5 - c, S_6 - g, S_7 - q, S_8 - e, S_9 - d, S_{10} - r.$$

Step 4: Construct a tournament with n vertices $A_1, A_2, \cdots, A_n$ as follows: given $1 \leq i, j \leq n$, draw an arc from A_i to A_j when and only when the representative of $A_i \cup A_j$ in the SDR obtained in Step 3 is from A_i.

Thus, for our instance, the tournament with 5 vertices is shown in Fig. 10.5.2.

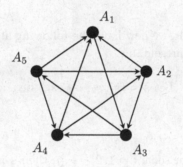

Fig. 10.5.2

Bang and Sharp showed that this tournament represents the given sequence $(s_1, s_2, \cdots, s_n)$.

. Note that Theorem 10.5.1 for tournaments bears some similarity with Theorem 2.5.3 (Erdös and Gallai Theorem) for graphic sequences. There is also a tournament analog of Theorem 2.5.5 (Havel and Hakimi Theorem) for graphic sequences which enables us to derive an algorithm for determining if a sequence is representable. To get to this, we first introduce a way to shorten a sequence.

Let $s = (s_1, s_2, \cdots, s_n)$ be a non-decreasing sequence of non-negative integers with $s_n \le n - 1$. Form

$$(s_1, s_2, \cdots, s_m, s_{m+1} - 1, s_{m+2} - 1, \cdots, s_{n-1} - 1),$$

where $m = s_n$, and let s' be the new sequence arranged in non-decreasing order.

For example, if $s = (1, 2, 3, 4, 4, 4, 5, 5)$, then $n = 8$ and $m = s_8 = 5$. Deleting the last term 's_8', keeping the first '5' terms, and subtracting '1' from each of the remaining $n - 1 - s_8$ terms, we have: $1, 2, 3, 4, 4, 3, 4$. Arranging them in non-decreasing order, we obtain

$$s' = (1, 2, 3, 3, 4, 4, 4).$$

We now state the following (see Problem 10.5.3):

Theorem 10.5.2. *Let $s = (s_1, s_2, \cdots, s_n)$ be a non-decreasing sequence of non-negative integers such that $s_n \le n - 1$. Then s is representable if and only if s' is so.*

$\square$

Based on this result, we now have the following algorithm for determining if a sequence is representable.

Algorithm 10.5.3. Let $s = (s_1, s_2, \cdots, s_n)$ be a sequence of n integers.

Step 1: Set $j := n$.

Step 2: If $s_i \ge j$ for some $i = 1, 2, \cdots, j$, then s is not representable.

Step 3: If there is a 'negative' term in the current sequence, then s is not representable.

Step 4: If the current sequence is (0), then s is representable.

Step 5: Arrange the current sequence in non-decreasing order.

Step 6: Delete the last term 's_j', keep the first 's_j' terms, and subtract '1' from each of the remaining $j - 1 - s_j$ terms. Let this new sequence be the current sequence.

Step 7: Reduce j by 1 and return to Step 2.

Example 10.5.4. Determine by Algorithm 10.5.3 if the following sequences are representable:

(i) $(0, 1, 1, 4, 4)$,
(ii) $(1, 1, 2, 2, 3)$,
(iii) $(0, 2, 2, 3, 3)$.

(i) Following the algorithm, we have:

0	1	1	4	4	Perform Step 6 with $(j = 5, s_5 = 4)$
0	1	1	4		$(j = 4, s_4 = 4)$

The sequence is not representable by Step 2.

(ii) Following the algorithm, we have:

1	1	2	2	3	Perform Step 6 with $(j = 5, s_5 = 3)$
1	1	2	1		Perform Step 4
1	1	1	2		Perform Step 6 with $(j = 4, s_4 = 2)$
1	1	0			Perform Step 4
0	1	1			Perform Step 6 with $(j = 3, s_3 = 1)$
0	0				Perform Step 6 $(j = 2, s_2 = 0)$
−1					

The sequence is not representable by Step 3. We also note that for this sequence,

$$\sum (s_i \mid 1 \le i \le 5) = 9 \ne \binom{5}{2},$$

which violates a condition in Landau's theorem.

(iii) Following the algorithm, we have:

0	2	2	3	3	Perform Step 6 with $(j = 5, s_5 = 3)$
0	2	2	2		Perform Step 6 with $(j = 4, s_4 = 2)$
0	2	1			Perform Step 4
0	1	2			Perform Step 6 with $(j = 3, s_3 = 2)$
0	1				Perform Step 6 with $(j = 2, s_2 = 1)$
0					

The sequence is representable by Step 4.

Remark 10.5.5. When applying Algorithm 10.5.3 manually, the procedure may end at the place where we know if the current sequence is representable. For instance, in Example 10.5.4(iii), the moment the current sequence $(0, 1, 2)$ is obtained (which is representable), we should know that the sequence $(0, 2, 2, 3, 3)$ is representable by Theorem 10.5.2.

Knowing that a sequence is representable by Algorithm 10.5.3, we can also reverse the procedure in the algorithm to construct a tournament representing the sequence.

Example 10.5.6. Construct a tournament representing the sequence $(0, 2, 2, 3, 3)$.

We reverse the procedure from $(0, 2, 2, 3, 3)$ to $(0, 1, 2)$ as shown in Example 10.5.4(iii).

First, we construct a tournament representing $(0, 1, 2)$, which is the transitive triple with vertex set $\{u, v, w\}$, as shown in Fig. 10.5.3(a).

Next, we extend it to construct a tournament representing $(0, 2, 2, 2)$. We add in a new vertex x and wish to make sure that the current score of x is '2'. From the above procedure, x should dominate w (its current score is 0) and u (its current score is 2), but be dominated by v (so that the score of v increases by 1) as shown in Fig. 10.5.3(b).

Finally, we extend the tournament representing $(0, 2, 2, 2)$ to construct a tournament representing $(0, 2, 2, 3, 3)$. We add in a last vertex y and wish to make sure that $s(y) = 3$. From the above procedure, y should dominate w ($s(w) = 0$) and two of the vertices with current score '2' (we choose, say u and x), but be dominated by v (so that the score of v increases by 1) as shown in Fig. 10.5.3(c). The resulting tournament is a desired one.

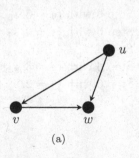

(a)

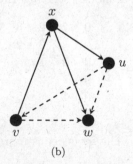

(b)

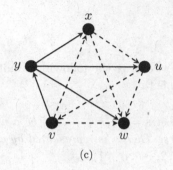

(c)

Fig. 10.5.3

What can be said about the score sequences of strong tournaments?

Theorem 10.5.7. *Let $s = (s_1, s_2, \cdots, s_n)$ be a non-decreasing sequence of non-negative integers. Then s is represented by a strong tournament if and only if*

(i) for any k with $1 \le k < n$, $\sum(s_i \mid 1 \le i \le k) > \binom{k}{2}$ and

(ii) $\sum(s_i \mid 1 \le i \le n) = \binom{n}{2}$.

Proof. ($\Rightarrow$) Let T be a strong tournament with $V(T) = \{v_1, v_2, \cdots, v_n\}$ that represents s. By Theorem 10.5.1, (ii) follows and $\sum_{i=1}^{k} s_i \ge \binom{k}{2}$ for each $k = 1, 2, \cdots, n-1$. Suppose that $\sum_{i=1}^{k} s_i = \binom{k}{2}$ for some k with $1 \le k \le n-1$. Consider the set $U = \{v_1, v_2, \cdots, v_k\}$. As $[U]$ itself is a tournament, $e([U]) = \binom{k}{2}$. Thus there is no arc from U to $\overline{U} = (V(T) \backslash U)$, and so T is not strong, a contradiction. Hence (i) follows.

($\Leftarrow$) By Theorem 10.5.1, s is the score sequence of some tournament T. Assume T is not strong. Then by Theorem 10.3.5, T is reducible. Let $\{A, B\}$ be a partition of $V(T)$ such that every vertex in A dominates every vertex in B. Let $B = \{u_1, u_2, \cdots, u_k\}$. Then observe that

$$\sum_{i=1}^{k} s_i \le \sum_{i=1}^{k} s(u_i) = \sum_{i=1}^{k} s_{[B]}(u_i) = \binom{k}{2},$$

a contradiction.

$\square$

To end this section, we state the following result for transitive tournaments (see Problem 10.5.4).

Theorem 10.5.8. *Let $s = (s_1, s_2, \cdots, s_n)$ be a non-decreasing sequence of non-negative integers. Then s is represented by a transitive tournament if and only if $s = (0, 1, 2, \cdots, n-1)$.*

$\square$

Exercise for Section 10.5

In this Exercise, we denote by T a tournament and by T_n a tournament of order $n \geq 2$.

1. Apply Theorem 10.5.1 to determine whether each of the following sequences is the score sequence of some tournament. If your answer is 'yes', then apply Bang and Sharp's method to construct such a tournament.

    (i) $(1, 1, 2, 3, 3)$;
    (ii) $(1, 1, 1, 2, 5, 5)$.

2. Consider the following four sequences of non-negative integers.

    $$a = (1, 1, 1, 3, 3, 6), \quad b = (1, 1, 2, 2, 4, 5),$$
    $$c = (0, 1, 2, 3, 4, 5), \quad d = (1, 2, 2, 2, 4, 4).$$

    (i) For each of them, determine whether it is the score sequence of some tournament. Give reason if your answer is 'no'; construct such a tournament if your answer is 'yes'.
    (ii) Which of them is the score sequence of a strong tournament? Justify your answer.

3. Prove Theorem 10.5.2.
4. Assume that $v(T) = n \geq 3$. Prove that the following statements are equivalent:

    (1) T is transitive.
    (2) $s(u) \neq s(v)$ for any two vertices u and v in T.

(3) The score sequence is given by $(0, 1, 2, \cdots, n-1)$.
(4) T has a unique Hamiltonian path.
(5) T is acyclic.
(6) T contains no C_3.

5. A tournament T is said to be **self-converse** if $T \cong \overrightarrow{T}$, where $\overrightarrow{T}$ is the converse of T (see Problem 10.2.25). Let $\boldsymbol{s} = (s_1, s_2, \cdots, s_n)$ be the score sequence of a tournament T. Show that T is self-converse if and only if $s_i + s_{n-i+1} = n-1$ for each $i = 1, 2, \cdots, n$.

10.6 Kings and Bases

Given a tournament T, no matter how large it is, by Theorem 10.4.1, we are guaranteed the existence of a vertex, from which we can reach all other vertices in T by a long (in fact, longest) but single path.

In contrast with this, we may ask a similar question, but from another extreme perspective, that is: Given any arbitrary tournament T, does there exist in T a vertex, from which we can reach all other vertices in T, each by some 'short' path?

What does 'short path' mean? The shortest ones are, of course, of length one. In this case, the question is: Does there exist a vertex in T from which we can reach all others by paths of length one (that is, single arcs)?

As shown in Fig. 10.6.1, clearly, this is possible when and only when T contains a source.

Source

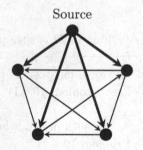

Fig. 10.6.1

Now, how 'short' are the paths we can expect if T contains no source?

Naturally, we ask: *Does there exist a vertex in T from which we can reach all others by paths of length at most two?'*

Example 10.6.1. Let T be the tournament of Fig. 10.6.2. Note that (i) T contains no source and (ii) from the vertex u, we can reach all others by paths of length at most two as indicated in bold.

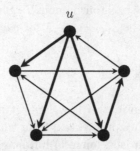

Fig. 10.6.2

Vertices of this sort are interesting, and we now give them a name.

Let D be a digraph and r, a positive integer. A vertex w in D is called an r-**king** if $d(w, x) \le r$ for each vertex x in D. Specially, a 2-king in D is also called a **king** in D.

We shall denote by $K_r(D)$ the set of all r-kings in D; and by $k_r(D)$ the number of r-kings in D, that is, $k_r(D) = |K_r(D)|$.

Remark 10.6.2.

(1) By definition, every r-king in D is also an $(r+1)$-king in D; that is, $K_r(D) \subseteq K_{r+1}(D)$.

(2) If D contains an r-king for some positive integer r, then D contains at most one source (see Problem 10.6.1).

In what follows, let T_n be a tournament, where $n \ge 2$. We have the following observations (see Problem 10.6.2).

Observation 10.6.3.

(1) A vertex w is a 1-king in T_n if and only if w is the (unique) source in T_n.

(2) $0 \leq k_1(T_n) \leq 1$, and $k_1(T_n) = 1$ if and only if T_n has a source.

The concept of 'king' was first introduced by Landau (1953) who also proved the existence of a king in any tournament. First of all, we have the following simple result.

Lemma 10.6.4. *Let u and v be any two vertices in T_n such that $s(u) \geq s(v)$. Then $d(u,v) \leq 2$.*

Proof. We may assume that $v \rightarrow u$. If there is a vertex in $O(u)$ which dominates v, then $d(u,v) = 2$, and we are done. Otherwise, v dominates every vertex in $O(u)$, which implies that $s(v) \geq |O(u)| + 1 > s(u)$, a contradiction.

$\square$

The following result, due to Landau, follows readily.

Corollary 10.6.5. *In T_n, any vertex with maximum score is a king, and so $k_2(T_n) \geq 1$.*

$\square$

Not every king in T_n is of maximum score. Indeed, the score of a king in some T_n could be as low as one (see Problem 10.6.3).

What can be said about the structure of T_n when $k_2(T_n) = 1$? Obviously, if T_n contains a source, then $k_2(T_n) = k_1(T_n) = 1$. Is the converse true?

Lemma 10.6.6. *Any vertex that is not the source in T_n is dominated by some king in T_n.*

Proof. Let u be a vertex in T_n which is not the source. Then $I(u)$ is not empty (see Fig. 10.6.3). Note that the subdigraph $[I(u)]$ of T_n is a tournament. Thus $[I(u)]$ contains a (local) king, say w. Observe that $w \rightarrow u \rightarrow O(u)$. Thus w is a king in T_n and $w \rightarrow u$.

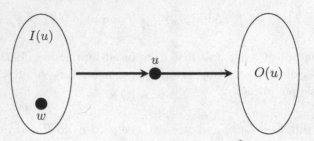

Fig. 10.6.3

$\square$

Corollary 10.6.7. $K_2(T_n) = \{w\}$ *if and only if w is the source in T_n.*

$\square$

Remark 10.6.8. It is interesting to note that Huang and Li (1987) showed that if T_n contains no source, then every vertex in T_n which is not a king is dominated by at least two kings in T_n. (See Problem X.8.)

The above corollary says that $k_2(T_n) = 1$ if and only if T_n contains a source. Does there exist a T_n such that $k_2(T_n) = 2$? By Lemma 10.6.6 and Corollary 10.6.7, we now have the following negative answer due to Moon (1962).

Theorem 10.6.9. *If T_n contains a source, then $k_2(T_n) = 1$; otherwise, $k_2(T_n) \geq 3$.*

$\square$

It is easy to construct T_n such that $k_2(T_n) = 3$ for each $n \geq 3$. A necessary condition for T_n to have the equality $k_2(T_n) = 3$ is given in Problem X.9.

Strong kings

In a round-robin tournament, one may want to find out who are the better players. Certainly, those who correspond to kings could be candidates.

However, as pointed out earlier, the score of a king in some tournament could be as low as one, and so saying that such a player is a better player may not be convincing. To have a better and fairer criterion, Ho and Chang (2003) imposed an additional condition on a king and introduced the notion of a strong king.

For every two vertices u, v in tournament T, let $b(u, v)$ denote the number of vertices x in T such that $u \to x \to v$ (see Fig. 10.6.4).

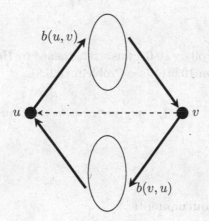

Fig. 10.6.4

A king w in T is called a **strong king** if $b(w, v) > b(v, w)$ for any vertex v with $v \to w$.

Question 10.6.10. *Find all the kings in the tournament of Fig. 10.6.5. Which of them are strong kings?*

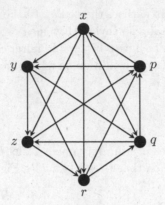

Fig. 10.6.5

The result in Corollary 10.6.5 was strengthened by Ho and Chang (2003) as shown in Theorem 10.6.11 (see Problem 10.6.8).

Theorem 10.6.11.   *In T_n, any vertex with maximum score is a strong king.*

$\square$

Multipartite tournaments

A digraph W is called an m-**partite tournament**, where $m \geq 2$, if $V(W)$ can be partitioned into m subsets $V_1, V_2, \cdots, V_m$ such that there is exactly one arc between u and v for any two vertices u and v in distinct V_i and V_j. We call $V_1, V_2, \cdots, V_m$ the **partite sets** of W. If $|V_i| = p_i$ for each $i = 1, 2, \cdots, m$, such an m-partite tournament W is denoted more precisely by $W(p_1, p_2, \cdots, p_m)$. Note that it is a digraph obtained by assigning a direction to each edge of a complete m-partite graph $K(p_1, p_2, \cdots, p_m)$ (see Problem III.24). A 2-partite tournament is better known as a **bipartite** tournament. An m-partite tournament, where $m \geq 2$, is also called a **multipartite** tournament.

Thus a tournament T is an n-partite tournament $W(p_1, p_2, \cdots, p_n)$, where $p_1 = p_2 = \cdots = p_n = 1$. Corollary 10.6.5 tells us that $K_2(T_n) \neq \emptyset$. It was pointed out in Remark 10.6.2 that if D is a digraph such that

$K_r(D) \neq \emptyset$ for some $r \geq 1$, then

$$D \text{ contains at most one source.} \tag{*}$$

Suppose that W is a multipartite tournament satisfying (*). Is it true that $K_2(W) \neq \emptyset$? Figure 10.6.6 shows an m-partite tournament W with $m \geq 2$, where (i) $[V_3 \cup V_4 \cup \cdots \cup V_m]$ is an arbitrary tournament of order $m-2$ and (ii) for any two sets A, B of vertices, '$A \to B$' signifies that every vertex of A dominates every vertex of B. It is easy to check that W satisfies (*) but $K_3(W) = \emptyset$.

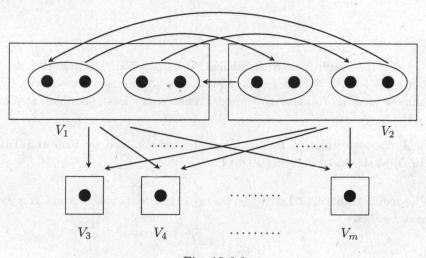

Fig. 10.6.6

The above example was in fact given by Gutin (1986) who further obtained the following result.

Theorem 10.6.12. *Let $W = W(p_1, p_2, \cdots, p_m)$ be a multipartite tournament, where $m \geq 2$. Then*

(i) $k_4(W) \geq 1$;
(ii) $k_3(W) \geq 1$ if $p_i \leq 3$ for each $i = 1, 2, \cdots, m$.

□

Theorem 10.6.12(i) was discovered independently by Petrovic and Thomassen (1991). Indeed, they proved a stronger result shown in

Theorem 10.6.13. For convenience, given an m-partitie tournament W with partite sets $V_1, V_2, \cdots, V_m$, we let

$$M_i = \{w \in V_i \mid s(w) \geq s(x) \text{ for each } x \in V_i\}$$

for $i = 1, 2, \cdots, m$.

Theorem 10.6.13. *Let W be an m-partite tournament, where $m \geq 2$. If T_m is a tournament contained in $[M_1 \cup M_2 \cup \cdots \cup M_m]$, then $K_2(T_m) \subseteq K_4(W)$.*

□

Let W be a multipartite tournament. It follows from Theorem 10.6.12(i) that $k_4(W) = 1$ if W contains a unique source. A question arises naturally. Suppose that W contains no source. What is the least possible value of $k_4(W)$?

A complete solution to this question was obtained by Koh and Tan (1995) as shown in Theorem 10.6.14.

Theorem 10.6.14. *Let W be an m-partite tournament which has no source. Then*

$$k_4(W) \geq \begin{cases} 4, & \text{if } m = 2 \\ 3, & \text{if } m \geq 3. \end{cases}$$

□

The structures of W for which the above equalities hold were also determined.

Bases

Corollary 10.6.5 guarantees the existence of a 2-king in any tournament. Theorem 10.6.12 ensures the existence of a 4-king in any multipartite tournament. These digraphs are connected with relatively large number of arcs. For a general digraph, which may be disconnected and sparse, looking for such a 'strong' vertex is clearly unrealistic. Thus the best we could hope for is to form a *team* of such vertices which can work together to reach the remaining vertices.

Let D be a digraph with vertex set V. Two vertices x and y are *adjacent* if either x is adjacent to y or y is adjacent to x. A subset B of V is said to be **independent** if no two vertices in B are adjacent. For a positive integer k, a subset B of V is called a k-**base** if (i) B is independent and (ii) for each z in $V \backslash B$, there exists a vertex x in B such that $d(x, z) \leq k$. Clearly, every k-base is a $(k+1)$-base in D.

Example 10.6.15.

(i) For the digraph of Fig. 10.6.7, the set $\{a, b\}$ is a 2-base.

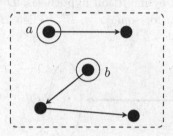

Fig. 10.6.7

(ii) As shown in Fig. 10.6.8, the cycle C_6 has a 1-base while C_5 has a 2-base but no 1-base. Indeed, it is easy to see that every even cycle has a 1-base while every odd cycle has a 2-base but no 1-base.

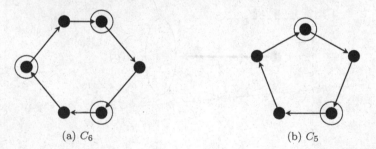

(a) C_6 (b) C_5

Fig. 10.6.8

A question arises naturally: Does any digraph always contain a 2-base?

Let $B \subseteq V$. We call B a k-**kernel** in D if it is a k-base in $\overrightarrow{D}$, the converse of D (see Problem 10.2.25). Chvátal and Lovász (1974) gave a positive answer to the above question using the term 'quasi-kernel' (that is, 2-kernel). Thus, we have:

Theorem 10.6.16.   *Every digraph contains a 2-base.*

Proof. The following proof is by induction on the order n of a digraph. The result is obvious for $n = 1, 2$. Now, let D be a digraph of order $n \geq 3$. Let $v \in V$ and $D' = D - (\{v\} \cup O(v))$.

If $V(D') = \emptyset$, then $B = \{v\}$ is a required 2-base.

If $V(D') \neq \emptyset$, then by induction hypothesis, D' has a 2-base, say B'.

Case (1). $x \rightarrow v$ for some $x \in B'$. Then B' is a 2-base of D (see Fig. 10.6.9(a)).

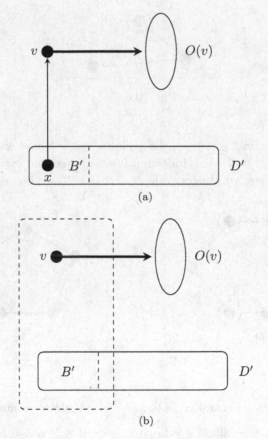

(a)

(b)

Fig. 10.6.9

Case (2). $x \to v$ for **no** $x \in B'$. Let $B = B' \cup \{v\}$. Then B is a required 2-base of D (see Fig. 10.6.9(b)).

The proof is thus complete.

$\square$

As an extension of the above theorem, Jacob and Meyniel (1996) proved the following result.

Theorem 10.6.17. *Let D be a digraph. If D contains no 1-base, then D contains at least three 2-bases.*

$\square$

To end this section, we summarize the above results on the existence of 2-kings in tournaments, 4-kings in multipartite tournaments and 2-bases in general digraphs below for comparison.

(1) Tournament T

$$\text{Landau:} \quad k_2(T) \geq 1.$$
$$\text{Silverman and Moon:} \quad \text{No source in } T \Rightarrow k_2(T) \geq 3.$$

(2) Multipartite tournament W

$$\text{Gutin, Petrovic-Thomassen:} \quad k_4(W) \geq 1$$
$$\text{Koh-Tan:} \quad \text{No source in } W \Rightarrow k_4(W) \geq 3.$$

(3) Digraph D.

Chvátal-Lovász: Number of 2-bases in D is at least 1

Jacob-Meyniel: No 1-base in $D \Rightarrow$ number of 2-bases in D is at least 3.

Exercise for Section 10.6

In this Exercise, we denote by T a tournament and by T_n a tournament of order $n \geq 2$.

1. Let D be a digraph which contains an r-king for some positive integer r. Show that D contains at most one source.

2. Prove the following statements:

   (i) A vertex w is a 1-king in T_n if and only if w is the (unique) source in T_n.

   (ii) $0 \leq k_1(T_n) \leq 1$, and $k_1(T_n) = 1$ if and only if T_n has a source.

3. For each $n \geq 2$, construct a T_n which contains a king w with $s(w) = 1$.

4. Assume that T_n is regular. How many kings does T_n have? Evaluate also its diameter.

5. Assume that $v(T) \geq 3$.

   (a) Show that a vertex w in T is an r-king, where $r \geq 2$, if and only if every arc incident to w is contained in a k-cycle, where $k \leq r + 1$.

   (b) Prove that the following statements are equivalent:

      (i) $\text{diam}(T) \leq r$;

      (ii) every vertex in T is an r-king;

      (iii) every arc is contained in a k-cycle, where $k \leq r + 1$.

6. Show that there is no tournament of order 4 in which every vertex is a king.

7. Construct

   (i) a tournament of order 5 in which every vertex is a king;

   (ii) a tournament of order 6 in which every vertex is a king;

   (iii) a tournament of order $n \geq 7$ in which every vertex is a king.

8. Show that any vertex with maximum score in T is a strong king. Is the converse true?

9. (a) Show that a vertex w in T is a strong king if and only if w dominates x for any vertex x in T with $s(x) > s(w)$.

   (b) Assume that $n \geq 3$ is odd and T_n contains exactly k strong kings. Show that $k \neq n - 1$.

   (c) Assume that T_n has exactly k strong kings, where $3 \leq k \leq n$. Show that T_n is a subtournament of a tournament T^* of order $n + 1$ that has also exactly k strong kings.

                              (Chen, Chang, Cheng & Wang (2008))

10. Let $T(X, Y)$ be a bipartite tournament with two partite sets X and Y. For simplicity, we denote $T(X, Y)$ by W. Assume that $d^+(u) = d^+(v)$ and $d^-(u) = d^-(v)$ for all u, v in $V(W)$. Show that

    (i) $|X| = |Y| = 2k$ for some positive integer k and

    (ii) $d^+(w) = d^-(w)$ for each w in $V(W)$.

11. Let $T(X, Y)$ be a bipartite tournament with two partite sets X and Y. For simplicity, we denote $T(X, Y)$ by W.

    (a) Let w be a vertex in W. Show that w is a 2-king if and only if w is the unique source in W.

    (b) Let w be a vertex in X. Show that w is a 3-king if and only if the following conditions are satisfied:

       (i) $O(w) \backslash O(x) \neq \emptyset$ for each $x \in X \backslash \{w\}$ and
       (ii) Y contains no source.

(Petrovic (1997))

    (c) Assume that W contains no source. Let z be a vertex in X such that $s(z) \geq s(x)$ for each x in X. Show that z is a 4-king in W.

(Goddard, Kubicki, Oellermann and Tian (1991))

12. Consider the digraph D shown in Fig. 10.6.10.

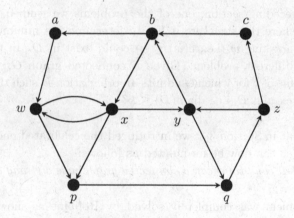

Fig. 10.6.10

Does there exist a 1-base in D? Find two 2-bases B in D with $|B| = 2$.

13. Let D be a digraph.

    (a) Show that if w is a source in D, then w is contained in any k-base of D.

    (b) Let B be a 1-base of D and A, a subset of $V(D)$ such that $B \subset A$. Show that A can never be independent.

(c) Let B be a 1-base of D and A, a subset of $V(D)$ such that $A \subset B$. Show that there exists x in $V(D) \backslash A$ such that $d(v, x) \neq 1$ for all $v \in A$.

(d) D is said to be **symmetric** if $xy \in E(D)$ whenever $yx \in E(D)$. Show that every symmetric D contains a 1-base.

14. Show that every acyclic digraph contains a unique 1-base.

10.7 Optimal Orientations of Graphs

Let G be a connected graph. Recall that (see Section 3.3) an **orientation** of G is a digraph obtained from G by assigning to each edge in G a direction. Thus, a *tournament* is an orientation of a complete graph. An orientation D of G is said to be **strong** if D is a strong digraph.

In the preceding section, one of the problems we studied is: Given a digraph D where the structure is fixed, determine the minimum value of k for which a k-king w (i.e., $e(w) \leq k$) could exist in D. In this section, we study a different problem: Given a connected graph G, what is the minimum value of k for which G admits an orientation D such that $e(v) \leq k$ for each vertex v in D, i.e., $\mathrm{diam}(D) \leq k$?

Recall that in Section 3.3, we introduced the celebrated one-way street problem which can now be formulated as follows:

Under what conditions can a connected graph have a strong orientation?

This problem was completely solved by Robbins as shown in Theorem 3.3.11, which states that **a connected multigraph has a strong orientation if and only if it contains no bridges.**

An efficient procedure for designing a strong orientation of a bridgeless connected graph based on the method of depth-first search, found by Roberts, was presented in Algorithm 3.3.12.

Discussion 10.7.1. As shown in Example 3.3.13, by applying Algorithm 3.3.12 on the graph $G(= P_3 \,\square\, P_3)$ of Fig. 10.7.1, we obtained three of its strong orientations as duplicated in Fig. 10.7.2.

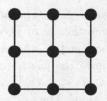

Fig. 10.7.1

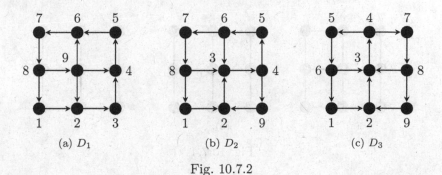

(a) D_1 (b) D_2 (c) D_3

Fig. 10.7.2

You may verify that $\operatorname{diam}(D_1) = 8$, $\operatorname{diam}(D_2) = 7$ and $\operatorname{diam}(D_3) = 6$. Can you find an orientation D of G such that $\operatorname{diam}(D) = 5$?

By trial and error, such an orientation D^* is shown in Fig. 10.7.3. It is noted that this orientation can never be obtained by applying Algorithm 3.3.12.

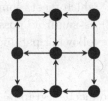

Fig. 10.7.3: $\operatorname{diam}(D^*) = 5$

For any strong orientation D of the above G, it is obvious that $\operatorname{diam}(D) \geq \operatorname{diam}(G) = 4$. Does there exist an orientation D of G such that $\operatorname{diam}(D) = 4$?

We claim that the answer is 'No'. The argument is by contradiction. Suppose there exists an orientation D of G such that $\mathrm{diam}(D) = 4$. As shown in Fig. 10.7.4(a), by symmetry, we may assume that $u \to v$ in D. Since $d(v, u) \leq 4$, we must have the partial orientation of G as shown in Fig. 10.7.4(b). As $d(y, z) \leq 4$, $y \to z$ as shown in Fig. 10.7.4(c). But then $d(z, v) \geq 5$ in D, a contradiction. Thus G admits no orientation D with $\mathrm{diam}(D) = 4$.

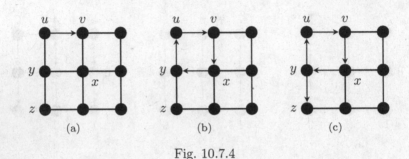

Fig. 10.7.4

By applying Algorithm 3.3.12 on a bridgeless connected graph G, we are sure to obtain a *strong* orientation D of G. However, for certain purposes, having a strong orientation may not be good enough; we may wish to design a strong orientation D which meets some other optimal criteria, for instance, having its $\mathrm{diam}(D)$ as small as possible. The above discussion motivates us to introduce the following parameter for G.

Let G be a bridgeless connected graph. Then G has its own non-empty family of strong orientations. Each strong orientation D of G has its $\mathrm{diam}(D)$. We wish to design a strong orientation D^* of G such that $\mathrm{diam}(D^*) \leq \mathrm{diam}(D)$ for any other strong orientation D of G. We call the *value* of $\mathrm{diam}(D^*)$ the **orientation number** of G, and call D^* an **optimal orientation** of G. Thus, the orientation number of the graph G of Fig. 10.7.1 is '5', and one of its optimal orientations is shown in Fig. 10.7.3.

Let G be a bridgeless connected graph. The **orientation number** of G, denoted by $\overrightarrow{d}(G)$, is defined by

$$\overrightarrow{d}(G) = \min\{\operatorname{diam}(D) \mid D \text{ is a strong orientation of } G\}.$$

An orientation D^* of G is said to be **optimal** if $\operatorname{diam}(D^*) = \overrightarrow{d}(G)$.

Example 10.7.2.

(i) For $n \geq 3$, what is the value of $\overrightarrow{d}(C_n)$? Up to isomorphism, the cycle C_n has one and only one strong orientation, and this orientation is of diameter $n - 1$. Thus $\overrightarrow{d}(C_n) = n - 1$.

(ii) Let $G(= P_4 + O_1)$ be the graph of Fig. 10.7.5. Evaluate $\overrightarrow{d}(G)$.

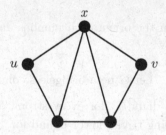

Fig. 10.7.5

We claim that $\overrightarrow{d}(G) = 3$. We first show that $\overrightarrow{d}(G) \geq 3$. Suppose this is not true. Then G admits an orientation D with $\operatorname{diam}(D) \leq 2$. This implies that $d(u, v) = 2$, and $u \to x \to v$ in D as shown in Fig. 10.7.6(a). But then $d(v, u)$ can never be two in D. Thus, $\overrightarrow{d}(G) \geq 3$.

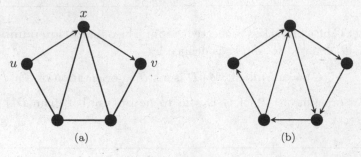

Fig. 10.7.6

Next, we show that $\overrightarrow{d}(G) \leq 3$. It suffices to design an orientation D^* of G with diam$(D^*) = 3$. Such an optimal orientation is shown in Fig. 10.7.6(b). We thus conclude that $\overrightarrow{d}(G) = 3$, as claimed.

Remark 10.7.3.   We learn from Example 10.7.2 and Discussion 10.7.1 the following two points in showing that $\overrightarrow{d}(G) = k$:

(1) To show that $\overrightarrow{d}(G) \geq k$ for some constant k, we may first assume that $\overrightarrow{d}(G) < k$, and proceed to derive a contradiction.
(2) To show that $\overrightarrow{d}(G) \leq k$ for some constant k, it suffices to design an orientation D of G with diam$(D) \leq k$.

Some useful facts on the orientation number and optimal orientations of G are presented below.

Observation 10.7.4.   Let G be a bridgeless connected graph.

(i) As diam$(D) \geq$ diam(G) for every strong orientation D of G, we have the following trivial lower bound for $\overrightarrow{d}(G)$, that is, $\overrightarrow{d}(G) \geq$ diam(G).
(ii) If G has an orientation D such that diam$(D) =$ diam(G), then $\overrightarrow{d}(G) =$ diam(G), and D is an optimal orientation of G.
(iii) Let D be a strong orientation of G. If x, y are vertices in G such that $d(x, y) = 1$ in D, then $d(y, x) \geq 2$ in D. It follows that diam$(D) \geq 2$, and thus $\overrightarrow{d}(G) \geq 2$.
(iv) If H is a spanning connected subgraph of G, then $\overrightarrow{d}(G) \leq \overrightarrow{d}(H)$ (see Problem 10.7.3).

The problem of evaluating $\vec{d}(G)$ for a general bridgeless connected graph G is very difficult (see Chvátal and Thomassen (1978)). Goldberg (1966) evaluated the extreme value of the diameter of a strong digraph in terms of its order and size. Expressing in $\vec{d}(G)$, his results amount to the following, which excludes that case when G is a cycle:

Theorem 10.7.5. *Let G be a bridgeless connected (n, m)-graph, where $n \geq 4$ and $m > n$. Then*

$$\vec{d}(G) \geq \left\lceil \frac{2(n-1)}{m-n+1} \right\rceil.$$

$\square$

While it is hard to evaluate $\vec{d}(G)$ for a general graph G, there are families of well-structured graphs whose orientation numbers have been found. In what follows, we present two such families.

We have pointed out in Observation 10.7.4(iii) that $\vec{d}(G) \geq 2$ for any bridgeless connected graph G. Which graphs have their $\vec{d}(G)$ equal to '2'? By Observation 10.7.4(iv), the complete graphs are likely candidates. Indeed, we have the following result due to Plesník (1975), Boesch and Tindell (1980) and Maurer (1980) independently. (See also Problem 10.6.7.)

Theorem 10.7.6. *For $n \geq 3$,*

$$\vec{d}(K_n) = \begin{cases} 2 \ \text{if } n \neq 4; \\ 3 \ \text{if } n = 4. \end{cases}$$

Proof. The orientations of complete graph are tournaments. It is obvious that $\vec{d}(K_3) = 2$. For $n = 4$, there are four non-isomorphic tournaments of order 4 as shown in Example 10.3.2. Among them, the one as shown in Fig. 10.7.7(a) is the only strong orientation, and it is of diameter 3. Thus $\vec{d}(K_4) = 3$.

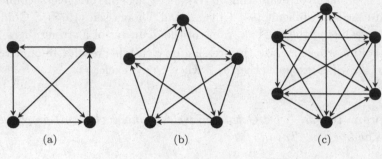

Fig. 10.7.7

Assume now that $n \geq 5$. The proof will be by induction on n with step '2'.

For the initial step, we consider the cases when $n = 5$ and $n = 6$. The orientations of K_5 and K_6 shown in Fig. 10.7.7(b) and (c) respectively are of diameter 2. Thus $\vec{d}(K_5) = \vec{d}(K_6) = 2$. Suppose $\vec{d}(K_k) = 2$ for $k \geq 5$. We shall show that $\vec{d}(K_{k+2}) = 2$.

View K_{k+2} as $K_k + K_2$, where $V(K_2) = \{u, v\}$, as shown in Fig. 10.7.8(a):

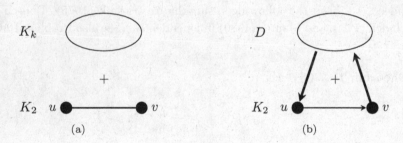

Fig. 10.7.8

As $\vec{d}(K_k) = 2$, K_k admits an orientation D with $\mathrm{diam}(D) = 2$. Design an orientation D' of K_{k+2} as follows:

(1) keep D for K_k,
(2) assign $u \to v$,
(3) assign $v \to x$ for all $x \in V(D)$,

(4) assign $x \rightarrow u$ for all $x \in V(D)$.

The orientation D' is shown in Fig. 10.7.8(b). It is straightforward to check that $\text{diam}(D') = 2$, and the proof is complete.

□

We now proceed to consider the family of complete bipartite graphs $K(p, q)$, where $q \geq p \geq 2$. Note that $\text{diam}(K(p, q)) = 2$.

Boesch and Tindell (1980) first observed that $\overrightarrow{d}(K(p,q)) \geq 3$ and $\overrightarrow{d}(K(p,p)) = 3$. Soltés (1986) and, independently, Gutin (1989) determined $\overrightarrow{d}(K(p,q))$ completely as shown in Theorem 10.7.7:

Theorem 10.7.7. *For $q \geq p \geq 2$,*

$$\overrightarrow{d}(K(p,q)) = \begin{cases} 3 \text{ if } q \leq \binom{p}{\lfloor \frac{p}{2} \rfloor}; \\ 4 \text{ otherwise.} \end{cases}$$

To prove Theorem 10.7.7, we make use of the following celebrated result in Combinatorics:

Lemma 10.7.8. (Sperner's Lemma) *Let p be a positive integer and let S be a collection of subsets of the set $\{1, 2, \cdots, p\}$ such that for any two members A, B in S, $A \nsubseteq B$. Then*

$$|S| \leq \binom{p}{\lfloor \frac{p}{2} \rfloor}$$

with equality occurring if and only if all of the members in S have the same cardinality.

□

Proof of Theorem 10.7.7. Let (X, Y) be the bipartition of $K(p, q)$, where $|X| = p$ and $|Y| = q$. We first prove the following:

(1) $\overrightarrow{d}(K(p,q)) \geq 3$.

Let D be an orientation of $K(p, q)$, and let u in X and v in Y be such that $u \rightarrow v$, say, in D. Then, as $K(p, q)$ is bipartite,

$d(v, u) \geq 3$ in D. This shows that $\text{diam}(D) \geq 3$ for any orientation D of $K(p, q)$. Thus (1) follows by definition.

We next prove the following:

(2) If $\vec{d}(K(p, q)) = 3$, then $q \leq \begin{pmatrix} p \\ \lfloor \frac{p}{2} \rfloor \end{pmatrix}$.

Let D be an orientation of $K(p, q)$ with $\text{diam}(D) = 3$. Observe that for any two vertices u, v in Y, as $\text{diam}(D) = 3$, we have $d(u, v) = 2$ in D, i.e., $O(u) \cap I(v) \neq \emptyset$, or equivalently, $O(u) \not\subseteq O(v)$.

Let $\mathcal{S} = \{O(u) \mid u \in Y\}$. Then $\mathcal{S}$ is a collection of subsets of X such that $O(u) \not\subseteq O(v)$ for any two vertices u, v in Y. By Lemma 10.7.8, and noting that $|\mathcal{S}| = |Y| = q$ and $|X| = p$, we have:

$$q \leq \begin{pmatrix} p \\ \lfloor \frac{p}{2} \rfloor \end{pmatrix}.$$

Finally, we leave it to the readers to design an orientation D of $K(p, q)$ with $\text{diam}(D) = 3$ when $q \leq \begin{pmatrix} p \\ \lfloor \frac{p}{2} \rfloor \end{pmatrix}$ and another orientation D' of $K(p, q)$ with $\text{diam}(D') = 4$ when $q > \begin{pmatrix} p \\ \lfloor \frac{p}{2} \rfloor \end{pmatrix}$. (See Problem 10.7.4.)

$\square$

Note: The original one-way street problem discussed in Section 3.3 discusses under what conditions a connected multigraph would have a strong orientation. The study of optimal orientations took this discussion further as we sought optimal orientations of a multigraph that minimizes the resulting digraph's diameter.

The Gossip Problem attributed to Boyd by Hajnal et al. (1972) is another instance where the study of optimal orientations is relevant. The problem is stated as follows:

The Gossip Problem. There are n ladies, and each of them knows an item of scandal which is not known to any of the others. They communicate by telephone, and whenever two ladies make a call, they pass on to each other, as much scandal as they know at that time. How many calls are needed before all ladies know all the scandal?

Another application for optimal orientations is communications in interconnecting networks. Initially each processor in the network has an item of information. The 'gossiping' process is implemented so that every item of information is finally distributed to every other processor of the system. At each stage, a processor can communicate with all its neighbors. The gossiping time is the minimum number of stages to complete the distribution.

If both adjacent processors can pass information to each other in the same stage, this communication model is usually referred to as Full-Duplex Δ-Port (F_*). Then the gossiping time is diam(G). On the other hand, if communication between two adjacent processors is only one way at a time, the communication model is called Half-Duplex Δ-Port (H_*). In that case, we have $\overrightarrow{d}(G)$ as an upper bound of the gossiping time.

The reader may also refer to Koh and Tay (2002) for a survey in the area of optimal orientations.

Exercise for Section 10.7

1. Let D be an orientation of a complete bipartite graph. Show that if D is not acyclic, then D contains a 4-cycle.

2. Let G be a bridgeless connected graph with girth g. Show that $\overrightarrow{d}(G) \geq g - 1$.

3. Let H be a spanning connected subgraph of a bridgeless connected graph G. Show that $\overrightarrow{d}(G) \leq \overrightarrow{d}(H)$.

4. Design (i) an orientation D of $K(p,q)$ with diam$(D) = 3$ when $q \leq \binom{p}{\lfloor \frac{p}{2} \rfloor}$ and (ii) an orientation D of $K(p,q)$ with diam$(D) = 4$ when $q > \binom{p}{\lfloor \frac{p}{2} \rfloor}$.

5. The fan of order $n \geq 4$, denoted by F_n, is defined as the join $P_{n-1} + O_1$. Show that

   (i) $\overrightarrow{d}(F_4) = \overrightarrow{d}(F_5) = 3$ (for $n = 5$, see Example 10.7.2(ii)) and
   (ii) $\overrightarrow{d}(F_n) = 4$ if $n \geq 6$.

6. Let $W_n \ (= C_{n-1} + O_1)$ be the wheel of order $n \geq 4$. Show that

   (i) $\overrightarrow{d}(W_4) = \overrightarrow{d}(W_5) = 3$ and
   (ii) $\overrightarrow{d}(W_n) = 4$ if $n \geq 6$.

7. Show that

   (i) $\vec{d}(C_4 + O_2) = 2$ and

  (ii) $\vec{d}(C_n + O_k) = 3$ for $n \geq 3$ and $k \geq 2$ such that $(n, k) \neq (4, 2)$.

(Ng (1999))

8. Let G_n be a tree of order n. Show that $\vec{d}(G_n + O_k) = 3$ for $n \geq 3$ and $k \geq 2$.

(Lee (2002))

9. Let G be the graph of order 8 shown in Fig. 10.7.9.

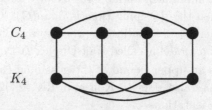

Fig. 10.7.9

Show that $\vec{d}(G) = 5$.

(Ng (2004) showed that the orientation number '5' remains unchanged if '4' (in C_4 and K_4) is replaced by any '$n \geq 3$' in the above structure of G.)

10. Let G be the graph shown in Fig. 10.7.10.

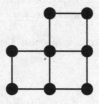

Fig. 10.7.10

   (i) Does there exist an orientation D of G such that diam$(D) \leq 5$?

  (ii) Design an orientation D^* of G such that diam$(D^*) = 6$.

 (iii) Find the value of $\vec{d}(G)$.

11. Evaluate the orientation number of the graph $P_2 \square P_k$, where $k \geq 3$.

(For the orientation number of the general $P_m \,\Box\, P_n$, see Roberts and Xu (1988), (1989), (1992) and (1994).)

12. Show that $\overrightarrow{d}\,(C_6 \,\Box\, P_2) = 5$.

(For the orientation number of the general $C_{2n} \,\Box\, P_k$, see Koh and Tay (1997).)

13. For $p \geq 3$, let B_p denote the spanning subgraph of the complete bipartite graph $K(p,p)$ obtained by deleting a perfect matching. Thus $B_3 \cong C_6$ and B_4 is shown in Fig. 10.7.11, which is isomorphic to the 3-cube graph Q_3.

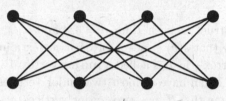

Fig. 10.7.11

Verify that $\overrightarrow{d}\,(B_3) = \overrightarrow{d}\,(B_4) = 5$.

(Goh (2006) showed that $\overrightarrow{d}\,(B_p) = 3$ for $p \geq 5$. Indeed, he proved more generally that by suitably removing almost $\frac{p^2}{2}$ edges from $K(p,p)$, the orientation number of the resulting spanning subgraph remains as '3'.)

10.8 Problem Set X

In this Exercise, we denote by T a tournament and by T_n a tournament of order $n \geq 2$.

1. Construct a digraph D such that (i) $d^+(v) = 2$ and $d^-(v) = 2$ for each vertex v in D and (ii) D contains no even cycle.

(Lovász conjectured in 1975 that if $d^+(v) \geq 2$ and $d^-(v) \geq 2$ for each v in D, then D contains an **even** cycle; see Koh (1976).)

2. Let $(s_1, s_2, \cdots, s_n)$ be the score sequence of T_n in non-decreasing order.

   (i) Show that, for each $k = 1, \cdots, n$,

   $$\frac{1}{2}(k-1) \leq s_k \leq \frac{1}{2}(n+k-2).$$

(ii) Show that if $s_n - s_1 < \frac{n}{2}$, then T_n is a strong tournament.

(iii) Is the converse of (ii) true?

3. Assume that $n \geq 4$. Show that the vertex set of T_n can be partitioned into two non-empty subsets A and B such that every vertex in A dominates a vertex in B, and every vertex in B is dominated by a vertex in A. Is the result still valid for $n = 3$?

4. For each $n \geq 3$, construct a T_n which contains a **unique** Hamiltonian cycle.

5. Suppose T is of order $n \geq 3$ and contains a **unique** Hamiltonian cycle.

   (a) Construct all such T for $n = 3, 4$ and 5.

   (b) Show that $s(v) \leq n - 2$ for each vertex v in T.

   (c) A vertex v in T is called a **major** vertex if $s(v) = n - 2$. Show that T can have at most two major vertices when $n \geq 4$.

   (d) Assume that T has two major vertices u, v and $u \to v$. We call u the first major vertex and v the second major vertex. Let $C = v_1 v_2 \cdots v_n v_1$ denote the unique Hamiltonian cycle in T, where $n \geq 4$. Suppose that v_n is the second major vertex of T. Show that

      (i) $v_{n-1} \to v_1$ and

      (ii) the tournament $T' = T - v_n$ also contains a **unique** Hamiltonian cycle.

                                                          (Moon (1982))

6. We say that T_n is **doubly regular** if for all x, y, u, v in $V(T_n)$ with $x \neq y$ and $u \neq v$, $|O(x) \cap O(y)| = |O(u) \cap O(v)|$. Show that if T_n is doubly regular, then T_n is regular and $n \equiv 3 \pmod 4$.

7. Let $(s_1, s_2, \cdots, s_n)$ be the score sequence of T_n. Evaluate

   (i) the number of transitive triples in T_n and

   (ii) the number of cyclic triples in T_n.

8. A vertex in T is said to be **normal** if it is not a king. For a vertex v in T, define

$$I^*(v) = \{x \in I(v) \mid x \text{ dominates every vertex in } O(v)\}.$$

Show that

   (i) a vertex w in T is a king if and only if $I^*(w) = \emptyset$;

   (ii) for each normal vertex v in T, there exists a king u in T such that $d(v, u) \geq 3$;

(iii) if T has no source, then every normal vertex is dominated by at least two kings.

(Huang and Li (1987))

9. (a) Assume that $n \geq 3$. Show that if $k_2(T_n) = 3$, then

   (i) $[K_2(T_n)]$ forms a 3-cycle and

   (ii) every vertex not in $K_2(T_n)$ is dominated by at least two vertices in $K_2(T_n)$.

   (b) For each $n \geq 3$, construct a T_n such that $k_2(T_n) = 3$.

   (c) Is the converse of (a) true?

10. Given T with $v(T) \geq 2$, show that there exists a tournament T^* such that $K_2(T^*) = T$ if and only if T contains no source.

(Reid (1982))

11. Let H be a subtournament of T such that $K_2(T) \subseteq V(H)$. Show that $K_2(H) \subseteq K_2(T)$.

(Bridgland and Reid (1984))

12. Let T be of order $n \geq 3$, and u, v be vertices in T such that $u \to v$. Let $D = T - uv$ be the subdigraph of T obtained by deleting uv. Show that D contains a king if and only if $d^-(u) + d^-(v) \geq 1$ in D.

(Yu, Di and Lin (2006))

13. Let T be of order $n \geq 3$, and u, v and w be vertices in T such that $u \to v$ and $v \to w$. Let $D = T - \{uv, vw\}$ be the subdigraph of T obtained by deleting uv and vw. Show that D contains a king if and only if $d^-(u) + d^-(v) \geq 1$ and $d^-(v) + d^-(w) \geq 1$ in D, not counting uw.

(Yu (2006))

Remark: The problem of characterizing $D = T - \{e, f\}$ that has a king for any two arbitrary arcs e and f has not been settled.

14. For each $k \geq 2$, construct a Hamiltonian bipartite tournament $W(X, Y)$ such that $|X| = |Y| = k$ and diam$(W(X, Y)) = 2k - 1$.

15. For integers k and n with $1 \leq k \leq n$, a tournament is called an (n, k)-**tournament** if it is of order n and contains exactly k kings. Show that an (n, k)-tournament always exists for all $1 \leq k \leq n$ except for the case when $n = k = 4$.

(Maurer (1980) and Reid (1982))

16. Show that, up to isomorphism, there exists one and only one (n, k)-tournament (see Problem X.15) for each of the following cases:

$(1,1), (2,1), (3,1), (3,3)$ and $(5,4)$.

Remark: Tay (2008) showed that for any (n, k) other than the above, there exist at least two non-isomorphic (n, k)-tournaments.

17. Let T be of order $n \geq 3$. Let w be a strong king in T and X be any set of vertices x such that $x \to w$ in T. Prove that w is still a strong king in the subtournament $T - X$ of T.

$$\text{(Ho and Chang (2003))}$$

18. For integers k and n with $1 \leq k \leq n$, a tournament is called an $(n, k)^s$-**tournament** if it is of order n and contains exactly k strong kings.

    (i) Show that for integers k and n with $1 \leq k \leq n$, an $(n, k)^s$-tournament always exists with the following exceptions: (i) n is odd and $k = n - 1$ and (ii) n is even and $k = n$.

    (ii) Show that, up to isomorphism, there exists one and only one $(n, k)^s$-tournament for each of the following cases:

    $$(1, 1), (2, 1), (3, 1), (3, 3), (4, 2), (4, 3), (5, 3), (5, 5),$$

    $$(6, 5) \text{ and } (7, 5).$$

Remark: Chen et al. (2008) showed that for any (n, k) other than the above, there exist at least two non-isomorphic $(n, k)^s$-tournaments.

19. Let D be a digraph and u a vertex in D. Show that u is either contained in a 2-base or dominated by a vertex in a 2-base of D.

$$\text{(Heard and Huang (2008))}$$

20. Let D be a digraph.

    (a) Let u be a vertex in D which is not a source. Show that D contains a 2-base B such that $u \notin B$.

    (b) Show that D has a unique 2-base if and only if D has a source and every vertex that is not a source is dominated by a source in D.

    (c) Show that if D has only one 2-base, say B, then B is a 1-base and B consists of the sources in D.

    (d) Assume that D contains no sources and has only two 2-bases P and Q. Show that $P \cap Q = \emptyset$.

    (See Gutin, Koh, Tay and Yeo (2004) and also Bowser and Cable (2005).)

21. (i) Let G be a bridgeless connected graph with $\text{diam}(G) = 2$. Show that $\overrightarrow{d}(G) \leq 6$. (Consider the case that some edge of G is contained in no C_3.)

(ii) Let H be the Petersen graph. Shown that $\overrightarrow{d}(H) = 6$.

(Chvátal and Thomassen (1978))

22. Let G be a graph obtained by adding two edges to link two disjoint K_5's in an arbitrary way as shown in Fig. 10.8.1:

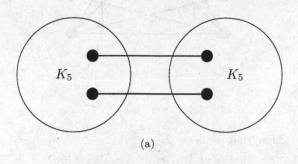

(a)

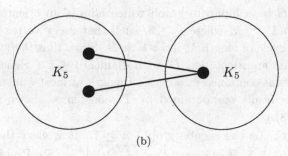

(b)

Fig. 10.8.1

(i) Show that for any orientation D of G, $\operatorname{diam}(D) \geq 4$.

(ii) Construct one such G such that $\overrightarrow{d}(G) = 4$.

(See Koh and Ng (2005) for more general results.)

23. (i) Let $G = K(2, 2, 2)$ as shown in Fig. 10.8.2.

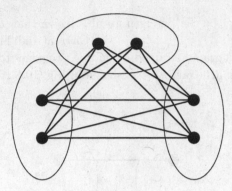

Fig. 10.8.2

Show that $\overrightarrow{d}(G) = 2$.

(ii) Show that $\overrightarrow{d}(K(p_1, p_2, \cdots, p_n)) \leq 3$ for $n \geq 3$.

(iii) Show that $\overrightarrow{d}(K(p, p, \cdots, p)) = 2$ for $n \geq 3$ and $p \geq 2$.

(See Plesník (1985), Gutin (1989), Koh and Tan (1996a), Koh and Tan (1996b).)

24. Let G be a bipartite graph which admits an orientation D with diam$(D) \leq k$, where $k \geq 3$, such that every vertex is contained in a cycle of length at most k in D. Show that the graph $G \,\square\, P_2$ admits an orientation D' with diam$(D') \leq k + 1$ such that every vertex is contained in a cycle of length at most k in D'.

(This result was obtained by Thomassen as shown in McCanna (1988).)

25. Let Q_n be the n-cube graph, $n \geq 1$. It is clear that $\overrightarrow{d}(Q_2) = \overrightarrow{d}(C_4) = 3$. It is known that $\overrightarrow{d}(Q_3) = 5$. (See Problem 10.7.13.) Applying the result in Problem X.24, or otherwise, show that $\overrightarrow{d}(Q_n) = n$ for $n \geq 4$.

(See McCanna (1988). For more general and further results along this direction, see Koh and Tay (2002).)

Bibliography

M. Adabi, E. Targhi, N. Jafari Rad and M. Moradi (2012), Properties of independent Roman domination in graphs. *Australasian J. Combinatorics* **52**, 11-18.

R.B. Allan and R. Laskar (1978), On domination and independent domination numbers of a graph. *Discrete Math.* **23**, 73-76.

A. Amahashi (1985), On factors with all degrees odd. *Graphs and Combinatorics* **1**, 111-114.

B. Andrásfai (1962), Über ein extremalproblem der graphentheorie. *Acta Math. Acad. Sci. Hungar.* **13**, 443-455.

P. Ang (2013), Some aspects of Roman domination in graphs. *Honours Project*, Department of Math., National University of Singapore.

F. Bäbler (1938), Über die Zerlegung regulärer Streckenkomplexe ungerader Ordnung. *Comment. Math. Helvetici.* **10**, 275-287.

C.M. Bang and H. Sharp (1979), Score vectors of tournaments. *J. Comb. Theory B* **26**, 81-84.

B. Bagheri, M. Jannesari and B. Omoomi (2011), Relations between metric dimension and domination number of graphs. (Preprint)

C. Berge (1957), Two theorems in graph theory. *Proc. Nat. Acad. Sci. USA* **43**, 842-844.

C. Berge (1962), *Theory of Graphs and Its Applications*. Methuen, London.

C. Berge (1973), *Graphs and Hypergraphs*. North-Holland, Amsterdam.

F. Boesch and R. Tindell (1980), Robbin's theorem for mixed multigraphs. *Amer. Math. Monthly* **87**, 716-719.

B. Bollobás and E.J. Cockyane (1979), Graph-theoretic parameters concering domination, independence, and irredundance. *J. Graph Theory* **3**, 241-249.

J.A. Bondy (1991), A graph reconstructor's manual, in: Surveys in combinatorics, Proceedings of the 13th British Combinatorial Conference, Guildford, UK, London Mathematical Society, Lecture Note Series **166**, 221-252.

J.A. Bondy and V. Chvátal (1976), A method in graph theory. *Discrete Math.* **15**, 111-136.

J.A. Bondy and F.Y. Halberstam (1986), Parity theorems for paths and cycles in graphs. *J. Graph Theory* **10**, 107-115.

J.A. Bondy and U.S.R. Murty (1976), *Graph Theory with Applications*. Elsevier North-Holland, New York.

R.C. Bose (1963), Strongly regular graphs, partial geometries, and partially balanced designs. *Pacific J. Math.* **13**, 389-419.

S. Bowser and C. Cable (2005), Digraphs with unique minimal king sets. *Discrete Math.* **305**, 347-349.

B. Brešar, P. Dorbec, W. Goddard, B.L. Hartnell, M.A. Henning, S. Klavžar and D. Rall (2012), Vizing's conjecture: A survey and recent results. *J. Graph Theory* **69**, 46-76.

M.F. Bridgland and K.B. Reid (1984), Stability of kings in tournaments, in J.A. Bondy and U.S.R. Murty (eds.), *Progress in Graph Theory*. Academic Press, New York, 117-128.

R.C. Brigham, P.Z. Chinn and R.D. Dutton (1988), Vertex domination-critical graphs. *Networks* **18**, 173-179.

R.C. Brigham and R.D. Dutton (1990), Bounds on the domination number of a graph. *Q. J. Math. Oxf. II. Ser.* **41**, 269-275.

R.L. Brooks (1941), On colouring the nodes of a network. *Proc. Cambridge Philosophical Society, Math. Phy. Sci.* **37**, 194-197.

P. Camion (1959), Chemins et circuits hamiltoniens des graphes complets. *Comptes Rendus de l'Académie des Sciences de Paris*, **249** 2151-2152.

J. Cao, M. Shi, M.Y. Sohn and X. Yuan (2008), Domination in graphs with minimum degree six. *J. Applied Mathematics and Informatics* **26(5-6)**, 1085-1100.

Y. Caro (1979), New results on the independence number. *Technical Report*, Tel-Aviv University.

Y. Caro and Y. Roditty (1990), A note on the k-domination number of a graph. *Internat. J. Math. Sci.* **13**, 205-206.

D. Cartwright and F. Harary (1968), On the coloring of signed graphs. *Elem. Math.* **23**, 85-89.

A. Cayley (1889), A theorem on trees. *Quart. J. Math.* **23**, 376-378.

E.W. Chambers, B. Kinnersley, N. Prince and D.B. West (2009), Extremal problems for Roman domination. *SIAM J. Discrete Math.* **23**, 1575-1586.

G. Chartrand and F. Harary (1967), Planar permutation graphs. *Annales de l'institut Henri Poincaré (B) Probabilités et Statistiques* **3(4)**, 433-438.

G. Chartrand and F. Harary (1968), Graphs with prescribed connectivities, in P. Erdös and G. Katona (eds.), *Theory of Graphs*. Akadémiai Kiadó, Budapest, 61-63.

G. Chartrand and S.F. Kapoor (1969), The cube of every connected graph is 1-hamiltonian. *J. Res. Nat. Bur. Standards Sect. B* **73**, 47-48.

G. Chartrand and O.R. Oellermann (1993), *Applied and Algorithmic Graph Theory*. McGraw-Hill, New York.

G. Chartrand, L. Eroh, M.A. Johnson and O.R. Oellermann (2000), Resolvability in graphs and the metric dimension of a graph. *Discrete Applied Math.* **105**, 99-113.

M. Chellali, N. Jafari Rad and L. Volkmann (2012), Some results on Roman domination edge critical graphs. *AKCE J. Graphs Comb.* **9**, 195-203.

A.H. Chen, J.M. Chang, Y. Cheng and Y.L. Wang (2008), The existence and uniqueness of strong kings in tournaments. *Discrete Math.* **308**, 2629-2633.

N. Christofides (1976), Worst-case analysis of a new heuristic for the traveling salesman problem. Report 388, Gradudate School of Industrial Administration, CMU.

V. Chungphaisan (1976), Factors of graphs and degree sequences. *Nanta Math.* **9**, 41-49.

V. Chvátal (1972), On Hamilton's ideals. *J. Comb. Theory B* **12**, 163-168.

V. Chvátal (1975), A combinatorial theorem in plane geometry. *J. Comb. Theory B* **18**, 39-41.

V. Chvátal and P. Erdös (1972), A note on Hamiltonian circuits. *Discrete Math.* **2**, 111-113.

V. Chvátal and L. Lovász (1974), Every directed graph has a semi-kernel. *Hypergraph Seminar, Lecture Notes in Math.* **441**, Springer-Verlag, Berlin, 175.

V. Chvátal and C. Thomassen (1978), Distances in orientations of graphs. *J. Comb. Theory B* **24**, 61-75.

E.J. Cockayne, P.A. Dreyer Jr., S.M. Hedetniemi and S.T. Hedetniemi (2004), Roman domination in graphs. *Discrete Math.* **278**, 11-22.

E.J. Cockayne, O. Favaron, C.M. Mynhardt and J. Puech (2000), A characterization of (γ, i)-trees. *J. Graph Theory* **34**, 277-292.

E.J. Cockayne and S.T. Hedetniemi (1974), Independence graphs. *Congr. Numer.* **10**, 471-491.

C. Delorme, M. Maheo, H. Thuillier, K.M. Koh and H.K. Teo (1980), Cycles with a chord are graceful. *J. Graph Theory* **4**, 409-415.

E.W. Dijkstra (1959), A note on two problems in connexion with graphs. *Numerische Mathematik* **1**, 269-271.

G.A. Dirac (1952a), Some theorems on abstract graphs. *Proceedings of the London Math. Soc.* 3rd Ser. **2**, 69-81.

G.A. Dirac (1952b), A property of 4-chromatic graphs and some remarks on critical graphs. *J. London Math. Soc.* **27**, 85-92.

G.A. Dirac (1960), In abstrakten Graphen vorhandene vollständige 4-Graphen und ihre Unterteilungen. *Math. Nachr.* **22**, 61-85.

M. Dorfling, W. Goddard, M.A. Henning and C.M. Mynhardt (2006), Construction of trees and graphs with equal domination parameters. *Discrete Math.* **306**, 2647-2654.

J. Edmonds (1965), The Chinese postman problem. *Operations Research* **13** Suppl. 1, 373.

J. Edmonds and E.L. Johnson (1973), Matching, Euler tours and the Chinese postman. *Mathematical Programming* **5**, 88-124.

E. Egerváry (1931), On combinatorial properties of matrices. *Matematikai és Fizikai Lapok* **38**, 16-28 (in Hungarian).

P. Erdös and T. Gallai (1960), Graphs with prescribed degrees of vertices. *Mat. Lapok* **11**, 264-274 (in Hungarian).

L. Euler (1736), The solution of a problem relating to the geometry of position. *Commentarii Academiae Scientiarum Imperialis Petropolitanae* **8**, 128-140.

G-H. Fan (1984), New sufficient conditions for cycles in graphs. *J. Comb. Theory B* **37**, 221-227.

O. Favaron, M. Maeho and J.F. Saclé (1991), On the residue of a graph. *J. Graph Theory* **15**, 39-64.

S. Fisk (1978), A short proof of Chvátal's watchman theorem. *J. Comb. Theory B* **24**, 374.

H. Fleischner (1974), The square of every two-connected graph is Hamiltonian. *J. Comb. Theory B* **16**, 29-34.

H. Fleischner (1989), Elementary proofs of (relatively) recent characterizations of Eulerian graphs. *Discrete Applied Math.* **24**, 115-119.

S. Fortin (1996), The graph isomorphism problem. *Technical Report TR96-20*, University of Alberta.

R. Frucht (1979), Graceful numbering of wheels and related graphs. *Ann. N.Y. Acad. Sci.* **319**, 219-229.

R. Frucht and J.A. Gallian (1988), Labeling prisms. *Ars Combin.* **26**, 69-82.

X. Fu, Y. Yang and B. Jiang (2009), Roman domination in regular graphs. *Discrete Math.* **309(6)**, 1528-1537.

T. Gallai (1959), Uber extreme Punkt- und Kantenmengen. *Ann. Univ. Sci. Budapest. Eötvös Sect. Math.* **2**, 133-138.

J.A. Gallian (2013), A dynamic survey of graph labeling. *Electron. J. Comb.* #DS6.

A. Ghouila-Houri (1960), Une condition suffisante d'existence d'un circuit hamiltonien. *C.R. Acad. Sci. Paris* **25**, 495-497.

W.D. Goddard, G. Kubicki, O. Oellermann and S.L. Tian (1991), On multipartite tournaments. *J. Comb. Theory B* **52**, 284-300.

H.C. Goh (2006), Strong orientations of spanning subgraphs of a complete bipartite graph. *M.Sc. Thesis*, Department of Math., National University of Singapore.

M.K. Goldberg (1966), The diameter of a strong connected graph. *Dokl. Akad. Nauk.* **170**, 767-769 (in Russian).

S.W. Golomb (1972), How to number a graph, in R.C. Read (ed.), *Graph Theory and Computing.* Academic Press, New York, 23-37.

D. Goncalves, A. Pinlou, M. Rao and S. Thomassé (2011), The domination number of grids. *SIAM J. Discrete Math.* **25(3)**, 1443-1453.

S. Goodman and S. Hedetniemi (1974), Sufficient conditions for a graph to be Hamiltonian. *J. Comb. Theory B* **16**, 175-180.

G. Gutin (1986), The radii of n-partite tournaments. *Math. Notes* **40**, 743-744.

G. Gutin (1989), m-sources in complete multipartite graphs. *Ser. Fiz.-Mat. Navuk.* **5**, 101-106 (in Russian).

G. Gutin, K.M. Koh, E.G. Tay and A. Yeo (2004), On the number of quasi-kernels in digraphs. *J. Graph Theory* **46**, 48-56.

A. Hajnal, E.C. Milner and E. Szemeredi (1972), A cure for the telephone disease. *Canad. Math. Bull.* **15**, 447-450.

G. Hajós (1961), Über eine Konstruktion nicht n-färbbarer Graphen. *Wiss. Z. Martin-Luther-Univ. Halle-Wittenberg Math.-Natur. Reihe* **10**, 116-117.

S.L. Hakimi (1962), On the realizability of a set of integers as degres of the vertices of a graph. *SIAM J. Appl. Math.* **10**, 496-506.

P. Hall (1935), On representatives of subsets. *J. London Math. Soc.* **10(1)**, 26-30.

A. Hansberg, N. Jafari Rad and L. Volkmann (2013), Vertex and edge critical Roman domination in graphs. *Util. Math.* **92**, 73-88.

P. Hansen (1975), Degrés et nombre de stabilité d'un graphe. *Cah. Centre Etudes Rech. Operation.* **17**, 213-220.

J. Harant and I. Schiermeyer (2001), On the independence number of a graph in terms of order and size. *Discrete Math.* **232**, 131-138.

J. Harant and I. Schiermeyer (2006), A lower bound on the independence number of a graph in terms of degrees. *Discuss. Math. Graph Theory* **26**, 431-437.

F. Harary (1962), The maximum connectivity of a graph. *Proc. Nat. Acad. Sci. USA* **48**, 1142-1146.

F. Harary, J.P. Hayes and H-J Wu (1988), A survey of the theory of hypercube graphs. *Computers & Mathematics with Applications* **15(4)**, 277-289.

F. Harary and M. Livingston (1986), Characterization of trees with equal domination and independent domination numbers. *Congr. Numer.* **55**, 121-150.

F. Harary and R.A. Melter (1976), On the metric dimension of a graph. *Ars Combin.* **2**, 191-195.

F. Harary and L. Moser (1966), The theory of round robin tournaments. *Amer. Math. Monthly* **73(3)**, 231-246.

B.L. Hartnell and D.F. Rall (1991), On Vizing's conjecture. *Congr. Numer.* **82**, 87-96.

V. Havel (1955), A remark on the existence of finite graphs (Czech.). *Casopis Pest. Mat.* **80**, 477-480.

T.W. Haynes, S. Hedetniemi and P. Slater (1998a), *Fundamentals of Domination in Graphs*. Marcel Dekker, New York.

T.W. Haynes, S. Hedetniemi and P. Slater (1998b), *Domination in Graphs: Advanced Topics*. Marcel Dekker, New York.

S. Heard and J. Huang (2008), Disjoint quasi-kernels in digraphs. *J. Graph Theory* **58(3)**, 251-260.

S. Hedetniemi and S. Mitchell (1977), Independent domination in trees. Proc. Eighth S.E. Conf. on Combinatorics, Graph Theory, and Computing, Baton Rouge, Los Angeles.

T.Y. Ho and J.M. Chang (2003), Sorting a sequence of strong kings in a tournament. *Inform. Process. Lett.* **87**, 317-320.

C. Hoede and H. Kuiper (1987), All wheels are graceful. *Util. Math.* **14**, 311.

R. Honsberger (1976), *Mathematical Gems II*. Mathematical Association of America, 104-110.

J. Huang and W. Li (1987), Toppling kings in a tournament by introducing new kings. *J. Graph Theory* **11**, 7-11.

H. Jacob and H. Meyniel (1996), About quasi-kernels in a digraph. *Discrete Math.* **154**, 279-280.

F. Jaeger and C. Payan (1972), Relations du ty Nordhaus-Gaddum pour le nombre d'absorption d'un graphe simple. *C.R. Acad. Sci. Paris A* **274**, 728-730.

S.R. Jayaram (1997), Minimal dominating sets of cardinality two in a graph. *Indian J. Pure appl. Math.* **28(1)**, 43-46.

J.J. Karaganis (1968), On the cube of a graph. *Canad. Math. Bull.* **11**, 295-296.

A.P. Kazemi (2012), Roman domination and Mycielski's structure in graphs. *Ars Combin.* **106**, 277-287.

J. Kobler, U. Schoning and J. Toran (1993), *The Graph Isomorphism Problem - Its Structural Complexity.* Springer-Verlag, Birkhäuser.

K.M. Koh (1976), Even circuits in digraphs and Lovász's conjecture. *Bull. Malaysian Math. Soc.* **7(3)**, 25-30.

K.M. Koh (1977), On the stability number of a graph. *Nanta Math.* **10**, 53-57.

K.M. Koh and B.H. Goh (1990), Two classes of chromatically unique graphs. *Discrete Math.* **82**, 13-24.

K.M. Koh and K.L. Ng (2005), The orientation number of two complete graphs with linkages. *Discrete Math.* **295**, 91-106.

K.M. Koh, D. G. Rogers and C.K. Lim (1979), On graceful graphs: sum of graphs. *Research Report 78*, College of Graduate Studies, Nanyang University.

K.M. Koh, D.G. Rogers, H.K. Teo and K.Y. Yap (1980), Graceful graphs: some further results and problems. *Congr. Numer.* **29**, 559-571.

K.M. Koh and B.P. Tan (1995), Kings in multipartite tournaments. *Discrete Math.* **147**, 171-183.

K.M. Koh and B.P. Tan (1996a), The minimum diameter of orientations of complete multipartite graphs. *Graphs and Combinatorics* **12**, 333-339.

K.M. Koh and B.P. Tan (1996b), The diamater of an orientation of a complete multipartite graph. *Discrete Math.* **149**, 131-139.

K.M. Koh and B.P. Tan (1996c), Number of 4-kings in bipartite tournaments with no 3-kings. *Discrete Math.* **154**, 281-287.

K.M. Koh and E.G. Tay (1997), On optimal orientations of cartesian products of even cycles and paths. *Networks* **30**, 1-7.

K.M. Koh and E.G. Tay (2002), Optimal orientations of graphs and digraphs: a survey. *Graphs and Combinatorics* **18(4)**, 745-756.

K.M. Koh and K.L. Teo (1989), The 2-Hamiltonian cubes of graphs. *J. Graph Theory* **13**, 737-747.

D. König (1931), Graphs and matrices. *Matematikai és Fizikai Lapok* **38**, 116-119 (in Hungarian).

J.B. Jr. Kruskal (1956), On the shortest spanning subtree of a graph and the traveling salesman problem. *Proc. Amer. Math. Soc.* **7**, 48-50.

K. Kuratowski (1930), Sur le problème des courbes gauches en topologie. *Fund. Math.* **15**, 271-283 (in French).

M.K. Kwan (1960), Programming method using odd or even points. *Acta Mathematica Sinica* **10**, 263-266 (in Chinese).

H.G. Landau (1953), On dominance relations and the structure of animal societies. III. The condition for a score structure. *Bulletin of Mathematical Biophysics* **15(2)**, 143-148.

C.E. Larson and R. Pepper (2014), Three bounds on the independence number of a graph. *Bulletin of the Institute of Combinatorics and Its Applications* **70**, 86-96.

S.L. Lee (2002), On the orientation number of graphs. *M.Sc. Thesis*, Department of Math., National University of Singapore.

L. Lovász (1975), Three short proofs in graph theory. *J. Comb. Theory B* **19**, 269-271.

C. Löwenstein and D. Rautenbach (2008), Domination in graphs of minimum degree at least two and large girth. *Graphs and Combinatorics* **24**, 37-46.

D.X. Ma and X.G. Chen (2004), A note on connected bipartite graphs having independent domination number half their order. *Appl. Math. Lett.* **17**, 959-962.

M. Maheo (1980), Strongly graceful graphs. *Discrete Math.* **29**, 39-46.

S.B. Maurer (1980), The king chicken theorems. *Math. Mag.* **53**, 67-80.

J.E. McCanna (1988), Orientations of the n-cube with minimum diameter. *Discrete Math.* **68**, 309-310.

W. McCuaig and B. Shepherd (1989), Domination in graphs with minimum degree two. *J. Graph Theory* **13**, 749-762.

T.A. McKee (1984), Recharacterizing Eulerian: intimations of new duality. *Discrete Math.* **51**, 237-242.

K. Menger (1927), Zur allgemeinen Kurventheorie. *Fund. Math.* **10**, 96-115.

M. Meyniel (1973), Une condition suffisante dexistence dun circuit Hamiltonien dans un graphe orient. *J. Comb. Theory B* **14**, 137-147.

B.P. Mobaraky and S.M. Sheikholeslami (2008), Bounds on Roman domination numbers of a graph. *Mat. Vesnik* **60**, 247-253.

J.W. Moon (1962), Solution to problem 463. *Math. Mag.* **35**, 189.

J.W. Moon (1966), On subtournaments of a tournament. *Canad. Math. Bull.* **9(3)**, 297-301.

J.W. Moon (1982), The number of tournaments with a unique spanning cycle. *J. Graph Theory* **6**, 303-308.

U.S.R. Murty (1969), Extermal nonseparable graphs of diameter 2, in F. Harary (ed.), *Proof Techniques in Graph Theory*. Academic Press, New York, 111-118.

J. Mycielski (1955), Sur le coloriage des graphes. *Colloq. Math.* **3**, 161-162.

K.L. Ng (1999), Orientations of graphs with minimum diameter. *M.Sc. Thesis*, Department of Math., National University of Singapore.

K.L. Ng (2004), A new direction in the study of the orientation number of a graph. *Ph.D. Thesis*, Department of Math., National University of Singapore.

E.A. Nordhaus and J.W. Gaddum (1956), On complementary graphs. *Amer. Math. Monthly* **63**, 175-177.

R.J. Nowakowski and D.F. Rall (1996), Associative graph products and their independence, domination and coloring numbers. *Discuss. Math. Graph Theory* **16**, 53-79.

O. Ore (1960), Note on Hamilton circuits. *Amer. Math. Monthly* **67**, 55.

O. Ore (1962), Theory of Graphs, in: *Amer. Math. Soc. Colloq. Publ.* **38**. Amer. Math. Soc., Providence, Rhode Island, 206-212.

J. Petersen (1891), Die Theorie der regul'aren Graph. *Acta Math.* **15**, 193-220.

V. Petrovic (1997), Kings in bipartite tournaments. *Discrete Math.* **173**, 187-196.

V. Petrovic and C. Thomassen (1991), Kings in k-partite tournaments. *Discrete Math.* **98**, 237-238.

J. Plesník (1975), Diametrically critical tournaments. *Casopis Pest. Mat.* **100**, 361-370.

J. Plesník (1985), Remarks on diameters of orientations of graphs. *Acta Math. Univ. Comenian.* **46/47**, 225-236.

R.C. Prim (1957), Shortest connection networks and some generalizations. *Bell System Technical Journal* **36**, 1389-1401.

L. Rédei (1934), Ein kombinatorischer Satz. *Acta Litteraria Szeged* **7**, 39-43.

B.A. Reed (1996), Paths, stars and the number three. *Combin. Probab. Comput.* **5**, 277-295.

B.A. Reed (1997), ω, Δ and χ. *J. Graph Theory* **27**, 177-212.

K.B. Reid (1982), Every vertex a king. *Discrete Math.* **38**, 93-98.

H.E. Robbins (1939), A theorem on graphs, with an application to a problem in traffic control. *Amer. Math. Monthly* **46**, 281-283.

F.S. Roberts (1978), The one-way street problem, *Graph Theory and its Applications to Problems of Society*, CBMS-NSF Regional Conference Series in Applied Mathematics **29**, Society for Industrial and Applied Mathematics (SIAM), Philadelphia, 7-14.

F.S. Roberts and Y. Xu (1988), On the optimal strongly connected orientations of city street graphs I: Large grids. *SIAM J. Discrete Math.* **1**, 199-222.

F.S. Roberts and Y. Xu (1989), On the optimal strongly connected orientations of city street graphs II: Two east-west avenues or north-south streets. *Networks* **19**, 221-233.

F.S. Roberts and Y. Xu (1992), On the optimal strongly connected orientations of city street graphs III: Three east-west avenues or north-south streets. *Networks* **22**, 109-143.

F.S. Roberts and Y. Xu (1994), On the optimal strongly connected orientations of city street graphs IV: Four east-west avenues or north-south streets. *Discrete Applied Math.* **49**, 331-356.

N. Robertson, D.P. Sanders, P. Seymour and R. Thomas (1997), The four-colour theorem. *J. Comb. Theory B* **70(1)**, 2-44.

A. Rosa (1966), On certain valuations of the vertices of a graph, *Theory of Graphs, Internat. Symposium, Rome, July 1966,* Gordon and Breach, N.Y. and Dunod Paris, 349-355.

D. Seinsche (1974), On a property of the class of n-colorable graphs. *J. Comb. Theory B* **16(2)**, 191-193.

M. Sekanina (1960), On an ordering of the set of vertices of a connected graph. *Spisy Přírod. Fak. Univ. Brno* **412**, 137-142.

H. Shank (1979), Some parity results on binary vector spaces. *Ars Combin.* **8**, 107-108.

P.J. Slater (1975), Leaves of trees. *Congr. Numer.* **14**, 549-559.

M.Y. Sohn and X. Yuan (2009), Domination in graphs of minimum degree four. *J. Korean. Math. Soc.* **46(4)**, 759-773.

L. Soltés (1986), Orientations of graphs minimizing the radius or the diameter. *Math. Solvaca* **36**, 289-296.

B.M. Stewart (1969), On a theorem of Nordhaus and Gaddum. *J. Comb. Theory* **6**, 217-218.

D. Sumner (1976), 1-factors and antifactor sets. *J. London Math. Soc.* **s2-13(2)**, 351-359.

D. Sumner and P. Blitch (1983), Domination critical graphs. *J. Comb. Theory B* **34**, 65-76.

G.Y. Tay (2008), Kings in tournaments and their generalizations. *Honours Project*, Department of Math., National University of Singapore.

S. Toida (1973), Properties of an Euler graph. *J. Franklin Inst.* **295**, 343-345.

M. Truszczyński (1981), Some results on uniquely colorable graphs. *Soloquia Math. Soc. János Bolyai* **37**, 733-746.

W.T. Tutte (1947), The factorization of linear graphs. *J. London Math. Soc.* **22**, 107-111.

S.M. Ulam (1960), *A collection of mathematical problems*. Wiley Interscience, New York.

O. Veblen (1912), An application of modular equations in analysis situs. *Annals of Mathematics, Second Series* **14(1)**, 86-94.

V.G. Vizing (1965), A bound on the external stability number of a graph. *Dokl. Akad. Nauk.* **164**, 729-731.

V.G. Vizing (1968), Some unsolved problems in graph theory. *Usephi Ma. Nauk.* **23(6(144))**, 117-134.

C. Wang (2008), The independent domination number of maximal triangle-free graphs. *Australasian J. Combinatorics* **42**, 129-136.

V.K. Wei (1981), A lower bound on the stability number of a simple graph. *Bell Laboratories Tech. Memorandum 81-11217-9* Murray Hill, New Jersey.

D.J.A. Welsh and M.B. Powell (1967), An upper bound for the chromatic number of a graph and its application to timetabling problems. *The Computer Journal* **10(1)**, 85-86.

H. Whitney (1932), Congruent graphs and the connectivity of graphs. *Amer. J. Math.* **54**, 150-168.

R. Wilson (2002), *Four Colors Suffice*, Penguin Books, London.

D.R. Woodall (1972), Suficient conditions for cycles in digraphs. *Proc. London Math. Soc.* **24**, 739-755.

H.M. Xing, L. Sun and X.G. Chen (2006), Domination in graphs of minimum degree five. *Graphs and Combinatorics* **22**, 127-143.

B. Xu (1993) 度补图的直径, *J. East China Jiaotong University* **1**, 94-98 (in Chinese).

S. Xu (1990), The size of uniquely colorable graphs. *J. Comb. Theory B* **50(2)**, 319-320.

I.G. Yero and J.A. Rodriguez-Velazquez (2013), Roman domination in Cartesian product graphs and strong product graphs. *Applicable Analysis and Discrete Math.* **7(2)**, 262-274.

Y. Yu, J. Di and M. Lin (2006), Kings in tournaments. *Math Medley* **33(1)**, 28-32.

Y. Yu (2006), Kings in tournaments (2). *Math Medley* **33(2)**, 15-18.

Index

Printed in the United States
By Bookmasters